2/36

Developments in Geomathematics 4

SAMPLING OF PARTICULATE MATERIALS

Further titles in this series

1. F.P. AGTERBERG
Geomathematics

2. M. DAVID
Geostatistical Ore Reserve Estimation

3. S. TWOMEY
Introduction to the Mathematics of Inversion in Remote Sensing and Indirect Measurements

Developments in Geomathematics 4

SAMPLING OF PARTICULATE MATERIALS
THEORY AND PRACTICE

PIERRE M. GY
Consulting Engineer, Cannes, France

ELSEVIER SCIENTIFIC PUBLISHING COMPANY
Amsterdam — Oxford — New York   1982

ELSEVIER SCIENTIFIC PUBLISHING COMPANY
1 Molenwerf
P.O. Box 330, 1000 AH Amsterdam, The Netherlands

Distributors for the United States and Canada:

ELSEVIER SCIENCE PUBLISHING COMPANY INC.
52, Vanderbilt Avenue
New York, N.Y. 10017

With 110 illustrations and 48 tables.

First edition 1979
Second revised edition 1982

Library of Congress Cataloging in Publication Data

```
Gy, Pierre.
   Sampling of particulate materials.

   (Developments in geomathematics ; 4)
   Bibliography: p.
   Includes index.
   1. Ores--Sampling and estimation.  2. Bulk solids--
Sampling.  I. Title.  II. Series.
TN560.G9        622'.1                      79-16075
```

ISBN 0-444-42079-7 (Vol. 4)
ISBN 0-444-41609-9 (Series)

© Elsevier Scientific Publishing Company, 1979.
All rights reserved. No part of this publication may be reproduced, stored in a retrieval system or transmitted in any form or by any means, electronic, mechanical, photocopying, recording or otherwise, without the prior written permission of the publisher, Elsevier Scientific Publishing Company, P.O. Box 330, 1000 AH Amsterdam, The Netherlands

Printed in The Netherlands

CONTENTS

	Page
INTRODUCTION	1
HISTORICAL SUMMARY	3
FIRST PART - ANALYSIS OF THE PROBLEM	7

Chapter 1 : DEFINITION OF BASIC TERMS AND NOTATIONS

1.1. Definition of basic terms	11
1.1.1. Material to be sampled - sampling	11
1.1.2. Analysis - Chemical and physical components	13
1.2. Statistical definitions and notations	15
1.3. Estimators - Estimates - Errors - Biases	15
1.4. Domains and their extent	16
1.5. Properties of a selective process	16
1.5.1. A priori qualities	16
1.5.2. A posteriori qualities	17
1.5.3. Search for an objective	18
1.6. Specific notations	19

Chapter 2 : LOGICAL APPROACH

2.1. First part - Analysis of the problem	23
2.2. Second part - Continuous model of the increment sampling process	24
2.3. Third part - From the continuous model to the discrete reality	25
2.4. Fourth part - Discrete model of the increment sampling process	25
2.5. Fifth part - Splitting processes	27
2.6. Sixth part - Preparation	27
2.7. Seventh part - Resolution of sampling problems	27
2.8. Eighth part - Problems associated with commercial sampling	27
2.9. Ninth part - Automatic sampling plants	28

Chapter 3 : PART OF SAMPLING IN QUALITY CONTROL

3.1. Introduction	29
3.2. Possibility of direct estimation of a critical content	29
3.3. Possibility of direct extraction of the analysis sample	29
3.4. Preparation stages - Preparation errors	30
3.5. Successive estimators and eventual estimate of the critical content	30
3.6. Breaking up of the overall estimation error	30

Chapter 4 : SAMPLING PROCESSES

4.1. Probabilistic and non-probabilistic sampling processes	33
4.2. Comments on non-probabilistic sampling processes	33
4.3. Probabilistic processes - Notion of movable and unmovable batches	33
4.4. Sampling of unmovable lots	34
4.5. Sampling of movable lots	34
4.6. Analysis of the increment sampling process	35
4.7. Analysis of the splitting process	36
4.8. Comparison of the increment sampling and splitting processes	37
4.9. Field of application of the increment sampling and splitting processes	38

Chapter 5 : MODELS OF THE INCREMENT SAMPLING PROCESS

5.1. What is a model and what is the use of it ? 39
   5.1.1. Model of an object 39
   5.1.2. Model of a process 39
   5.1.3. Qualities of a model or of a group of models 39
5.2. Introduction to a group of models of the sampling processes .. 40
   5.2.1. Continuous model 41
   5.2.2. Discrete model 42
5.3. Continuous model 42
   5.3.1. Continuous model of the material to be sampled 42
   5.3.2. Degenerate models of the lot L 43
   5.3.3. General continuous model of the lot L 47
   5.3.4. Punctual, extended and fragmental functions 47
   5.3.5. Continuous model of a selection process 49
   5.3.6. Continuous model of the increment sampling process ... 49
5.4. Discrete model 50
   5.4.1. Discrete model of the lot L 50
   5.4.2. Discrete model of a selection process 50
5.5. Objectives pursued when developing selection models 51
5.6. Resolution of sampling problems 51

SECOND PART - CONTINUOUS MODEL OF THE INCREMENT SAMPLING PROCESS 53

Chapter 6 : HETEROGENEITY OF A CONTINUOUS SET

6.1. Introduction 55
6.2. Definition and properties of a homogeneous material 55
6.3. Description of a heterogeneous material 56
6.4. Definition of the variogram 56
6.5. General properties of the variogram 57
   6.5.1. Breaking up of the variogram $v_f(\theta)$ 58
   6.5.2. Properties of $v_{f1}(\theta)$ 58
   6.5.3. Properties of $v_{f2}(\theta)$ 59
   6.5.4. Properties of $v_{f3}(\theta)$ 59
   6.5.5. Properties of $v_f(\theta)$ 60
   6.5.6. Particular non-periodic variograms 61
6.6. Experimental determination of the variogram - Logical approach .. 63
   6.6.1. Definition of the discrete variogram 63
   6.6.2. Definition of the experimental variogram 64
   6.6.3. Definition of the corrected variogram 65
   6.6.4. Definition of the model variogram 65
   6.6.5. Recapitulation 66
6.7. Interpretation of the results of a variographic experiment 66
   6.7.1. Estimation of v'_{f1} 67
   6.7.2. Estimation of $v'_{f2+3}(\theta)$ 67
   6.7.3. Checking the existence of a periodic term 68
   6.7.4. Analysis of a non-periodic variogram 68
   6.7.5. Analysis of a periodic variogram 69
   6.7.6. Objective definition of the components of $f(t)$ 72
   6.7.7. Variographic parameters 73
6.8. Particular case of zero-dimensional lots 73
6.9. Conclusions 74

Chapter 7 : REFERENCE SELECTION SCHEMES

7.1. Introduction 75
   7.1.1. Punctual sample 75
   7.1.2. Definition of a random selection 75

7.2. Systematic selection with random positioning	76
7.2.1. Definition	76
7.2.2. Properties of the number Q of increments	76
7.3. Random stratified selection	79
7.3.1. Definition	79
7.3.2. Properties of the number Q of increments	79
7.4. Random selection	82
7.4.1. Definition	82
7.4.2. Number Q of increments	82
7.5. Full description of a selection scheme	82
7.6. Comparison of the three selection schemes	82
7.7. Field of application of the three selection schemes	83

Chapter 8 : DEVELOPMENT OF THE CONTINUOUS SELECTION MODEL
              CONTINUOUS SELECTION ERROR CE

8.1. Introduction	85
8.2. Definition of the model	85
8.2.1. Definition of the lot L	85
8.2.2. Definition of the material to be sampled	86
8.2.3. Definition of the selecting process	86
8.2.4. Definition of the increment delimitation process	86
8.2.5. Definition of the sample S	86
8.2.6. Definition of the continuous selection error CE	87
8.3. Distribution of the weight M_S of active components in the sample S	87
8.3.1. Distribution law of M_S	87
8.3.2. Mean of M_S	88
8.3.3. Relative variance of the sample weight M_S	91
8.4. Distribution of the weight A_S of critical component in the sample S	93
8.4.1. Distribution law of A_S	93
8.4.2. Mean of A_S	93
8.4.3. Relative variance of A_S	94
8.5. Correlation between the distributions of M_S and A_S	94
8.6. Distribution of the critical content a_S of the sample S - Introduction	95
8.6.1. Review of the relevant factors	95
8.6.2. State of the material to be sampled	96
8.6.3. Selecting conditions	96
8.6.4. Cases studied in the following sections	96
8.6.5. Sources of selection bias	97
8.7. Distribution of the critical content a_S of the sample S - Case No.1	98
8.7.1. Distribution law of a_S	98
8.7.2. Moments of a_S - Introduction	98
8.7.3. Moments of a_S - First approximation	100
8.7.4. Moments of a_S - Second approximation	100
8.8. Distribution of the critical content a_S of the sample S - Case No.2	101
8.8.1. Mean of a_S	101
8.8.2. Variance of a_S	101
8.9. Distribution of the critical content a_S of the sample S - Case No.3	101
8.10. Distribution of the critical content a_S of the sample S - Case No.4	102
8.10.1. General formulas	102
8.10.2. Practical formulas	102
8.11. Distribution of the critical content a_S of the sample S - Case No.5	103
8.12. Distribution of the critical content a_S of the sample S - Case No.6	104

Chapter 9 : BREAKING UP OF THE CONTINUOUS SELECTION ERROR CE

9.1. Introduction	105
9.2. Definition of the weighting error and of the quality fluctuation error	106

9.3. Analysis of the quality fluctuation error QE	107
9.4. Moments of the quality fluctuation error QE	107
9.4.1. Mean of QE	107
9.4.2. Variance of QE	107
9.5. Recapitulation	108

Chapter 10 : SHORT-RANGE QUALITY FLUCTUATION ERROR QE_1

10.1. Definition	109
10.2. Mean of QE_1	109
10.3. Variance of QE_1	109
10.4. Cancelling and minimizing of QE_1	110
10.5. Further analysis of the short-range quality fluctuation error QE_1	110

Chapter 11 : LONG-RANGE QUALITY FLUCTUATION ERROR QE_2

11.1. Definition	111
11.2. Mean of QE_2	111
11.3. Variance of QE_2	111
11.4. Comparison of the three selection schemes	112
11.5. Cancelling and minimizing of $\sigma^2(QE_2)$	112

Chapter 12 : PERIODIC QUALITY FLUCTUATION ERROR QE_3

12.1. Definition	113
12.2. Occurrence of periodic fluctuations	114
12.3. Moments of QE_3	115
12.3.1. Most general case	116
12.3.2. First particular case	116
12.3.3. Second particular case	116
12.3.4. Third particular case	117
12.4. Comparison of the three selection schemes	118
12.5. Example	119
12.6. General conclusion concerning the choice of a selection scheme	119

Chapter 13 : WEIGHTING ERROR WE

13.1. Definition	121
13.2. Mean of WE	121
13.3. Variance of WE	121
13.3.1. Expression of the variance	121
13.3.2. Weighting fluctuation factor	122
13.3.3. Weighting variance and fluctuations of the rate-of-flow	122
13.3.4. Comparison of the three selection schemes	122
13.4. Constant tonnage sampling systems	122
13.4.1. The problem to be solved	123
13.4.2. Systems using a uniform speed cutter	123
13.4.3. Systems using a proportional speed cutter	124
13.5. Error resulting from the non-uniformity of the cutter speed from one increment to the next	124
13.6. Particular rate of flow functions	124
13.6.1. Crenellated rate-of-flow functions	125
13.6.2. Long-wave variations	125
13.6.3. Variogram of the rate-of-flow functions	126
13.7. Conclusions	127

Chapter 14 : PRACTICAL IMPLEMENTATION OF THE CONTINUOUS MODEL VARIOGRAPHIC EXPERIMENT

14.1. Introduction	129

14.2. Organization of a variographic experiment 129
   14.2.1. Planning of the experiment 129
   14.2.2. Practical considerations 130
14.3. Analysis of a simple periodic variogram 131
   14.3.1. Organization of the variographic experiment 131
   14.3.2. Analysis of the first series with h_1 = 2 seconds 132
   14.3.3. Analysis of the second series with h_2 = 1 minute 133
   14.3.4. Analysis of the third series with h_3 = 20 minutes 133
14.4. Analysis of a non-periodic variogram 140

Chapter 15 : PRACTICAL IMPLEMENTATION OF THE CONTINUOUS MODEL
ERROR ESTIMATION

15.1. Introduction 143
15.2. Example No.1 - Lead ore 143
   15.2.1. The material to be sampled 143
   15.2.2. Relative variance of the sample weight M_S 144
   15.2.3. Relative variance of the weight A_S of critical component 144
   15.2.4. Variance of the continuous selection error CE 145
   15.2.5. Mean of the continuous selection error CE 146
   15.2.6. Confidence interval of the Pb content of the sample S 146
   15.2.7. Breaking up of the total variance $\sigma^2(CE)$ 146
   15.2.8. Interpretation of the results 147
15.3. Example No.2 - Feed to the blending system of a cement plant 148
   15.3.1. The material to be sampled 148
   15.3.2. Relative variance of the sample weight M_S 148
   15.3.3. Relative variance of the weight A_S of CaO in the sample 149
   15.3.4. Variance of the continuous selection error CE 150
   15.3.5. Mean of the continuous selection error CE 150
   15.3.6. Confidence interval of the CaO content 150
   15.3.7. Breaking up of the total variance $\sigma^2(CE)$ 151
   15.3.8. Interpretation of the results 152

THIRD PART - FROM THE CONTINUOUS MODEL TO THE DISCRETE REALITY
MATERIALIZATION OF THE PUNCTUAL INCREMENTS 155

Chapter 16 : COMPONENTS OF THE MATERIALIZATION ERROR ME

16.1. Introduction 157
16.2. Increment sampling of flowing streams 158
   16.2.1. The one-dimensional model 158
   16.2.2. The punctual increment I_P 158
   16.2.3. The model extended increment I_E 158
   16.2.4. The actual extended increment I'_E - Increment delimitation error 160
   16.2.5. The model fragmental increment I_F 162
   16.2.6. The actual fragmental increment I'_F - Increment extraction error 163
16.3. Recapitulation 164
16.4. Selection, extraction and sampling probabilities 165
   16.4.1. Density of selection probability 165
   16.4.2. Selection probability 165
   16.4.3. Extraction probability 165
   16.4.4. Sampling probability 166

Chapter 17 : INCREMENT DELIMITATION ERROR DE

17.1. Definition 167
17.2. Falling stream sampling - correctness conditions involving the cutter geometry 167

17.2.1. Straight-path cross-stream cutters - correct geometry	167
17.2.2. Straight-path cross-stream cutters - deviations from correctness	168
17.2.3. Straight-path cutters of other types - deviations from correctness	169
17.2.4. Circular-path cutters - correct geometry	171
17.2.5. Circular-path cutters - deviations from correctness	171
17.3. Falling stream sampling - Correctness conditions involving the cutter speed	173
17.3.1. Achievement of the uniformity of the cutter velocity during its travel across the stream	173
17.3.2. Achievement of the uniformity of the cutter velocity from one increment to the next	175
17.4. Falling stream sampling - Correctness conditions involving the general lay-out	175
17.4.1. Correct lay-out of the sampling device	175
17.4.2. Incorrect lay-out : part of the stream escapes sampling	176
17.4.3. Incorrect lay-out : part of the stream enters the cutter in idle position	177
17.4.4. Incorrect lay-out : reciprocating sampler with a non-idle reversing position	177
17.4.5. Sampling dry materials containing fines	177
17.5. Falling stream sampling - Recapitulation of the conditions of delimitation correctness	179
17.6. Stopped belt sampling	180
17.7. Two-dimensional sampling	180
17.8. Cost of correct delimitation	182

Chapter 18 : INCREMENT EXTRACTION ERROR EE

18.1. Definition	183
18.2. Analysis of the rebounding rule	183
18.2.1. Definitions and notations	183
18.2.2. Chronology of the fragment F	184
18.2.3. Chronology of the leading edge C_L	184
18.2.4. Chronology of the trailing edge C_T	185
18.2.5. Collision between the fragment F and parts of the cutter C	185
18.2.6. Respective positions of F and C_L at time t_{F3}	186
18.2.7. Collision between F and C_L	187
18.2.8. Respective positions of F and C_T at time t_{F3}	188
18.2.9. Rebounding rule and model increment	189
18.3. Conditions of extraction correctness involving the material to be sampled	189
18.3.1. Fragments do not fall one by one	190
18.3.2. Fragments are liable to spin	190
18.3.3. Fragments do not fall in a vertical plane	190
18.3.4. Recapitulation	191
18.4. Conditions of extraction correctness involving the cutter characteristics	191
18.4.1. Straightness of the cutter edges	191
18.4.2. Thickness of the cutter edges	191
18.4.3. Inclination of the cutter edges	192
18.4.4. Cutter width and velocity - qualitative theoretical approach	194
18.4.5. Cutter width and velocity - rules of extraction correctness	198
18.4.6. Increment integrity	202
18.5. Cutter width and velocity - Experimental determination of the critical characteristics	204
18.5.1. Introduction and notations	204
18.5.2. Choice of an experimental method	205
18.5.3. General lay-out of the testing plant	206

18.5.4. Critical value of the ratio W/d_M . . . 207
18.5.5. Critical cutter speed when $W = W_o$ . . . 208
18.5.6. Critical cutter speed when $W = n\, W_o$ . . . 208
18.5.7. Acknowledgments . . . 210
18.6. Recapitulation . . . 210
18.7. Cost of extraction correctness . . . 211

FOURTH PART - DISCRETE MODEL OF THE INCREMENT SAMPLING PROCESS . . . 213

Chapter 19 : HETEROGENEITY OF A DISCRETE SET

19.1. Introduction and notations . . . 215
19.2. Definition and properties of homogeneous and heterogeneous materials . . . 216
19.3. Characterizing the heterogeneity of a discrete set . . . 217
   19.3.1. Heterogeneity carried by a particle . . . 217
   19.3.2. Heterogeneity carried by a group of particles . . . 218
   19.3.3. Constitution heterogeneity and distribution heterogeneity of the lot L . . . 218
   19.3.4. Relationship between constitution and distribution heterogeneity . . . 219
   19.3.5. Properties of the constitution and distribution heterogeneities . . . 219
19.4. General expression of the distribution heterogeneity . . . 225

Chapter 20 : DEVELOPMENT OF THE DISCRETE SELECTION MODEL

20.1. Introduction and notations . . . 227
20.2. Distributions of N_{Zm}, N_{Sk}, M_{Sk} and A_{Sk} . . . 228
   20.2.1. Distribution of N_{Zm} frequency of unit U_m in the set Z . . . 228
   20.2.2. Distribution of N_{Sk} number of units in the sample S_k . . . 228
   20.2.3. Distribution of M_{Sk} weight of S_k . . . 229
   20.2.4. Distribution of A_{Sk} weight of critical component in S_k . . . 229
20.3. Distribution of a_S critical content of the sample S - Incorrect selection . . . 229
   20.3.1. Distribution law of the critical content a_S . . . 229
   20.3.2. Moments of a_S - Introduction . . . 230
   20.3.3. Moments of a_S - First and second approximations . . . 231
20.4. Practical implementation of formulas involving sums $\sum_i$ . . . 233
   20.4.1. Introduction . . . 233
   20.4.2. Practical estimation of the sums extended to the N_L particles of the lot L . . . 233
   20.4.3. Properties of the selection probability $P_{\alpha\beta}$ and of the sampling ratio $\tau_{\alpha\beta}$ . . . 235
20.5. Moments of N_S, M_S, A_S and a_S - Correct selection . . . 236
   20.5.1. Number of units N_S in the sample S . . . 236
   20.5.2. Weight M_S of the sample S . . . 236
   20.5.3. Weight A_S of critical component in the sample S . . . 236
   20.5.4. Critical content a_S of the sample S . . . 237
   20.5.5. Expression of the sums $\sum_i$ extended to the N_L particles of the lot . . . 238
20.6. Moments of N_S, M_S, A_S and a_S - Correct selection - Uniform weighting . . . 238
   20.6.1. Moments of N_S . . . 238
   20.6.2. Moments of M_S . . . 238
   20.6.3. Moments of A_S . . . 239
   20.6.4. Moments of a_S . . . 239

20.6.5. Moments of the sampling error SE	239
20.6.6. Remark	
20.7. Practical implementation of the results of the discrete model	239

**Chapter 21 : LINKING UP OF THE CONTINUOUS AND DISCRETE MODELS
FUNDAMENTAL ERROR FE - GROUPING AND SEGREGATION ERROR GE**

21.1. Introduction and notations	241
21.2. Moments of $SE = QE_1$ according to the continuous and discrete models	242
21.2.1. Continuous perspective	242
21.2.2. Discrete perspective	242
21.2.3. Comparison of the results of the continuous and discrete models	244
21.3. Analysis of the short-range quality fluctuation error QE_1	245
21.3.1. Incidence of the increment size on the variance $\sigma^2(QE_1)$	245
21.3.2. Definition of the fundamental error FE	246
21.3.3. Definition of the grouping and segregation error GE	247
21.3.4. Breaking up of QE_1	247
21.4. Cancelling and minimizing of the fundamental error FE	248
21.4.1. Cancelling of FE	248
21.4.2. Minimizing of FE	248
21.5. Cancelling and minimizing of the grouping and segregation error GE	249
21.5.1. Cancelling of GE	249
21.5.2. Minimizing of GE	250

**Chapter 22 : PRACTICAL IMPLEMENTATION OF THE THEORETICAL RESULTS -
CORRECT SELECTION**

22.1. Introduction	251
22.2. Estimation of the moments of the fundamental error FE - Introduction of Y and Z	251
22.3. Estimation of Y and Z - Method No.1	252
22.3.1. The critical component is a mineral	252
22.3.2. The critical component is a size fraction	254
22.4. Estimation of Y and Z - Method No.2	254
22.4.1. Principle of the method	255
22.4.2. Estimation of $Y_1 = Z_1$	255
22.4.3. Estimation of Z_2	256
22.4.4. Simplified expressions of Z and $\sigma^2(FE)$	258
22.4.5. Conclusions concerning the second method	258
22.5. Estimation of Y and Z - Method No.3	258
22.6. Properties and practical estimation of c, ℓ, f, g and d	259
22.6.1. Mineralogical composition factor c	259
22.6.2. Liberation factor ℓ	262
22.6.3. Particle shape factor f	263
22.6.4. Size range factor g	263
22.6.5. Maximum particle diameter d	263
22.6.6. Sampling slide rule	264
22.7. Resolution of sampling problems involving the fundamental variance	264
22.7.1. Estimation of the fundamental variance - Example 1	264
22.7.2. Estimation of the minimum sample weight - Example 2	265
22.7.3. Estimation of the maximum particle size - Examples 3 and 4	266
22.8. Practical application of methods No.1, 2 and 3	267
22.8.1. Example 5 : application of methods No. 1 and 2	267
22.8.2. Example 6 : application of method No.3	276
22.8.3. Example 7 : application of method No.3	278
22.9. Recapitulation and conclusions	279

Chapter 23 : PRACTICAL IMPLEMENTATION OF THE THEORETICAL RESULTS - INCORRECT SELECTION

23.1. Introduction	281
23.2. Incorrect extraction curve	283
23.3. Practical determination of the curve of incorrect extraction	284
23.3.1. Method No.1	284
23.3.2. Method No.2	284
23.3.3. Method No.3	285
23.3.4. Method No.4	286
23.4. Examples	286
23.4.1. Example No.1	286
23.4.2. Example No.2	287
23.4.3. Example No.3	288
23.5. Conclusions	288

FIFTH PART - SPLITTING PROCESS   289

Chapter 24 : SPLITTING METHODS AND DEVICES

24.1. Introduction	291
24.2. True and degenerate splitting processes	292
24.3. Coning and quartering	293
24.4. Riffling	294
24.4.1. Description	294
24.4.2. Possibility of an operating bias	295
24.4.3. Practical conclusions	296
24.5. Fractional shovelling	297
24.5.1. Procedure	297
24.5.2. Possibility of a bias	298
24.5.3. Practical conclusions	299
24.6. Sectorial splitters	300
24.6.1. Revolving feeder sectorial splitter	300
24.6.2. Stationary feeder sectorial splitter	301

Chapter 25 : MODEL OF THE SPLITTING PROCESS - SPLITTING ERRORS

25.1. Linking up with the existing models	305
25.2. Moments of the continuous selection error CE	306
25.3. Minimizing of $\sigma^2(CE)$	307
25.4. Delimitation error DE	308
25.5. Extraction error EE	309

Chapter 26 : PRACTICAL IMPLEMENTATION OF SPLITTING PROCESSES - EXAMPLE - REDUCTION OF DRILL CORE SAMPLES

26.1. Introduction	311
26.2. Core sample reduction methodology	311
26.2.1. Longitudinal core halving	311
26.2.2. Drying	312
26.2.3. Size reduction	312
26.2.4. Sampling	312
26.2.5. Mixing	312
26.3. Selection of a core sample reduction scheme	313
26.3.1. Sample reduction diagram	313
26.3.2. Safety rule	314
26.3.3. Core sample L	315
26.3.4. Primary comminution stage	316
26.3.5. Primary sampling stage	316

26.3.6. Secondary comminution stage	316
26.3.7. Secondary sampling stage	316
26.3.8. Tertiary comminution stage	316
26.3.9. Tertiary sampling stage	317
26.4. Examples	317
26.4.1. Safe reduction scheme	318
26.4.2. Unsafe reduction scheme	319
26.5. Recommendations	321

SIXTH PART - LOT AND SAMPLE PREPARATION 323

Chapter 27 : PREPARATION ERRORS

27.1. Introduction	325
27.2. Errors resulting from contamination	325
27.2.1. Contamination by dust	325
27.2.2. Contamination by material present in the sampling circuit and equipment	326
27.2.3. Contamination by abrasion	326
27.2.4. Contamination by corrosion	326
27.2.5. Contamination by salting	327
27.3. Errors resulting from losses	327
27.3.1. Loss of fines as dust	327
27.3.2. Loss of material remaining in the preparation or sampling circuit	327
27.3.3. Loss of certain fractions of the sample	327
27.3.4. Deliberate loss of fractions of the sample	328
27.4. Errors resulting from alteration of the chemical composition	328
27.4.1. Errors by addition or fixation	328
27.4.2. Errors by substraction or elimination	329
27.5. Errors resulting from alteration of the physical composition	330
27.5.1. Addition or creation of critical component	330
27.5.2. Substraction or destruction of critical component	331
27.6. Errors resulting from unintentional mistakes	331
27.7. Errors resulting from frauding or sabotage	332
27.8. Conclusions	332

SEVENTH PART - RESOLUTION OF SAMPLING PROBLEMS 333

Chapter 28 : RECAPITULATION OF THE SAMPLING ERRORS

28.1. Analysis of the overall estimation error	335
28.1.1. Breaking up of the overall estimation error OE	335
28.1.2. Breaking up of the total sampling error TE	335
28.1.3. Breaking up of the total sampling error TE_n arising at stage n	336
28.1.4. Breaking up of the sampling error SE	336
28.1.5. Breaking up of the selection error CE	336
28.1.6. Breaking up of the quality fluctuation error QE	336
28.1.7. Breaking up of the short-range quality fluctuation error QE_1	337
28.1.8. Breaking up of the materialization error ME	337
28.1.9. Breaking up of the total preparation error PE	337
28.1.10. Recapitulation	338
28.2. Fundamental error FE	340
28.2.1. Definition	340
28.2.2. Properties of the mean	340
28.2.3. Properties of the variance	340
28.2.4. Cancellation of the fundamental error FE	342
28.2.5. Minimization of the fundamental error FE	342
28.3. Grouping and segregation error GE	342
28.3.1. Definition	342
28.3.2. Properties of the mean	342

28.3.3. Properties of the variance — 342
28.3.4. Cancellation of the grouping and segregation error GE — 343
28.3.5. Minimization of the grouping and segregation error GE — 343
28.4. Long-range quality fluctuation error QE_2 — 343
  28.4.1. Definition — 343
  28.4.2. Properties of the mean — 344
  28.4.3. Properties of the variance — 344
  28.4.4. Cancellation of the long-range quality fluctuation error QE_2 — 344
  28.4.5. Minimization of the long-range quality fluctuation error QE_2 — 344
28.5. Periodic quality fluctuation error QE_3 — 345
  28.5.1. Definition — 345
  28.5.2. Properties of the mean — 345
  28.5.3. Properties of the variance — 345
  28.5.4. Cancellation of the periodic quality fluctuation error QE_3 — 345
  28.5.5. Minimization of the periodic quality fluctuation error QE_3 — 346
28.6. Weighting error WE — 346
  28.6.1. Definition — 346
  28.6.2. Properties of the mean — 346
  28.6.3. Properties of the variance — 346
  28.6.4. Cancellation of the weighting error WE — 346
  28.6.5. Minimization of the weighting error WE — 347
28.7. Increment delimitation error DE — 347
  28.7.1. Definition — 347
  28.7.2. Properties of the increment delimitation error DE — 347
  28.7.3. Cancellation of the increment delimitation error DE — 347
28.8. Increment extraction error EE — 348
  28.8.1. Definition — 348
  28.8.2. Properties of the increment extraction error EE — 348
  28.8.3. Cancellation of the increment extraction error EE — 348
28.9. Preparation errors PE — 349
  28.9.1. Definition — 349
  28.9.2. Properties of the preparation errors PE — 350
  28.9.3. Elimination of the preparation errors PE — 350
28.10. Conclusions — 350

Chapter 29 : SOLVABLE AND UNSOLVABLE SAMPLING PROBLEMS

29.1. Definitions — 351
29.2. Representativeness and cost — 351
  29.2.1. Notion of acceptable representativeness standard — 352
  29.2.2. Notion of acceptable cost — 353
  29.2.3. Tentative conclusion — 355
29.3. Sampling of three-dimensional objects — 355
  29.3.1. Definition - Examples — 355
  29.3.2. Theoretical and practical solvability — 356
  29.3.3. Possible solutions — 357
29.4. Sampling of two-dimensional objects — 357
  29.4.1. Definition - Examples — 357
  29.4.2. Theoretical and practical solvability — 358
  29.4.3. Other possible solutions — 358
29.5. Sampling of one-dimensional stationary objects — 358
  29.5.1. Definition - Examples — 358
  29.5.2. Theoretical and practical solvability — 358
  29.5.3. Other possible solutions — 359
29.6. Sampling of one-dimensional flowing streams — 359
29.7. Sampling of zero-dimensional objects — 359
  29.7.1. Definition - Examples — 359
  29.7.2. Theoretical and practical solvability — 359
  29.7.3. Further processing of the primary sample — 360
29.8. Sampling of small or valuable objects — 361
  29.8.1. Definition — 361

    29.8.2. Theoretical and practical solvability   361
    29.8.3. Japanese slab-cakes   361
29.9. Conclusions   363

EIGHTH PART - PROBLEMS ASSOCIATED WITH COMMERCIAL SAMPLING   365

Chapter 30 : NOTION OF EQUITY

30.1. Introduction - Definition   367
30.2. Properties of the settlement price assumed to be a linear function of the critical content   368
30.3. Properties of the settlement price assumed to be a non-linear function of the critical content   370
30.4. Relative importance of bias and random error in commercial sampling   371
    30.4.1. Relative unimportance of random errors in commercial sampling   372
30.5. Conclusions - Recommendations   374
30.6. Equity - Louis-le Débonnaire's splitting method   376

Chapter 31 : TESTING THE AGREEMENT BETWEEN TWO SERIES OF INDEPENDENT ESTIMATES OF A SAME CHARACTERISTIC - DISCREPANCIES BETWEEN SELLER AND BUYER

31.1. Introduction   379
31.2. Notations and definitions   380
    31.2.1. Positive and negative tests   380
    31.2.2. Certainty, presumption and uncertainty   380
    31.2.3. Agreement and disagreement between two series of estimates   381
    31.2.4. Kinds of risk and level of risk   381
31.3. Testing the hypothesis $H \equiv \{D = 0\}$ 381
31.4. Testing the hypothesis $H' \equiv \{D' = |D| - D_A = 0\}$ 383
31.5. Practical implementation of the test   385
31.6. Graphical implementation of the results of the test   386
31.7. Examples   388
31.8. Average number of trials necessary to show up a systematic difference   389

Chapter 32 : TESTING THE AGREEMENT BETWEEN AN ESTIMATE AND THE TRUE VALUE - CHECK OF THE SAMPLING BIAS

32.1. Introduction - Notations   391
32.2. Method No.1 - Absolute method involving a synthetic lot   392
32.3. Method No.2 - Relative method involving a reference method or device   393
32.4. Method No.3 - Relative method involving a comparison between sample and sampling reject   394
    32.4.1. Variant No.1   394
    32.4.2. Variant No.2   395
    32.4.3. Interest of this method   395
32.5. Example of application of method No.2   395
    32.5.1. Critical inspection of the sampler   395
    32.5.2. Experimental testing   396
    32.5.3. Conclusions   398
32.6. Accuracy vs. reproducibility - Example of application of method No.3   400
32.7. Critical inspection of a sampling device vs. experimental testing of accuracy   402

NINTH PART - AUTOMATIC SAMPLING PLANTS   405

Chapter 33 : DESIGN OF AUTOMATIC SAMPLING PLANTS

33.1. Introduction   407
33.2. Achievement of sampling accuracy   407
33.3. Achievement of sampling reproducibility   408
33.4. Allotment of the expendable sampling variance   409

33.5. Sampling large tonnages of coarse materials   410
33.6. Conclusions   412

Chapter 34 : TYPICAL FLOW-SHEETS OF AUTOMATIC SAMPLING PLANTS

34.1. Introduction   413
34.2. Sampling of a bauxite ore   414
34.3. Sampling of a nickel ore   416
34.4. Sampling of an iron ore   418
34.5. Sampling of the raw mix fed to a cement factory   420
34.6. Commercial sampling of uranium concentrates   422

REFERENCES   425

INDEX   427

INTRODUCTION

The failure of mining or metallurgical undertakings can nearly always be traced back to the confusion between "specimens" on the basis of which no sane financial decision should ever be made and "samples" known to be representative of the object to be valued (orebody, shipment of ores or concentrates, etc..) within the limits of a certain confidence interval that can be estimated and relied upon. In other words, the failure of what is aptly called a mining or metallurgical "venture" can nearly always be attributed to unaccountable sampling errors.

Sampling problems arising in the mineral industries belong to two categories :

- sampling of compact solids such as mineral deposits, metal ingots, etc..
- sampling of particulate solids such as cores extracted from mineral deposits, run-of-mine ores fed to processing plants after crushing or grinding, concentrates or minerals shipped by a producer to a consumer, etc..

The sampling of compact solids and more specifically mineral deposits is covered by the science known as "Geostatistics". The fundamentals of this science, established by Krige, Sichel, de Wijs were developed by Matheron and his team (references in appendix). Worked out in France too, Matheron's theories are slowly but steadily gaining acceptance in English speaking countries around the world thanks to an increasing teaching and to technical textbooks such as Michel David's "Geostatistical Ore Reserve Estimation" (1977).

The sampling of particulate solids such as ores and concentrates has remained the poor relation of the family for a long time and as far as we know, the present theory seems to be the first and only comprehensive theory ever published on this ill-known subject in the English language. Its only competitor is the textbook we have published in French (1975).

This theory, however, has been taught in Europe (including England) and South-America for a number of years. Such a situation is the obvious result of the difficulties of scientific communication between people speaking different languages. It justifies our effort to present this synthesis of a quarter century of research and practical experience as a consultant, trouble-shooter, judicial expert and teacher.

The particulate materials currently submitted to sampling are mainly of mineral and vegetable origin. The theoretical part of this book was indeed worked out for minerals and all our examples are borrowed from the mining, metallurgical and cement industries but when analysing the fundamentals of this theory one comes to the conclusion that it is applicable to any particulate or discrete material and more specifically to :

- all particulate materials of mineral origin, whether the interstitial fluid is air, water or any gas or liquid,
- metallurgical by-products such as granulated drosses or slags,
- raw materials and products of cement, glass or ceramic industries,
- raw materials and products of chemical and pharmaceutical industries,
- raw materials and products of vegetable origin such as grains, seeds, roots, tubers, stems, fruits, flours, etc ..
- raw materials and products of food and paper industries,
- some products of animal origin,
- miscellaneous by-products such as industrial or household refuses, etc ...

The sampling of vegetables such as cereals, oil seeds, sugar beets, etc .. is very primitive and might be considerably improved for the benefit of all parties involved. This is not due to a theoretical gap : as already mentioned, the theory of vegetable sampling can very easily be adapted from the general theory presented here but the responsibles of the trade of vegetable commodities still have to become conscious of the economical importance of sampling. The progress observed in the theory of mineral sampling has however roused some interest, at least in France and the situation is changing rapidly.

A sizeable fraction of this theory is also applicable to one-phase media such as compact solids or liquids and for instance :

- melted and flowing metallurgical products and by-products,
- solid metallurgical products in the form of bars, plates, ingots, etc ..
- all kinds of liquids or solutions, either clear or containing fine solid particles such as those handled in chemical, hydrometallurgical, pharmaceutical industries, in water processing plants, etc ..
- all kinds of industrial or town effluents to be controlled in order to protect the environment, etc ...

Huge fields of application of the results of this theory remain to be prospected and developed.

HISTORICAL SUMMARY

His diploma in his pocket, the young geologist, mining engineer, metallurgist or analyst is thrown head on against the realities of his professional life. When confronted with his first sampling problem he is often surprised to realize that his masters have given him only vague, ingenuous or even contradictory indications. Bibliography if consulted is disappointing : he may find for instance in the U.S. Bureau of Mines "Dictionary of Mining, Mineral and related terms" (1968) the following statement quoted from Walter H. Weed:"Honest sampling requires good judgment and practical experience" and wonder whether it does not require a bit more than that.

Papers and books have been written in French but who reads French nowadays ? Historically the first concern of authors on sampling has been to establish a relationship between the minimum sample weight and the diameter of the coarsest particle in the material to be sampled. Vezin (1865) was reported to use a relationship of the form :

$M_S \geq k\, d^3$ with :

M_S : sample weight
d : diameter of the coarsest fragment
k : a universal constant

This formula was officially proposed by Brunton (1895) in a paper based on very sound homothetic considerations. But Richards (1909) disposed of Brunton's formula under the astounding pretext that the resulting sample weights were way over those in actual use. Without any theoretical justification he enforced the formula :

$M_S \geq k\, d^2$

which was to remain the golden rule during more than 50 years and still today keeps being used in a number of mines and processing plants. In Richards' formula, k varies from one material to the next according to charts where the notions of homogeneity and grade are confusingly muddled up. Probably looking for a compromise between Brunton and Richards, Demond and Halferdahl (1922) proposed the formula :

$M_S \geq k\, d^\alpha$

with k and α to be experimentally determined. During the 1930's a number of authors working specifically in the field of coal sampling carried out a huge amount of experimental work without the slighest attempt to give the problem a theoretical basis. Apart from the very reasonable considerations of geometrical similarity developed by Vezin and Brunton, sampling has not yet relinquished the lower level of empiricism. Not a single step has been made so far towards a better understanding of the subject. After 1940, with the rapid development of mathematical statistics, various authors attempted to snatch a ready-made solution out of specialized books and the binomial distribution was retained as a model by those who were ready to identify a batch of ore with a population of black and white balls. Therefrom, Hassialis (1945) proposed a multinomial model that is statistically correct but, despite the worldwide distribution of the handbook containing his formulas, we are still looking 30 years after its publication for someone who has been fortunate enough to get workable results out of Hassialis formulas. Following the same line, Becker (1964) developed a sound theory based on the same model but failed to propose practical formulas.

Thus far, all authors speak in the singular of THE sampling error, implicitly assuming that this error is unique and follows a simple law. It was under the same assumption that we presented (1953) the "Equiprobable sampling model" resulting in the formula :

$$M_S \geq \frac{C d^3}{\sigma_0^2} \quad \text{with}$$

σ_0^2 : variance of the tolerated sampling error,
C : a constant characterizing the material to be sampled. According to its definition as a product of several factors, C can be easily estimated in every particular case.

Later on we gave (1955) precise rules and charts allowing a quick estimation of C. Then, we devised for the same purpose a circular calculator (1956) replaced a few years later by a slide rule (1963). Both the circular calculator and the slide rule allow an easy estimation of the minimum sample weight.

The equiprobable model could be regarded as a faithful representation of reality but a part of it only. In fact, we very soon found that a still greater part of reality was left in the shade. So far theoreticians had taken into account one and only one of the various errors that may take place in the course of sampling. This error is very important indeed inasmuch as it is the only one never to cancel out, even when the sampling is assumed to be carried out in an ideal way. But this fundamental error is far from being unique and is seldom the most important. In a series of attempts to analyse the total sampling error, we published (1956 to 1966) a number of articles dealing with one aspect of the problem or the other.

During the same period, G. Matheron and his team developed (1965-1969-1970-1973) and matured the theory of geostatistics covering the sampling of mineral deposits. When we became conscious of the ambivalence of particulate materials and of the necessity of representing these by two models :

- a <u>continuous model</u> taking into account the continuous nature of the space or time variability of its characteristics, and

- a <u>discrete model</u> taking into account the discrete nature of the population of fragments submitted to sampling, the problem was solved. In a first essay (1967) we presented the various errors tied to the continuous and discrete models. This was followed (1971) by a study of errors tied to the sampling operation and not taken into account by the models. These two documents, however, are to be considered as progress reports. They had to be completed and reorganized : we had gathered most pieces of the jig-saw puzzle but some of them were still missing and the general representation of the subject remained hazy. Our next step was a thorough analysis of the concept of heterogeneity (1972) followed by a study of the notions of sampling accuracy (1972) and sampling correctness (1972). The discrete selection model (1973) is a generalization of the equiprobable model formerly proposed (1953 and 1964).

A synthesis was then necessary in order to incorporate the results of our latest studies (1975). The present book is a matured, revised and reorganized English edition of this synthesis

March 21 st, 1975
May 3 rd, 1979

Dr. Pierre Gy
Consulting Engineer
Résidences de Luynes,
14, Avenue Jean-de-Noailles
06400 CANNES
FRANCE

FIRST PART

ANALYSIS OF THE PROBLEM

Sampling was the art and becomes the science of representing a batch of matter by a fraction of it on which quality estimations are carried out by proxy. The weight of the analysis or assay sample may be as small as one gram or a few decigrams whilst that of the batch may reach several hundred thousand tons. This gives an idea of the frequently undervalued importance of sampling. Sampling and analysis are complementary but whereas the random nature of assaying is usually well understood, that of sampling is too often neglected and too many people from Presidents of Mining Companies to Equipment Manufacturers and sampling Operators regard it as another mechanical operation such as for instance crushing or screening. This amounts to forgetting that sampling is a selection, the purpose of which is to replace the quantity to be estimated (a given characteristic of the lot) by an estimator (the corresponding characteristic of the sample), the difference between the former and the latter being what we shall call the "sampling error".

The choice of a crusher or of a screen can be made according to mere mechanical or cost considerations but that of a sampler must be founded above all on its aptitude for avoiding or reducing sampling errors : all other qualities are but of secondary importance.

Most objects hold in themselves the proof of their value and this can be estimated by measurements or analyses carried out on the object itself but it is not true of a sample. The economical importance of a sample is not related to its material value and its value as a sample, its "representativeness", leaves no material trace : one can always tell a false diamond from a genuine one but the only way to tell a representative sample from a specimen without value is to perform a critical inspection of the conditions under which it has been extracted from the lot it is supposed to represent.

By disregarding these pieces of evidence a number of equipment manufacturers still design, advertize and eventually sell sampling devices transgressing the most elementary rules and mining companies still buy, install and use these, without being aware that their interests are jeopardized.

In our studies and especially in this book we have endeavoured to :
 - analyse and explain the sampling error generating mechanisms,
 - disclose the parameters of the material to be sampled and the variables of the sampling method or device on which those errors are depending,

- show how to suppress these errors whenever possible and how to minimize them whenever it is not,

- express the mean and variance of the sampling errors that can be estimated,in terms of characteristics of the material to be sampled and of the sampling device,

- define the qualities a sampler can and must have in order to deliver representative samples,

- state the requirements in order for a sampling to be "correct", "unbiased", "accurate", "reproducible" or "represertative", each of these qualities receiving a clear scientific definition.

Our approach to the theory of sampling is that of a Cartesian analysis. We endeavoured to reduce a complex problem to a sum of simple ones. Sampling belonging t the calculus of probabilities and to mathematical statistics, it was therefore unavoidable to use the mathematical language and notations : the time when "honest sampling" was made of "good judgment" and "practical experience" is over. Our purpose is to speak of risks and confidence intervals, of variances and variograms. But we always made a point of explaining our logical approach and of translating our mathematical results into practical conclusions expressed in common language.

The sampling of minerals is usually acknowledged as economically important. The world production of solid mineral commodities is in excess of 60 billion dollars and represents more than two percents of the world gross yield. In the chain of mineral production and transformation, quality control is omnipresent : prospecting, geological or geochemical surveys, mining, processing or extractive metallurgy, as well as chemical, physico-chemical or mechanical analysis. The purpose of such controls may be commercial, technical or merely administrative but all rely on sampling and when this sampling is biased,it is liable or likely to result in losses of money, technical disturbances or false information.

Sampling errors have numerous causes that we shall disclose in due time but they nearly always have detrimental consequences : consider for instance a 150,000 ton shipment of iron ore worth, say, one million dollars. At the end of a complex sampling procedure,this huge tonnage is eventually reduced to a few assay portions of one gram each. Now, we can vouch that mechanical samplers introducing a 20 % relative bias are still in use here and there. We were recently confronted, in a commercial operation involving tin concentrates, with a 9 % bias. A 5 % bias is far from exceptional. One can easily imagine the effect of such biases on the estimation of the commercial value of a one-million dollar shipment. Since they are not always conspicuous, biases very often escape detection except to the eye of an experienced specialist. A good size book could be written on the "historical" sampling errors and its readers could meditate upon the huge amounts of money that have been lost or squandered as the result of sampling errors. All those who have had some experience in that line know well enough that sampling is undoubtedly the trickiest and most treacherous of all operations taking place in the mining and metallurgical industries. Unfortunately it has kept for a great length of time and still today remains in many countries, in the shade of more spectacular techniques. Its laws are still widely ignored and with a handful of exceptions its teaching is primitive if not non-existent.

The first part of this book consists of a general analysis of the problem. It is made of five chapters :
    Chapter 1 : Definition of basic terms and notations.
    Chapter 2 : Logical approach.
    Chapter 3 : Part of sampling in quality control.
    Chapter 4 : Sampling processes.
    Chapter 5 : Models of the sampling processes.

CHAPTER 1

DEFINITION OF BASIC TERMS AND NOTATIONS

> *If you want to prevent war, define the meaning of your words before speaking.*
> Confucius

1.1. DEFINITION OF BASIC TERMS :

Frequent misunderstandings take place in scientific discussions, due very often to the fact that interlocutors give different meanings to the words they use. When developing a theory such as this, the author is practically compelled to create new expressions, even new words and to give common terms a very precise scientific meaning. It is therefore his duty to give a clear definition of the vocabulary he is going to use. In this section we shall only define the basic language of sampling. A number of more specific definitions will be given in due course.

1.1.1. Material to be sampled - Sampling :

Lot L : A batch of matter, the composition of which is to be estimated. This may refer to a 200,000 ton shipment as well as to the 50 grams contained in a small jar.

Particle or fragment F (we shall use both words with the same meaning, irrespective of the size) : A compact unit assumed, during a selective operation, to be indivisible.

Group of particles G : A set of usually neighbouring particles.

Domain (D) : a) A closed volume containing a lot, a sub-lot, a fraction or a group of particles.
b) Generally speaking, a closed portion of a three-, two- or one-dimensional geometrical space.

Domain (T) : A closed fraction of a one-dimensional temporal space.

Increment I : A group of particles extracted from the lot in a single operation of the sampling device.

Sample S : A part of the lot, often obtained by reunion of several increments or fractions of the lot, and meant for representing it in further operations. A sample is not just any part of the lot : its extraction must respect certain rules that the theory of sampling intends to establish.

Specimen : A part of the lot obtained without respecting these rules. A specimen should never be used for representing the lot and should be labelled as such.

Sampling reject R : Complement of the sample. By definition : R = L - S .

Primary sample S_1 , secondary sample S_2 , n th stage sample S_n : Sampling is often carried out by progressive stages : a primary sample is extracted from the lot, a secondary sample from the primary sample, etc .. This genealogy is illustrated by the following family tree :

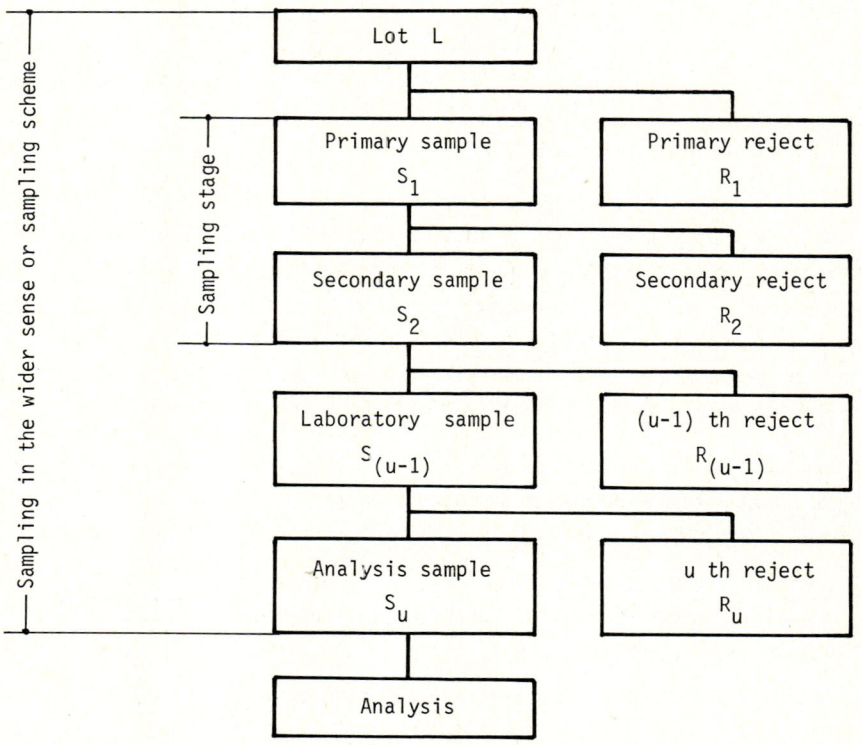

Laboratory sample : In a progressive sampling scheme, sample of convenient weight delivered to the laboratory where further preparation and analysis will take place. The laboratory sample is often the last sample but one $S_{(u-1)}$.

Analysis sample : Ultimate sample of the series, submitted as a whole to the analytical procedure. It is often directly extracted from the laboratory sample. In chemical analysis, it is the assay portion.

Twin samples : Two or more samples of equal bulk obtained in the same time and in similar conditions, often by splitting.

Sampling (in the proper sense) or sampling stage : A selective process implemented on a given batch (lot or sample) in order to reduce its bulk without altering (too much) its other characteristics.

Preparation or preparation stage : A sequence of non-selective operations such as transfer, crushing, grinding, pulverizing, drying, mixing, etc .. carried out on a batch (lot or sample) in order to bring it at the place and under the form convenient for the next processing stage (sampling and ultimately analysis).

Sampling (in the wider sense) or sampling scheme : A sequence of non-selective and selective operations (preparation and sampling stages) carried out on a lot and ending up with the extraction of one or several analysis samples.

Sampling and sample reduction : English speaking authors usually distinguish "sampling" (our primary sampling stage) from "sample reduction" (sequence of further preparation and sampling stages). We do not encourage such a distinction which seems to infer that there are two categories of problems belonging either to sampling or to sample reduction. There is only one theory of sampling and its results are applicable irrespective of the bulk of the object to be sampled.

1.1.2. Analysis - Chemical and Physical Components :

The purpose of sampling is to prepare one or several analysis samples which will be analysed. We shall take four types of analysis into consideration :

- Chemical analysis or assaying : estimation of the percentage (in weight) of one or several chemical components (metal, radical, mineral, etc..),
- Size analysis : estimation of the percentage (in weight) of one or several size fractions of a particulate solid,
- Moisture analysis : estimation of the percentage (in weight) of moisture in a wet solid,
- Determination of a pulp concentration : estimation of the percentage (in weight) of dry solids in a pulp.

Physical components : The domain occupied by the lot can be broken up into a certain number of sub-domains respectively occupied by a single "physical component" of the material. The sampling theory of particulate materials involves the proportions or percentages of physical components whereas assaying estimates the percentages of chemical components. Difficulties may arise for instance when the same metal (chemical component the proportion of which is estimated by assaying) is present in several minerals (physical components the proportion of which is involved in the theory of sampling). For the moment we shall distinguish :
- active physical components
- passive physical components.

Active and passive physical components : A physical component of the material is said to be "active" when it enters into the definition of the proportion to be estimated. It is said to be "passive" when it does not enter into this definition. According to the characteristic to be estimated, the active physical components are :

- a set of minerals (or corresponding components when dealing with non-mineral materials), when estimating the chemical or mineralogical composition of the lot,

- a set of size fractions when estimating the size analysis of the lot,

- the solid phase and the adsorbed moisture when estimating the moisture content of the lot,

- the solid phase and the liquid phase when estimating the solid concentration of a pulp, etc..

whilst the passive physical components are :

- the interstitial fluid (air, water, any gas or liquid) when estimating the chemical or mineralogical composition, or else the size distribution of a solid,

- the interstitial air (or gas) when estimating the moisture content of a wet solid,

One should observe that, when estimating the solid percentage of a pulp, there is no passive component since the interstitial water (or solution) plays an active part in the definition of this percentage.

<u>Critical component</u> : Chemical or physical component, the proportion of which is to be estimated. There may be several critical components : there are as many sampling problems as there are critical components. These problems must be solved independently from one another. If a minimum sample weight is to be estimated, the largest figure will be retained.

<u>Content</u> : We shall use the word "<u>content</u>" with the general meaning of "<u>proportion of a given active component</u>":

$$\text{Content} = \frac{\text{Weight of a given active component}}{\text{Weight of all active components}}$$

(synonyms : grade or percentage).

<u>Critical content</u> : Proportion of critical component. Usually denoted by "a".

<u>Analysis</u> : We shall use the word "<u>analysis</u>" with the general meaning of "<u>estimation of a critical content</u>". The analytical procedure directly provides :

- the size distribution,
- the moisture content,
- the solid concentration of a pulp,

but it seldom provides the mineralogical composition involved in the sampling theory when sampling prior to a chemical assay. In the practical applications of the theory, we shall therefore have to estimate the mineralogical composition of the sampled material on the basis of its chemical analysis and of mineralogical examinations. This point will be dealt with in due course.

1.2. STATISTICAL DEFINITIONS AND NOTATIONS :

If x denotes a random variable, then,

$m(x)$ = true unknown <u>mean</u> of the distribution of x

$\sigma^2(x)$ = true unknown <u>variance</u> of the distribution of x. Its square root $\sigma(x)$ is the <u>standard-deviation</u> of x

$r^2(x)$ = true unknown <u>mean square</u> of x. By definition :

$$r^2(x) = m(x^2) = m^2(x) + \sigma^2(x)$$

Its square root $r(x)$ is the <u>quadratic mean</u> of x.

x' = experimental estimate of x

$\bar{x}$ = experimental estimate of $m(x)$

$s(x)$ = experimental estimate of $\sigma(x)$

$\rho(x,x')$ = <u>coefficient of correlation</u> between the distributions of x and x'.

If y denotes the estimator of a physical characteristic with a non-zero mean,

$u(y)$ = <u>coefficient of variation</u> of y. According to Pearson's definition :

$$u(y) = \frac{\sigma(y)}{m(y)}$$

It is often referred to as "<u>relative standard deviation</u>" of y and its square $u^2(y)$ as "<u>relative variance</u>" of y.

1.3. ESTIMATORS - ESTIMATES - ERRORS - BIASES :

We shall use the following notations :

a_L = unknown critical content of the lot L

a_S = unknown critical content of the sample S. The content a_S is an <u>estimator</u> of a_L.

SE = relative selection or sampling error incurred when replacing a_L by its estimator a_S. By definition :

$$SE = \frac{(a_S - a_L)}{a_L}$$

a_{Su} = unknown critical content of the ultimate sample S_u analysed in totality,

a'_{Su} = result of S_u analysis. Then, a'_{Su} is an "<u>estimate</u>" of a_{Su} and the eventual estimate of a_L.

AE = relative analysis error incurred when replacing a_{Su} by its estimate a'_{Su} :

$$AE = \frac{(a'_{Su} - a_{Su})}{a_L}$$

(all relative errors are expressed with a_L on the denominator).

$B(a_S)$ = relative selection or sampling "<u>bias</u>" associated with the estimator a_S :

$$B(a_S) = m(SE) = \frac{m(a_S) - a_L}{a_L}$$

$b(a_S)$ or $b(SE)$ = corresponding <u>coefficient of bias</u>. By definition :

$$b(a_S) = \frac{|B(a_S)|}{u(a_S)} = b(SE) = \frac{|m(SE)|}{\sigma(SE)}$$

1.4. DOMAINS AND THEIR EXTENT :

We shall use the following notations :

(D) = a certain domain of a p-dimensional space (p undefined) and more specifically a certain domain of a three-dimensional space.

D = extent of (D).

(T) = a certain domain of a one-dimensional temporal space,

T = extent of (T).

(D') : (D) denoting a domain of the three-dimensional Cartesian space, (D') shall denote the projection of (D) in a two-dimensional space. (D') is a surface.

(D") : Projection of (D) in a one-dimensional space. (D") is a segment.

D' = extent of (D').

D" = extent of (D").

D(X) = a certain domain of extent D centred on point X

1.5. PROPERTIES OF A SELECTIVE PROCESS :

A simple or complex selective process can be appraised from two different standpoints :

- on the basis of the <u>selecting conditions</u> ("a priori" qualities),
- on the basis of the <u>selecting results</u> ("a posteriori" qualities).

1.5.1. A priori qualities :

A selective process or the resulting sample is said to be :

<u>Probabilistic</u> : when the selection is founded on the notion of selecting probability,

<u>Correct</u> : when, being probabilistic, the selecting chances are uniformly distributed,

<u>Incorrect</u> : when, remaining probabilistic, the latter condition is not fulfilled

<u>Non-probabilistic</u> : when the selection is not founded on the notion of selecting probability. It can be for instance deterministic or purposive.

<u>Deterministic</u> : when the selection is founded on the implementation of a rigid system, without intervention of a random element. To illustrate this point, the well known systematic sampling process can be either probabilistic or deterministic according as the systematic grid is positioned at random or not. Grab sampling is usually a deterministic selection process.

<u>Purposive</u> : when the selection is founded on the choice by the sampling operator of the elements of the lot to be retained as a sample. The "hammer and shovel method" is the prototype of a purposive selection process.

1.5.2. A posteriori qualities :

A selective process is error-generating. Its a posteriori appraisal must be based on the properties of the selection error already defined in the preceding section. A selective process is said to be :

Unbiased : when the mean of the selection error is zero : $m(SE) = 0$

Biased : when the mean of the selection error is non-zero : $B = m(SE) \neq 0$

Accurate : when the absolute value of the bias $|m(SE)|$ is not larger than a certain standard of accuracy $m_o(SE)$: $|m(SE)| \leq m_o(SE)$. The terminology does not seem to be definitely stabilized : let us quote J. Mandel (The analysis of experimental data) : "Regarding the concept of accuracy there exist two schools of thought. Many authors define accuracy as the more or less complete absence of bias. The second school of thought defines accuracy in terms of the total error". Mandel decides that "he will for the sake of consistency adopt the first view only". So will we do. But the reader should know that, though lacking consistency, the second school of thought is supported by the British Standards Institution from which we quote : (BS 1017 Sampling of coal and coke) "Accuracy is the measure of the ability of a method to provide accurate results, i.e. results which are precise and free from bias".

This is but an example of Confucius' wisdom.

Reproducible or precise : when the variance of the selection error is not larger than a certain standard of reproducibility $\sigma_o^2(SE)$:

$$\sigma^2(SE) \leq \sigma_o^2(SE)$$

Representative : when the mean square of the selection error is not larger than a certain standard of representativeness $r_o^2(SE)$:

$$r^2(SE) = m^2(SE) + \sigma^2(SE) \leq r_o^2(SE) \quad \text{or} \quad r^2(SE) = \{b^2(SE) + 1\}\sigma^2(SE) \leq r_o^2(SE)$$

From a practical standpoint, a selection is said to be representative when it is in the same time accurate and reproducible with :

$$r_o^2(SE) = m_o^2(SE) + \sigma_o^2(SE)$$

Exact : when the selection error is identically zero:

$SE \equiv 0$ which entails $m(SE) \equiv 0$ and $\sigma^2(SE) \equiv 0$

An exact selection is therefore characterized by a zero mean (it is unbiased) and by a zero variance (it is perfectly reproducible). Exactness appears as a consequence of certain hypotheses concerning the material to be sampled. It is never achieved in actual practice but the word had to be defined if only for avoiding the misuse of it.

Equitable : in commercial sampling, when the merchant value of the lot L as computed on the basis of the sample content a_S is a random variable, the mean of which is equal to the merchant value computed on the basis of the true content a_L.

1.5.3. Search for an objective :

The user of sampling equipment thinks (or rather should think) in terms of results : he wants accurate and/or reproducible results. Now the manufacturer thinks (or rather should think) in terms of sampling conditions : he designs and builds correct or incorrect devices. One of the purposes of a theory of sampling is therefore to conciliate the user's requirements with the manufacturer's possibilities. In other words, this amounts to disclosing relationships between the conditions and the results of sampling, between the a priori and the a posteriori qualities.

If exactness was not a limit case occurring only under theoretical hypotheses never met with in actual practice, it would probably be agreed upon as the best choice by all parties, with the possible exception of those who, for good reasons of their own, find a definite advantage in the fuzziness still often present in sampling : sampling is a very useful scapegoat when something goes wrong. But, fortunately or not, exactness is ruled out as a practical objective.

In commercial sampling, when a long term contract provides that large tonnages of the same commodity shall be shipped at more or less regular intervals by a same seller to a same buyer, equity is an obvious objective. The delicate notion of equity will be analysed in chapter 30 but we shall anticipate the conclusions of this chapter by stating that accuracy is the only warrant for equity. Our first concern should be to suppress all possible sources of bias. As the relative importance of random errors tends to become negligible in the long run, reproducibility stands only second in the priority order.

Now, when sampling for a technical control, in an automated processing plant for instance, the fluctuations of a given variable (e.g. the critical content of the feed) are more important than its true value : a constant bias would be perfectly acceptable. Then, reproducibility is more important than accuracy but this is an academic hypothesis : there is no such thing as a reproducible inaccurate sampling. We shall see in due time that inaccuracy always goes with lack of reproducibility, the best warrant of reproducibility as well as accuracy being correctness.

When sampling an isolated lot, either for commercial or for technical purposes, we need the closest possible estimate of the content of the lot. We must therefore look for representativeness or in other words for accuracy and reproducibility. Here again, correctness is the best warrant of success.

Anticipating again the conclusions of the next chapters, it is certainly not too early to state that whatever the sampling problem to be solved, the first and foremost quality to be achieved is correctness : all elements making up the lot to be sampled must have the same selecting probability.

1.6. SPECIFIC NOTATIONS :

We strove to adopt a consistent notation system throughout this book : as a general rule, a given notation will keep the same meaning save a few duly pointed out exceptions. We shall distinguish :

- the body of a symbol. It is written on the line. This may be a single capital latin or greek letter, a double capital latin letter or a single lower case latin or greek letter.
- the subscript(s) of the symbol. The body is very often completed by one or several subscripts : capital or lower case latin or greek letters, figures, etc..
- the superscript. It always denotes the corresponding power.

We shall give the meaning of the symbol bodies. The subscripts will always be defined whenever necessary.

Single capital latin letters :

A : mass (or weight) of critical component in a given object specified by the subscript.
B : bias
C : Chapter 18 : cutter and cutter edges
   Chapter 22 : sampling constant
D : domain and its extent. Chapter 31 : systematic difference.
E : chapter 5 : interstitial fluid,
F : fragment or particle
G : group of neighbouring fragments
H : Chapter 5 : thickness of a two-dimensional object,
   Chapter 18 : horizontal plane,
   Chapters 31 and 32 : hypothesis.
I : increment
J : reference increment
K : integer
L : lot or batch
M : mass (or weight) of active components in a given object specified by the subscript.
N : number of elements in a set.
O : origin of a system of co-ordinates.
P : probability
Q : number of increments in a sample
R : sampling reject
S : sample
T : time interval
U : unit or element of a discrete set.

V : velocity of a moving object (cutter, fragment, belt, etc..) specified by the subscript,
W : width of a cutter (chapter 18). Shifting mean of a variogram (chapter 6).
X : point in a p-dimensional space,
Y : Chapter 18 : point. Chapter 22 : auxiliary parameter.
Z : chapter 5 : section of a one-dimensional lot,
   chapter 18 : point. Chapter 22 : auxiliary parameter.

Double capital latin letters :

The second letter is either an E for error or an H for heterogeneity.
AE : analysis error
CE : continuous selection error
DE : increment delimitation error
EE : increment extraction error
FE : fundamental error
GE : grouping and segregation error
ME : materialization error
OE : overall estimation error
PE : preparation errors
QE : quality fluctuation error
SE : sampling or selection error
TE : total sampling error in the wider sense
VE : settlement error (commercial sampling)
WE : weighting error

CH : constitution heterogeneity
DH : distribution heterogeneity

Single lower case latin letters :

a : critical content
b : coefficient of bias
c : mineralogical composition factor
d : diameter of a fragment. In dx, dy, dz or dt, d is the differential symbol.
e : error
f : shape factor of a particle,
   in f(t) denotes a non-specified function of time.
g : size distribution factor of the material to be sampled
h : heterogeneity of a fragment or group of fragments. Chapter 14 : time interval
k : integer
l : liberation factor of the material to be sampled. To prevent any confusion between l and 1 we shall use the "script" type ℓ
m : mean of a random variable

n : chapter 18 : ratio of the actual cutter width to the minimum cutter width
p : dimension of a given space (p = 1, 2, 3). Chapter 30 : price of a commodity
r : quadratic mean of a random variable
s : experimental estimate of a standard deviation
t : time variable
u : coefficient of variation of a random variable
v : variogram or variographic parameter. Chapter 20 : volume of a particle.
w : discrete variogram
x : a random variable
x, y, z : co-ordinates of point X

Capital greek letters :

Γ : curves
Λ : family of planes or surfaces
Π : probability density
Σ : introduces a sum of terms

Lower case greek letters :

α : in $\alpha(t)$, rate of flow of critical component at time t
γ : grouping factor
δ : increase of a function
ε : infinitely small auxiliary variable
ζ : infinitely small auxiliary variable
η : infinitely small auxiliary variable
θ : time interval
λ : density of a fragment
μ : in $\mu(t)$, rate of flow of active components at time t
ν : number of degrees of freedom
ξ : segregation factor
π : circumference of a circle of unit diameter
ρ : coefficient of correlation
σ : true unknown value of a standard deviation
τ : sampling ratio
ϕ : auxiliary function
χ : auxiliary function
ψ : auxiliary variable
ω : set of samples

In any case, notations will always be defined at the beginning or in the course of all chapters.

CHAPTER 2

LOGICAL APPROACH

2.1. FIRST PART - ANALYSIS OF THE PROBLEM - CHAPTERS 1 TO 5

A theory of sampling is first and foremost a theory of the sampling errors and a study of their properties. The first part of this book begins with the definition of basic terms and notations (chapter 1) and with a synopsis of our logical approach (chapter 2). Sampling is a necessary preliminary step to practically all analytical procedures (chapter 3). Both sampling and analysis are error-generating with the consequence that the "Overall estimation error" OE is the sum of the "Total sampling error" TE and of the "Analytical error" AE.

$$OE = TE + AE$$

The subject of this work is a study of the total sampling error TE. Owing to the fact that it is usually impossible to extract the analysis sample directly from the lot to be evaluated, sampling, in the wider sense used in the preceding paragraphs, sampling is a multistage operation involving "Preparation stages" alternating with "selection stages" also referred to as "sampling stages" in the proper sense. All these operations are, potentially at least, error-generating. We shall therefore encounter "Preparation errors" PE and "Sampling errors" SE.

$$TE = TE_1 + TE_2 + \ldots + TE_n + \ldots$$

$$TE_n = PE_n + SE_n \quad \text{and} \quad TE = \sum_n (PE_n + SE_n) \quad \text{and} \quad OE = \sum_n (PE_n + SE_n) + AE$$

Chapters 4 to 26 are dedicated to a study of the components of the sampling error SE and Chapter 27 to a review of the preparation errors PE.

Probabilistic sampling schemes belong to one of the two following categories (chapter 4) :
- Increment sampling process (chapters 6 to 23),
- Splitting process (chapter 24 to 26).

Non-probabilistic sampling processes are not taken into consideration in this book. The sampling theory of particulate materials requires the development of two complementary models (chapter 5) :
- Continuous selection model,
- Discrete selection model.

The second part deals with the continuous model of the increment sampling process, the third part is dedicated to the delicate problem of matching the continuous model with the discrete reality, the fourth part to the discrete selection model.

2.2. SECOND PART - CONTINUOUS MODEL OF THE INCREMENT SAMPLING PROCESS - CHAPTERS 6

The results of the continuous model are applicable to any sampling problem :
- irrespective of the physical state of the material to be sampled : compact or particulate solids, liquids, pulps, gases, multi-phase systems, etc..
- irrespective of its origin : mineral, vegetable, animal, synthetic or mixed materials such as town refuses, etc..
- irrespective of the "critical component" taken into consideration : mineralogical constituents, size fractions, moisture, solid phase of a pulp, etc..
- irrespective of the mathematical model retained to represent the lot : three-, two-, one-dimensional geometrical spaces, one-dimensional time space (flowing stream

We shall however restrict our demonstrations to :
- particulate solids, whether the interstitial fluid is air, water or any other gas or liquid,
- materials of mineral origin, and more specifically ores and concentrates,
- an undefined critical component which can be, as the case may be, a given mineral (whether valuable or penalized impurity), a given size fraction (whether coarse or fine), moisture, etc..
- flowing streams of solids or pulps represented by a one-dimensional temporal mod or movable lots of solids or pulps.

The continuous model identifies the lot L to be sampled with the continuous set of all instants belonging to the time interval $(0, T_L)$ with 0 origin of the flow and T_L end of the flow. As far as its sampling is concerned, the lot L is complete defined by two functions :

- a <u>qualitative function</u> : the critical content $a(t)$ of the material flowing past the sampling point at instant t,
- a <u>quantitative function</u> : the rate of flow $\mu(t)$ at instant t.

The results of this one-dimensional model can readily be generalized to two- or three-dimensional objects and easily adapted to non-mineral materials. In fact, it has been used for materials of vegetable origin such as cereals or sugar beets. The practical problem set by the sampling of objects other than flowing streams is dea with in chapter 29.

All sampling errors result from one form or another of heterogeneity. The concep of heterogeneity of a continuous set is analysed and quantified by means of the variogram function (chapter 6). Sampling in the proper sense is a selection process and we present three reference selection schemes (chapter 7) :
- Systematic selection with random positioning,
- Random stratified selection,
- Random selection.

The continuous selection model is developed in chapter 8 with the purpose of expressing the moments (mean and variance) of the <u>continuous selection error CE</u> in terms of the characteristics of the stream to be sampled and of the selection process retained. When analysing the properties of the material making up the stream, we find (chapter 9) that the continuous selection error CE can be regarded as the sum of two complementary errors taking into account the properties of the two functions $a(t)$ and $\mu(t)$:
- <u>quality fluctuation error</u> QE
- <u>quantity fluctuation or weighting error</u> WE (studied in chapter 13).

Now, the quality function $a(t)$ can be broken up into a sum of a constant a_0 and three independent components $a_1(t)$, $a_2(t)$ and $a_3(t)$ each generating an independent error (chapter 9) :
- <u>short-range quality fluctuation error</u> QE_1 (studied in chapter 10)
- <u>long-range non-periodic quality fluctuation error</u> QE_2 (studied in chapter 11)
- <u>periodic quality fluctuation error</u> QE_3 (studied in chapter 12).

$$CE = QE + WE = QE_1 + QE_2 + QE_3 + WE$$

The last two chapters of this second part are dedicated to the practical implementation of the theoretical results. The variographic experiment is the key to the quality and quantity fluctuation analysis and to the experimental estimation of the moments of the error CE and of its components QE_1, QE_2, QE_3 and WE. Chapter 14 deals with the organization of a variographic experiment and with the delicate problem of interpretation of the variograms. Chapter 15 presents examples of error estimation.

2.3. THIRD PART - FROM THE CONTINUOUS MODEL TO THE DISCRETE REALITY - CHAPTERS 16-18

The continuous model is based on the selection of points on the time axis whilst reality is made of fragments and groups of fragments. A very thorough analysis is carried out in chapter 16. It shows that a <u>materialization error</u> ME adds to the continuous selection error CE. The materialization error is the sum of two and only two errors not accounted for by the continuous model :
- <u>increment delimitation error</u> DE (studied in chapter 17),
- <u>increment extraction error</u> EE (studied in chapter 18).

$$SE = CE + ME = CE + DE + EE$$

These errors are studied in great detail as they are liable to be the largest of all sampling errors.

2.4. FOURTH PART - DISCRETE MODEL OF THE INCREMENT SAMPLING PROCESS - CHAPTERS 19-23

The discrete model identifies the lot L to be sampled with a discrete set of units U_m that can be either individual particles or groups of particles. As far as its sampling is concerned, the lot L is completely defined by the finite set of the parameters a_m and M_m characterizing unit U_m :

- <u>critical content</u> a_m of unit U_m
- <u>weight</u> M_m of unit U_m

The concept of heterogeneity of a discrete set of units is analysed and its two forms are quantified by means of two parameters (chapter 19) :

- <u>Constitution heterogeneity</u> CH_L of the lot L, taking into account the intrinsic properties of the population of individual fragments or particles,
- <u>Distribution heterogeneity</u> DH_L, taking into account the properties of the spatial distribution of the particles throughout the lot L.

The properties of these two parameters are carefully reviewed with the purpose of linking up the model of heterogeneity to the discrete selection model and to the moments of the discrete selection error. The discrete selection model is then developed (chapter 20). The moments of the <u>discrete selection error SE</u> are expressed in terms of the characteristics a_m and M_m and of the selecting probability P_m of unit U_m, as well as in terms of constitution and distribution heterogeneities.

Chapter 21 shows how the discrete model can be linked up to the continuous mode at the level of the short-range quality fluctuation error QE_1 which, in the continuous model, takes globally into account all that is relevant to the discrete structure of the material. The error QE_1 is shown in this chapter to be the sum of two components :

- <u>fundamental error</u> FE (associated to the constitution of the material),
- <u>grouping and segregation error</u> GE (associated to the distribution of the mate

$$QE_1 = FE + GE$$

The last two chapters of the fourth part show how to implement the results of the discrete model in the two following cases :

- <u>correct selection</u> - computation of the fundamental error (chapter 22)
- <u>incorrect selection</u> - computation of the delimitation and extraction errors and more specifically biases (chapter 23).

The purpose of chapter 22 is especially to give the reader practical formulas making it possible to compute the fundamental variance in a matter of minutes and even of seconds when using the sampling slide rule devised by the author. That of chapter 23 is to show on examples the dangers of incorrect sampling devices and especially those that do not respect the requirements stated in chapters 17 and 18 in order for the increment delimitation and extraction to be correct and for the sampling devices to deliver unbiased samples.

A number of examples illustrate these two chapters.

2.5. FIFTH PART - SPLITTING PROCESSES - CHAPTERS 24 to 26

Splitting processes are restricted to objects small or valuable enough to stand the cost of handling for the sole purpose of their sampling. Chapter 24 is a review of the most popular splitting methods or devices. Chapter 25 shows that the theoretical part of splitting can be covered by means of the models already studied and precises that the splitting error usually amounts to the fundamental error FE and the grouping and segregation error GE, the other errors being present but negligible. As an example of practical implementation of splitting processes and of computation of splitting errors, chapter 26 gives a few pieces of advice on how to reduce drill core samples without incurring unacceptable errors.

2.6. SIXTH PART - PREPARATION - CHAPTER 27

The word "preparation" covers all non-selective operations undergone by the lot and the successive samples (transfer, comminution, mixing, drying, etc..). Chapter 27 reviews the six categories of "preparation errors" PE liable to take place :

- errors by contamination PE_C
- errors by loss PE_L
- errors by alteration of the chemical composition PE_A
- errors by alteration of the physical composition PE_P
- errors by unintentional mistakes or accidents PE_M
- errors by frauding or sabotage PE_F

The properties and examples of such errors are detailed in chapter 27.

2.7. SEVENTH PART - RESOLUTION OF SAMPLING PROBLEMS - CHAPTERS 28 AND 29

The seventh part begins (chapter 28) with a practical recapitulation of the various sampling errors and of their properties : how to suppress those that can be suppressed and how to minimize those that cannot. The notion of "solvable and unsolvable sampling problems" is the subject of chapter 29 where the concepts of acceptable sampling error and acceptable sampling cost are analysed. A sampling problem is said to be "solvable" if and only if the overall sampling error can be estimated.

2.8. EIGHTH PART - PROBLEMS ASSOCIATED WITH COMMERCIAL SAMPLING - CHAPTERS 30 to 32

Sampling errors may be particularly detrimental in commercial sampling. Chapter 30 introduces the concept of "equity", analyses its content and concludes with a few pieces of advice to sellers and buyers of mineral commodities. Chapter 31 is concerned with the testing of agreement between two series of independent estimates of a same characteristic, for instance the weight of metal contained in a given shipment or series of shipments, as estimated by seller and buyer. This test is devised so as to always lead to a practically certain conclusion (probability 99 % or risk of error 1 %) that either the estimates agree within the limits of a certain confidence interval (systematic difference smaller than a <u>certain standard of agreement</u> D_A) or disagree (systematic difference larger than D_A). The same test can be implemented outside the commercial field as well and for instance to check the agreement between the material fed to a processing plant and the material recovered in concentrates or lost in tailings. Chapter 32 deals with the controversial problem of testing for a sampling bias. From a theoretical standpoint, it is a particular case of the problem dealt with in the preceding chapter. The emphasis is put here on the practical side of the problem and on the dangers of certain standards which tend to maintain confusion between the absence of certainty of bias (state of uncertainty : no safe conclusion can be drawn) and the certainty of absence of bias (state of practical certainty : a safe conclusion can be drawn).

Though the results of chapters 31 and 32 are particularly useful in commercial sampling, they are also relevant to the problems of technical or accounting sampling (e.g. metallurgical balances in processing plants or smelters, detection of losses of valuables materials, etc..).

2.9. NINTH PART - AUTOMATIC SAMPLING PLANTS - CHAPTERS 33 AND 34

The last two chapters of this book have been written to the attention of designers of automatic sampling plants. Chapter 33 provides these with a few pieces of advice on how to solve a problem even though it may seem unsolvable or point to paradoxical solutions. Chapter 34 presents typical sampling flow-sheets.

CHAPTER 3

PART OF SAMPLING IN QUALITY CONTROL

3.1. INTRODUCTION

For technical, accounting or commercial purposes the average quality of batches of particulate materials has very often to be estimated. This quality is nearly always expressed as a series of proportions :

- metal or mineral content,
- proportion of one or several size fractions,
- moisture content of a wet solid
- solid concentration of a pulp, etc ..

These proportions, relevant to the problem, will be generally referred to as "<u>critical contents</u>". They are estimated by analysis.

3.2. POSSIBILITY OF DIRECT ESTIMATION OF A CRITICAL CONTENT

The estimation of a critical content can very seldom be directly carried out on the whole of the batch. A few exceptions can however be mentioned :

- radiometric estimation of a uranium content,
- size analysis of a batch small enough to be screened in totality,
- moisture estimation on a batch small enough to be dried in totality or by means of physical methods.

These direct analyses have in common to be cheap and non-destructive, hence their possible application to relatively large weights. Chemical assaying, for instance, costly and destructive as it is, is restricted to assay portions weighing one gram or a few decigrams (a little more for precious metals.

3.3. POSSIBILITY OF DIRECT EXTRACTION OF THE ANALYSIS SAMPLE

When a direct estimation of the critical content is impossible or uneconomical, the analytical procedure usually fixes at least the order of magnitude of the amount of material to be analysed : the analysis sample or assay portion. But here again, the direct extraction of such a sample is seldom possible. We shall however note the following exceptions :

- the lot is at the particle size required by chemical assaying and its weight does not exceed a few hundred times that of the assay portion (for instance the lot is a "laboratory sample").

- the batch is a primary sample made of increments extracted from the lot itself and is processed for size or moisture analysis in an automatic plant.

From these examples we can infer that a direct or one-stage sampling is possible when simultaneously the sampled material is in the physical state required by the analytical procedure and the weight of the lot does not exceed a few hundred times that of the analysis sample. When these conditions are not fulfilled the sampling cannot be directly carried out. Several sampling and preparation stages are required

3.4. PREPARATION STAGES - PREPARATION ERRORS

The lot and its successive samples are seldom in the convenient physical state required by the analytical procedure and where the next sampling stage is to be performed. Lot and samples must therefore be transferred, crushed, pulverized, dried, stored, mixed or reclaimed, etc ... All these non-selective operations are referred to as "preparation operations". A complete sampling scheme in the wider sense of the word is therefore a sequence of preparation stages alternating with selection or sampling (in the proper sense) stages.

During preparation, the lot and its descendants can be adulterated either by contamination, loss, alteration, etc .. Preparation stages are therefore potentially at least error-generating. Preparation errors will be reviewed in chapter 27 (sixth part).

3.5. SUCCESSIVE ESTIMATORS AND EVENTUAL ESTIMATE OF THE CRITICAL CONTENT OF THE LOT

We shall respectively call :

L and L', S_1 and S'_1, ... S_u and S'_u : the unprepared and prepared lot and successive samples.

a_L and $a_{L'}$, a_{S1} and $a_{S'1}$, ... a_{Su} and $a_{S'u}$: their respective critical contents.

In the course of the sampling scheme the critical content a_L which is to be estimated is successively substituted by the estimators $a_{L'}$, a_{S1}, $a_{S'1}$ ... $a_{S'u}$. The eventual estimator $a_{S'u}$ is the true critical content of the analysis sample actually submitted in totality to the analytical procedure. The result of this analysis is an estimate $a'_{S'u}$ of $a_{S'u}$ and will be retained as eventual estimate of the critical content a_L. Each step of this multi-stage operation is error-generating

3.6. BREAKING UP OF THE OVERALL ESTIMATION ERROR

We shall define the following relative errors :

PE_1, PE_2, ... : the <u>successive preparation errors</u> :

$$PE_1 = \frac{a_{L'} - a_L}{a_L} \qquad PE_2 = \frac{a_{S'1} - a_{S1}}{a_L} \qquad \text{etc ...}$$

SE_1, SE_2, ... : the <u>successive sampling errors</u> :

$$SE_1 = \frac{a_{S1} - a_{L'}}{a_L} \qquad SE_2 = \frac{a_{S2} - a_{S'1}}{a_L} \qquad \text{etc} \ldots$$

TE_1, TE_2, ... : the <u>successive total preparation and sampling errors</u> :

$$TE_1 = PE_1 + SE_1 \qquad TE_2 = PE_2 + SE_2 \qquad \text{etc} \ldots$$

AE : the <u>analysis error</u> defined as follows :

$$AE = \frac{a'_{S'u} - a_{S'u}}{a_L}$$

OE : the <u>overall estimation error</u> :

$$OE = \frac{a'_{S'u} - a_L}{a_L} = PE_1 + SE_1 + PE_2 + SE_2 + \ldots + PE_u + SE_u + AE$$

$$OE = \sum_n (PE_n + SE_n) + AE = \sum_n TE_n + AE \qquad \text{with} \quad n = 1, 2, \ldots u.$$

This book is mainly concerned with the study of all kinds of sampling errors SE. Chapter 27, as already mentioned, deals with the preparation errors PE. Analytical errors AE would by themselves justify a multi-volume textbook still to be written. We shall just mention that in chemical assaying the analyst usually extracts himself the assay portion S_u from the laboratory sample $S_{(u-1)}$. In the rare documents dealing with analytical accuracy and reproducibility (or repeatability) the last sampling stage is always considered as part of the analytical procedure with the consequence that the last stage sampling error is included in the analysis error AE of which it may be the largest part.

CHAPTER 4

SAMPLING PROCESSES

4.1. PROBABILISTIC AND NON-PROBABILISTIC SELECTING PROCESSES

Sampling is a selecting process. We shall distinguish two categories of selecting processes :

- <u>Probabilistic selecting processes</u> according to which all elements of the lot are submitted to the selection with a given probability density (continuous model) or a given probability (discrete model) of being selected.

- <u>Non-probabilistic selecting processes</u> according to which the selection is not founded on the notion of selecting probability. It can be for instance either <u>deterministic</u> (e.g. grab sampling) or <u>purposive</u> (e.g. hammer and shovel method).

4.2. COMMENTS ON NON-PROBABILISTIC PROCESSES

There does not seem to be any theoretical approach to non-probabilistic selecting processes. The errors incurred when sampling by the hammer and shovel method for instance can hardly be regarded as random errors with stationary properties. Experiments related in GY (1967 chapter 12) tend to show that non-probabilistic sampling errors are usually large enough to deprive the samples of any practical value. Such methods are always dangerous and it is somewhat astounding to note that they are still in use where they can be the most detrimental, namely in commercial sampling. Unfortunately Standards Organizations still mention or even recommend these, delaying the moment when such methods that belong to the past will be at last abandoned.

There is no place in this book for a study or even a simple description of non-probabilistic sampling methods. We would like, however, to caution the reader against them : they always introduce unaccountable errors. In other words, they deliver specimens not samples. They were developed when sampling was a primitive art practiced only by a limited number of initiates who failed to update their scientific knowledge. We definitely recommend to avoid them at any cost.

4.3. PROBABILISTIC PROCESSES - NOTION OF MOVABLE AND UNMOVABLE BATCHES

We shall therefore restrict our study to the probabilistic sampling processes. A batch of particulate material is said to be "<u>movable</u>" when it is small or valuable enough to be economically handled in totality for the sole purpose of its sampling. This definition can widely vary according to the social and economic conditions prevailing at the time and place of the sampling and also according to the mechanical help (especially mechanical shovels) at the sampler's disposal. This point will be developed in the fifth part when reviewing the splitting methods and devices (chapter 24). We shall now distinguish the sampling of unmovable lots from that of movable lots.

4.4. SAMPLING OF UNMOVABLE LOTS

The only probabilistic sampling process applicable to unmovable lots of particulate materials is the "increment sampling process" by which the lot is represented by a set of increments extracted from the lot according to a certain "selection scheme" and by means of a certain extracting device. The theoretical problem set by the increment sampling of three- and two-dimensional unmovable lots of particulate material is in no way distinct from that of three- and two-dimensional compact solids such as mineral deposits. This problem is dealt with by the science known as "Geostatistics" (see for instance David - 1977). The practical problem set by such batches is to be regarded as "unsolvable" as will be shown in chapter 29.

There remains the problem of the flowing streams of particulate materials that can be represented by a one-dimensional model. Although practically and economically very important, this problem had not been thoroughly studied before. There are three ways of sampling flowing streams of particulate materials :

- taking the whole of the stream part of the time,
- taking part of the stream all of the time,
- taking part of the stream part of the time.

The trial of devices taking part of the stream all or part of the time has been made several times by various authors since Warwick (1903) and they have always been found guilty of introducing uncontrollable biases. Such devices have practically disappeared from modern sampling plants. We shall just mention in chapters 18 and 23 examples of the heavy biases they are likely to introduce. As far as the sampling of unmovable lots is concerned we shall therefore restrict our study to the sampling of one-dimensional flowing streams by application of the increment process taking the whole of the stream part of the time (see section 4.6.).

4.5. SAMPLING OF MOVABLE LOTS

Movable lots can be sampled by application of one of the two following probabilistic processes :

- "increment process" already mentioned in the preceding section where the batch to be sampled (lot or sample obtained at an earlier sampling stage) is put under the form of a flowing stream and sampled as such (section 4.6.).

- "splitting process" where the batch is first partitioned into several fractions, one of which being retained as a sample (section 4.7.)

As a conclusion of sections 4.4. and 4.5. we shall retain that, whether the lot is movable or not, all probabilistic unbiased sampling devices or methods fall within the province of two sampling processes : increment sampling and splitting.

4.6. ANALYSIS OF THE INCREMENT SAMPLING PROCESS

A typical example of increment sampling is the sampling of a flowing stream at the discharge of a belt conveyor by means of a cross-stream cutter. The increment sampling process can be broken up into a sequence of four elementary and independent steps :

- Point selection : all points of the domain (D_L) occupied by the lot L in the p-dimensional space retained as a model (general definition) or all instants of the time domain (T_L) occupied by the lot L on the time axis (flowing streams) are submitted to a certain selection scheme (chapter 7) with a certain density of selecting probability. The selected points will be called the "punctual increments" (chapter 8). Their materialization is a two-step operation which will be studied in the third part of this book (chapters 16 to 18). It consists of the two following steps.

- Increment delimitation : moving through the domain (D_L) actually occupied by the lot L in the three-dimensional Euclidean space the sampling device delimits the geometrical boundaries of the domain (D_I) occupied by the "extended increments" around the punctual increments. This purely geometrical delimitation is not supposed to respect the fragments integrity. Increment delimitation errors will be studied in chapter 17.

- Increment extraction : The extended increments are volumes. We must now take the particulate structure of the material into consideration. When crossing the stream, the cutter extracts a certain number of fragments making up what we shall call the "fragmental increments" coinciding more or less with the set of fragments the centre of gravity of which falls within the boundaries of the extended increments. Increment extraction errors will be studied in chapter 18.

- Increment reunion : the sets that can be obtained by reunion of the punctual, extended or fragmental increments will be called respectively the "punctual sample", "extended sample" and "fragmental sample".

The second part of this book will be dedicated to a thorough study of the "continuous selection model" fitting the point selection step of the increment sampling process (chapters 6 to 15).

The third part will deal with the materialization of the punctual increments and with a study of the corresponding delimitation and extraction errors (chapters 16-18).

In the fourth part, we shall study the discrete selection model. This model leads to a better understanding of the results of the continuous selection model (chapters 19 to 23).

4.7. ANALYSIS OF THE SPLITTING PROCESS

Typical examples of the splitting process are :
- coning and quartering,
- alternate or fractional shovelling,
- riffling, etc, ...

The fifth part of this book will present a study of the splitting process (chapters 24 to 26).

The splitting process can be broken up into a sequence of four elementary and independent logical steps :

- Fraction delimitation : in its relative movement through the domain (D_L) occupied by the lot L, the sampling tool or device delimits the geometrical boundaries of the domains occupied by the "geometrical fractions" of the lot. This relative movement can be realized in three different ways :

•Stationary lot, moving tool : coning and quartering, alternate shovelling, etc .. (the geometrical boundaries of the fractions are but approximately defined),

•Moving lot, stationary device : riffle divider, revolving feeder sectorial splitters (geometrical boundaries defined with precision),

•Moving lot, moving device : stationary feeder sectorial splitters (geometrical boundaries defined with precision).

This purely geometrical operation does not take the fragment integrity into consideration. It corresponds to the delimitation step of the increment process.

- Fraction separation : the particulate structure of the sampled material is taken into account in this second step. The "material fractions" actually separated coincide more or less with the set of particles whose centre of gravity falls within the boundaries of the geometrical fractions. It corresponds to the extraction step of the increment process.

- Fraction dealing out : the material fractions are then dealt out according to a usually systematic scheme between a certain number of "potential samples" (two with coning and quartering, alternate shovelling, riffling, etc .., more than two with fractional shovelling, sectorial dividers, etc ..) obtained by reunion of several fractions (two with coning and quartering, more than two with all other splitting methods or devices). It corresponds to the reunion step of the increment sampling process.

- Sample selection : all potential samples are submitted to a selection scheme retaining one "sample" or several "twin-samples". This selection step may be regarded as probabilistic if and only if it is made at random. It corresponds to the point selection step of the increment process.

These four steps are, potentially at least, error-generating. They will be studied as such in the light of the two complementary models already developed for the study of the increment process.

4.8. COMPARISON OF THE INCREMENT AND SPLITTING PROCESSES

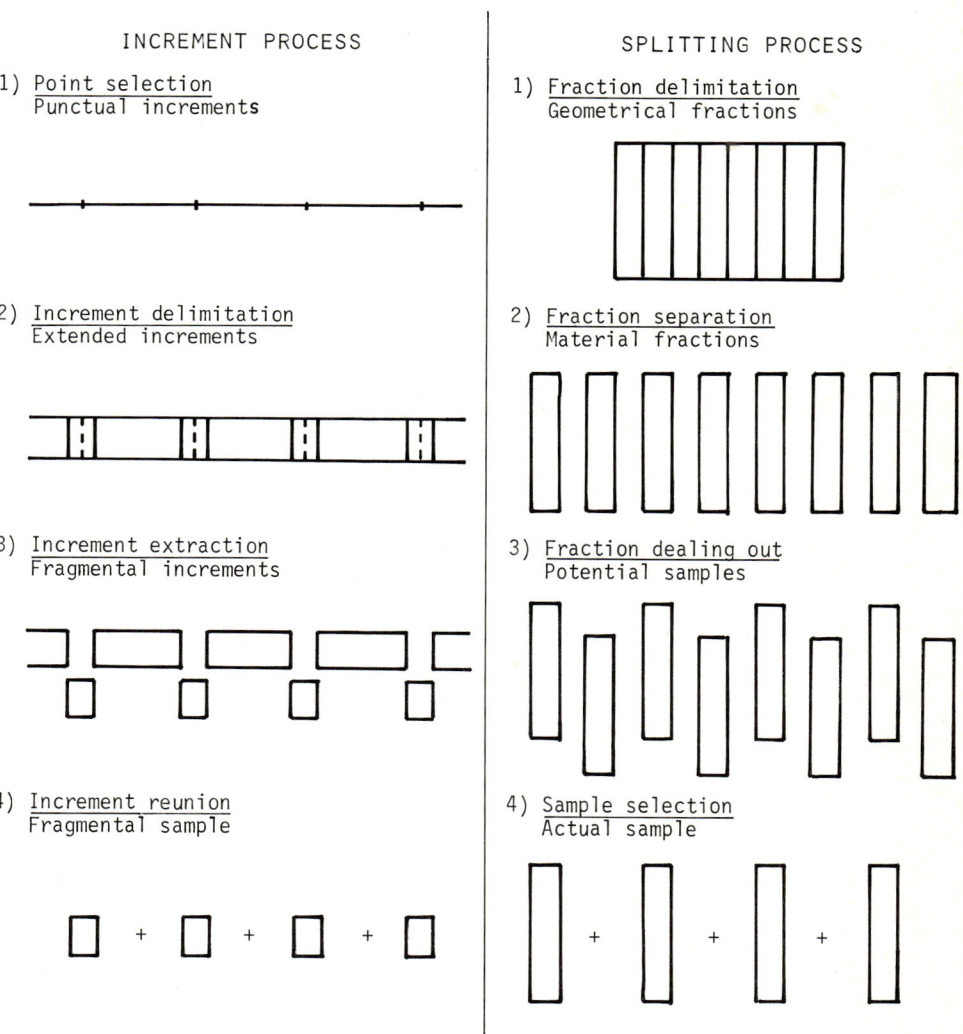

Fig. 4.1. Schematization of the increment and splitting sampling processes

Fig. 4.1. schematizes the four steps of both the increment and splitting processes. From a logical standpoint and also from a practical standpoint, the main difference between these processes is that the selection step takes place :

- Increment process : prior to the extraction step,
- Splitting process : after the separation step (equivalent to the extraction step of the increment process).

Due to this apparently unimportant property, splitting can be made equitable even when it is technically biased, whilst increment sampling is equitable if and only if it is technically unbiased. The notion of equity, so important in commercial sampling, is analysed in chapter 30.

4.9. FIELD OF APPLICATION OF THE INCREMENT AND SPLITTING SAMPLING PROCESSES

Increment sampling is always implemented to sample unmovable lots or flowing streams of particulate materials. The usual range of the sampling ratio is from $1/1000$ to $1/10$.

Splitting is of necessity restricted to the sampling of movable lots. The usual range of the splitting ratio is from $1/20$ to $1/2$. By repeated splitting into halves (alternate shovelling, riffling, etc ..) any sampling ratio can be obtained.

In mechanical and automated sampling plants, the first stages always implement increment sampling whereas the last stages nearly always implement splitting in order to obtain one or a certain number of twin laboratory samples that can be directly received in Mason jars.

CHAPTER 5

MODELS OF THE INCREMENT SAMPLING PROCESS

A scientific theory is usually built on one or several models but ...

5.1. WHAT IS A MODEL AND WHAT IS THE USE OF IT ?

A model is a more or less idealized representation of an often complex reality. One can model objects, phenomena and processes, these words being used in their wider sense.

When developing the theory of a selection process such as sampling, the purpose is to establish mathematical relationships between the "properties" (again in the wider sense of the word) of the object submitted to the selection, the properties of the phenomena involved in the process and one or several "appreciation factors" characterizing the "efficiency" of the process, in this case the moments of the sampling error.

5.1.1. Model of an object :

When an object is submitted to a given process, only a limited number of its properties are relevant to this process. The most efficient model devised to represent the object will therefore take into account all these relevant properties and forget the others.

According to the perspective in which the object is observed, we can devise several models of the same object corresponding to its "projections" on different planes just as we can take several photographs of the same object under different angles or through different lenses. Each photograph, each projection, each model brings into light a different aspect of the reality and helps to understand it better.

5.1.2. Model of a process :

Much in the same way, we shall retain the properties of the process which are relevant to the problem to be solved and forget the others. Here again, we shall devise several models matching the corresponding models of the object to be processed.

5.1.3. Qualities of a model or of a group of models :

A model providing a distorted picture of reality is not only useless, it is dangerous. The first quality of a model is therefore to provide an undistorted, unbiased picture of reality. If this picture is fragmentary, a group of fragmentary but undistorted models is necessary to cover all aspects of the reality.

The second quality of a model is to be simple enough to allow a straightforward mathematical processing. Following Descartes' philosophy, we shall therefore endeavour to break up any complex reality into a set or a sequence of simple logical elements covered by simple models. We have already given two examples of such an approach in sections 4.6. and 4.7.

The third quality expected of a group of models is to be linked to one another like the pieces of a jig-saw puzzle.

The fourth quality of a group of models is to cover all relevant aspects of the problem and to answer all questions : all pieces of the puzzle must be gathered, matched to one another and all gaps must be filled up.

5.2. INTRODUCTION TO A GROUP OF MODELS OF THE SAMPLING PROCESSES

Sampling is the first link of quality control, the final purpose of which is to estimate a certain property of the lot L, its critical content a_L. Sampling provides a sample S the critical content a_S of which is retained as an estimator of a_L. As far as the estimation of a_L is concerned, the appreciation factor of the sampling process is the difference $a_S - a_L$ or, which is more convenient, its relative counterpart the "sampling error" SE defined as follows :

$$SE = \frac{a_S - a_L}{a_L}$$

The purpose of a theory of sampling is to ascertain the properties of SE in terms of :
- the properties of the material to be sampled,
- the properties of the sampling process.

According to its definition, SE solely depends on a_L and a_S. Our approach will therefore consist in analysing how a_L can be mathematically expressed in terms of the properties of the material to be sampled and how a_S can be expressed in terms of both the properties of the material to be sampled and of the sampling process.

A lot of particulate material can be regarded as a set of N_{LF} fragments or particles F_i (i = 1, 2, .. N_{LF}) enclosed in a certain domain (D_L) and surrounded by an interstitial fluid E which can be any gas or liquid, for instance air or water, and which can play an active or passive part (section 1.1.2.). Such a lot L is schematized in fig. 5.1. We shall represent L in two different ways or in other words by means of two different and complementary models: the continuous and the discrete models.

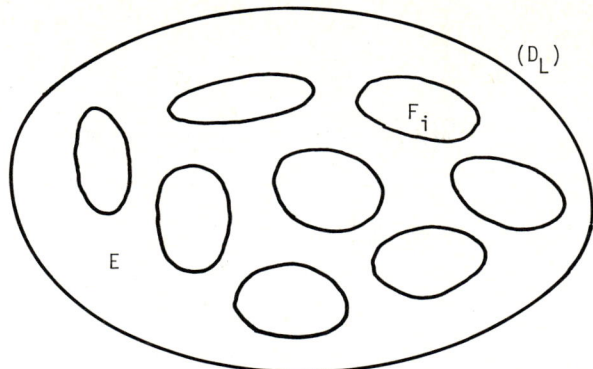

Fig. 5.1. Schematization of a lot L of particulate material.

5.2.1. Continuous model :

The lot L is regarded as the amount of matter contained within the boundaries of a certain domain (D_L) of a given space, assuming that these boundaries respect the integrity of the particles and that (D_L) contains all the N_{LF} fragments of L and none other. This envelope is nearly always an immaterial surface or line.

According to the continuous model, L is regarded as the continuous set of all points X (co-ordinates x, y, z) falling within the boundaries of the domain (D_L). The matter present at point X is conveniently described by means of various functions of its co-ordinates x, y, z, whether X belongs to a critical, to an active non-critical or to a passive component. The continuous model does not take into account the discrete nature of the material. We can therefore surmise that such a model will be very general and will be applicable to compact solids or to fluids as well as to particulate materials. This model is schematized in fig. 5.2.

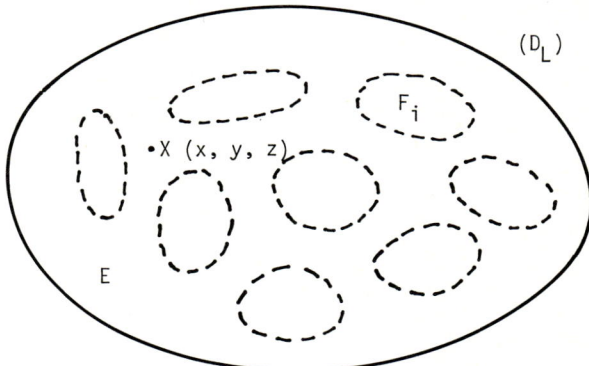

Fig. 5.2. Continuous model of the lot L - The dotted lines recall the boundaries of the fragments F_i.

5.2.2. Discrete model :

The lot L is regarded as the discrete set of the N_{LF} particles making up the lot L. The discrete model does not take into account the interstitial fluid E, with the consequence that it is valid only when this fluid plays a passive part, which excludes sampling for the estimation of the solid concentration of a pulp. The discrete model is schematized in fig. 5.3.

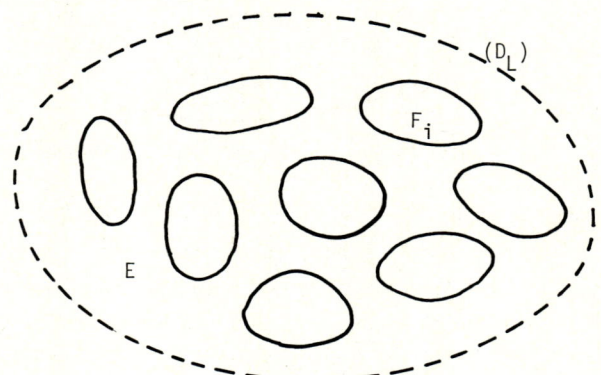

Fig. 5.3. Discrete model of the lot L - The dotted line recalls the boundaries of the domain (D_L).

5.3. CONTINUOUS MODEL

We must model the material to be sampled and the sampling process.

5.3.1. Continuous model of the material to be sampled :

In this continuous perspective, L is assimilated to the continuous set of all points X (x, y, z) falling within the boundaries of the domain (D_L). In order to complete this model, we must now describe the relevant properties of the matter present at point X.

Continuity is a mathematical concept : matter is essentially discontinuous, made as it is of discrete elements such as atoms, molecules, ions, etc.. to say nothing of other elementary particles. We must therefore surmise difficulties when trying to apply a continuous mathematical model to a physically discontinuous matter. From a practical standpoint, however, we shall forget the fundamental discontinuity of matter which is not relevant at the scale of our observations. We shall consider :

dX : the elementary volume centred at point X,
dM : the weight of active components present in dX,
dA : the weight of critical component present in dX.

Then, we shall define the following "punctual functions" characterized by the subscript P.

$\mu_P(x, y, z) = \lim_{dX \to 0} \frac{dM}{dX}$: weight of active components per unit volume at point X

$\alpha_P(x, y, z) = \lim_{dX \to 0} \frac{dA}{dX}$: weight of critical component per unit volume at point X

$a_P(x, y, z) = \lim_{dX \to 0} \frac{dA}{dM}$: critical content at point X : $a_P(x, y, z) = \frac{\alpha_P(x, y, z)}{\mu_P(x, y, z)}$

We shall use the following notations :

M_L : weight of active components in L or more simply weight of L. By definition :

$M_L = \iiint_{(D_L)} \mu_P(x, y, z) \, dx \, dy \, dz$

A_L : weight of critical component in L. By definition :

$A_L = \iiint_{(D_L)} \alpha_P(x, y, z) \, dx \, dy \, dz$

a_L : critical content of L : $a_L = \frac{A_L}{M_L} = \frac{\iiint_{(D_L)} \alpha_P(x, y, z) \, dx \, dy \, dz}{\iiint_{(D_L)} \mu_P(x, y, z) \, dx \, dy \, dz}$

The matter present at point X is - as far as its sampling is concerned - completely defined by any two of the three functions $\mu_P(x, y, z)$, $\alpha_P(x, y, z)$, $a_P(x, y, z)$ and by their mathematical relationship. This definition is necessary and sufficient. We shall admit that the functions $\mu_P(x, y, z)$ and $\alpha_P(x, y, z)$ are defined everywhere in the domain (D_L) and that they are integrable. This latter assumption is acceptable as long as one admits the additivity of weights, a proposition that might be questioned by atomic physicists working at atomic or sub-atomic scale but that we shall accept at the scale of mineral fragments. The model of L is therefore defined by the domain (D_L) and by the two functions :

$a_P(x, y, z)$: <u>critical content</u> at point X, or "<u>quality function</u>",

$\mu_P(x, y, z)$: specific gravity at point X (when X falls within an active component) or 0 (when X falls within a passive component). It is also called the "<u>weighting function</u>" since, according to the definition of a_L, the critical content of the lot is the weighted mean of $a_P(x, y, z)$ throughout the domain (D_L).

5.3.2. Degenerate models of the lot L :

Strictly speaking, all compact or particulate materials extend in a three-dimensional space and can be represented by a three-dimensional model but it is sometimes convenient to represent them by degenerate models such as two-, one- or even zero-dimensional models.

The three-dimensional model must be retained to represent all lots of irregular shape that cannot be reduced to one of the degenerate models. Fig. 5.4. represents an irregular pile of ore stored on the (x, y) horizontal plane.

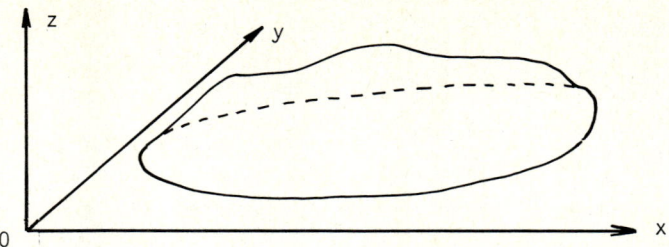
Fig. 5.4. *Irregular pile of ore represented by a three-dimensional model.*

5.3.2.1. Two-dimensional model : This model is convenient to represent lots, one of the dimensions of which (usually the thickness) is small and reasonably uniform. This covers for instance flat piles, some sedimentary mineral deposits or metallurgical products under the form of plates. Fig. 5.5. represents a flat pile of ore.

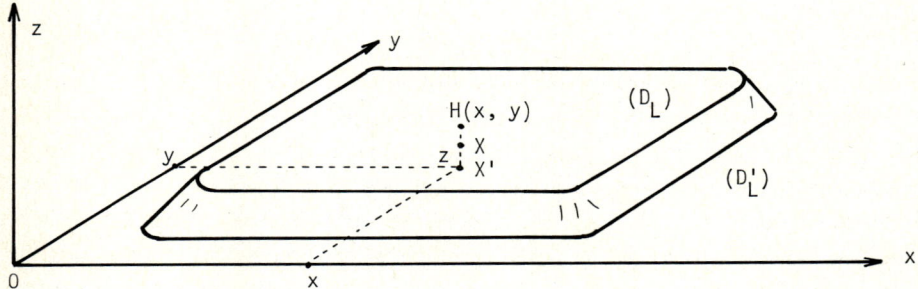
Fig. 5.5. *Flat pile of ore represented by a two-dimensional model*

The extension plane of the lot being retained as (x, y) plane, we shall call :

X' : the projection of X on this plane (co-ordinates x, y),
(D'_L) : the projection of (D_L) on this plane,
H(x, y) : the thickness of L at point X',
$\mu_p(x, y)$: the accumulation of active components at point X'. By definition :
$$\mu_p(x, y) = \int_{H(x,y)} \mu_p(x, y, z) \, dz$$
$\alpha_p(x, y)$: the accumulation of critical component at point X'. By definition :
$$\alpha_p(x, y) = \int_{H(x,y)} \alpha_p(x, y, z) \, dz = \int_{H(x,y)} \mu_p(x, y, z) \, a_p(x, y, z) \, dz$$
$a_p(x, y)$: the critical content at point X'. By definition :
$$a_p(x, y) = \frac{\alpha_p(x, y)}{\mu_p(x, y)}$$

It can be easily shown that :

$$a_L = \frac{\iint_{(D'_L)} a_p(x, y) \, \mu_p(x, y) \, dx \, dy}{\iint_{(D'_L)} \mu_p(x, y) \, dx \, dy}$$

From this expression, the co-ordinate z has completely disappeared : the problem is degenerate and can be solved in a two-dimensional space. It should be pointed out that no assumption has been made as to the relative size and uniformity of H(x,y) throughout (D_L') which means that the two-dimensional model may be used for lots of non-uniform thickness. We shall however restrict its use to "flat" lots.

5.3.2.2. One-dimensional model : This model is convenient to represent lots, the section of which is small and reasonably uniform. This covers for example elongated piles like those obtained in blending systems, metallurgical products under the form of bars, etc... Fig. 5.6. represents an elongated pile of ore.

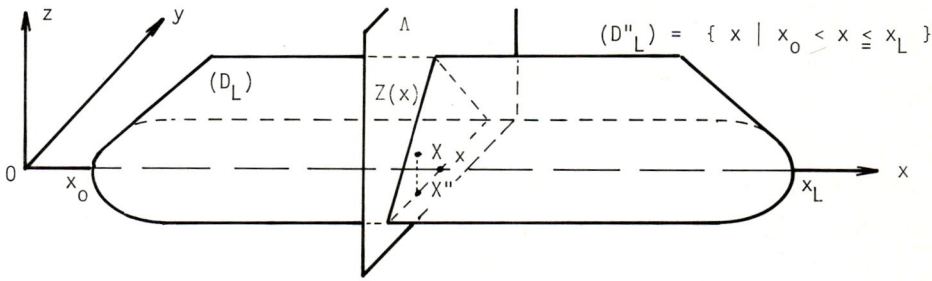

Fig. 5.6. Elongated pile of ore represented by a one-dimensional model.

The extension axis of the lot being retained as axis of the abscissae, we shall call :

X" : the projection of X on the Ox axis. Its abscissa is x.
(D_L'') : the projection of D_L on the Ox axis
Z(x) : the section of (D_L) by the Λ plane parallel to yOz at point X".
$\mu_P(x)$: the accumulation of active components at point X". By definition :
$\mu_P(x) = \iint_{Z(x)} \mu_P(x, y, z) \, dy \, dz$
$\alpha_P(x)$: the accumulation of critical component at point X". By definition :
$\alpha_P(x) = \iint_{Z(x)} a_P(x, y, z) \mu_P(x, y, z) \, dy \, dz$
$a_P(x)$: the critical content at point X". By definition :

$$a_P(x) = \frac{\alpha_P(x)}{\mu_P(x)}$$

It can be easily shown that :

$$a_L = \frac{\int_{(D_L'')} a_P(x) \mu_P(x) \, dx}{\int_{(D_L'')} \mu_P(x) \, dx}$$

From this expression, the co-ordinates y and z have completely disappeared : the problem is degenerate and can be solved in a one-dimensional space. Here again, we have made no assumption as to the relative size and uniformity of Z(x) throughout (D_L''). This means that the one-dimensional model may be used for lots with non-uniform section. We shall however restrict its use to "elongated lots".

5.3.2.3. One-dimensional temporal model - flowing streams

: The most important problem falling within the province of the one-dimensional model is certainly that of the flowing streams where the material, for instance deposited on a belt conveyor moves with a certain velocity V_B (B for belt) that we shall assume to be constant in time and uniform for all parts of the stream. We shall call :

t : time of passage of point X'' in front of a reference mark, the sampling point for instance,
t_0 : beginning of the flow of L
t_L : end of the flow of L
(T_L) : domain occupied by the lot L on the time axis :
$(T_L) \equiv \{t \mid t_0 < t \leq t_L\}$. By definition, its extent is : $T_L = t_L - t_0$

We can obviously write :

$$x - x_0 = V_B (t - t_0) \qquad dx = V_B \, dt \qquad D_L'' = V_B T_L$$

We shall now define :

$\mu_P(t)$: <u>rate of flow of active components</u> at time t : weight of active components flowing during the unit of time at time t. By definition :
$\mu_P(t) = V_B \, \mu_P(x)$
$\alpha_P(t)$: <u>rate of flow of critical component</u> at time t : weight of critical component flowing during the unit of time at time t. By definition :
$\alpha_P(t) = V_B \, \alpha_P(x)$
$a_P(t)$: <u>critical content</u> of the material flowing at instant t : critical content of the slice of matter flowing between t and $t + dt$. By definition :
$a_P(t) = \alpha_P(t)/\mu_P(t) = \alpha_P(x)/\mu_P(x) = a_P(x)$

The critical content a_L of the lot is :

$$a_L = \frac{\int_{(T_L)} a_P(t) \, \mu_P(t) \, dt}{\int_{(T_L)} \mu_P(t) \, dt}$$

5.3.2.4. Zero-dimensional model

: By convention, we shall call zero-dimensional lot, a lot which is naturally divided into a large number of "<u>units</u>" (usually transportation units) of reasonably uniform bulk. This model covers for instance units such as the contents of railwagons, trucks, shovelfuls (whether manual or mechanical drums, sacks, etc .. All units usually have the same nominal capacity and are more or less uniformly loaded. We shall now define :

N_{LU} : number of units in the lot L,
U_m : typical unit ($m = 1, 2, \ldots N_{LU}$)
a_m : critical content of U_m
M_m : net weight of U_m.

The lot L can be regarded :

- either as a sequence of units arranged in some natural order. This case falls within the province of the continuous model (second part of this book),
- or as a set of unarranged units. This case falls within the province of the discrete model (fourth part of this book).

In both cases, the critical content a_L of the lot L is :

$$a_L = \frac{\sum_m a_m M_m}{\sum_m M_m} \quad \text{with } m = 1, 2, \ldots N_{LU}$$

This formula is valid even when no assumption is made as to the uniformity of the weights M_m but the model is efficient only when the weights M_m are fairly uniform.

The lot L may be regarded as a zero-dimensional object only when its primary sampling consists in selecting a certain number of whole units. This representation is no longer valid when the primary sampling stage consists in extracting increments from each unit, irrespective of how these increments are extracted.

5.3.3. General continuous model of the lot L :

Without prejudice to the number p of dimensions of the model retained to describe the lot L (p = 1, 2, 3 and p ≠ 0), we shall adopt the following notations :

X : point of a p-dimensional space,

$\mu_p(X)$: weight of active components per unit of space (volume, surface, length or time) at point X,

$\alpha_p(X)$: weight of critical component per unit of space at point X,

$a_p(X)$: critical content of the matter present or projected at point X,

dX : elementary fraction of the considered space (dx dy dz, dx dy, dx, dt).

(D_L) : domain occupied by L in the space considered. Then :

$$a_L = \frac{\int_{(D_L)} a_p(X) \mu_p(X) \, dX}{\int_{(D_L)} \mu_p(X) \, dX}$$

According to the case, the integral may be triple, double, simple. This general expression is valid, irrespective of the value of p.

5.3.4. Punctual, extended and fragmental functions :

5.3.4.1. <u>Punctual functions</u> : We have just defined the punctual functions $\mu_p(X)$ $\alpha_p(X)$ and $a_p(X)$ describing the properties of the matter present in the elementary fraction dX of the space in question. Now, when trying to fill up the gap between theory and practice, between model and reality, we encounter two kinds of difficulties :

- the first difficulty arises from the abstract nature of the mathematical concept of extensionless point X or of durationless instant t.
- the second difficulty arises from the discrete nature of the particulate

materials object of this study.

In order to overcome these difficulties, we shall now define two new kinds of functions attached to the point X :
- the "extended functions" describing the properties of the matter present within the boundaries of a certain "extension domain" centred at point X.
- the "fragmental functions" describing the properties of the group of particles whose centre of gravity falls within the boundaries of the extension domain centred at point X.

5.3.4.2. Extended functions : The extended functions will be characterized by the subscript E. We shall define :

(D_E) : a certain domain of the p-dimensional space retained to represent the lot L. We shall call it the "extension domain". Its extent is D_E. We shall assume that D_E remains small in comparison with D_L and that (D_E) is isotropic. In other words, (D_E) is a sphere, a circle, a segment or a time interval. It is completely defined by its extent D_E.

$D_E(X)$: the extension domain centred at point X,

$\mu_E(X)$: the "extended function $\mu(X)$". It is defined as the mean of the punctual function $\mu_p(X)$ throughout the domain $D_E(X)$. By definition :

$$\mu_E(X) = \frac{1}{D_E} \int_{D_E(X)} \mu_p(X') \, dX'$$

$\alpha_E(X)$: the "extended function $\alpha(X)$". It is defined as the mean of the punctual function $\alpha_p(X)$ throughout the domain $D_E(X)$. By definition :

$$\alpha_E(X) = \frac{1}{D_E} \int_{D_E(X)} \alpha_p(X') \, dX'$$

$a_E(X)$: the "extended function $a(X)$". It is the critical content of the matter contained within the boundaries of the domain $D_E(X)$. By definition :

$$a_E(X) = \frac{\alpha_E(X)}{\mu_E(X)}$$

5.3.4.3. Fragmental functions : The fragmental functions will be characterized by the subscript F. The extended functions describe the properties of a certain amount of matter without regard to the particulate structure of the material. We shall now define the "fragmental functions" describing the properties of the group $G_E(X)$ of particles the centre of gravity of which falls within the boundaries of the extension domain $D_E(X)$. This definition is not arbitrary and we shall explain the reasons of this choice in due course (chapter 18 : rule of the centre of gravity). Fig.5.7. and 5.8. illustrate the definitions of the extended and fragmental functions. The hachured areas show the matter taken into account by these two functions. X is the centre of the extension domain $D_E(X)$.

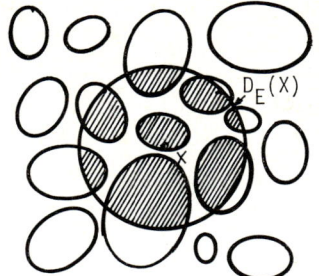

Fig. 5.7. *Extended functions*

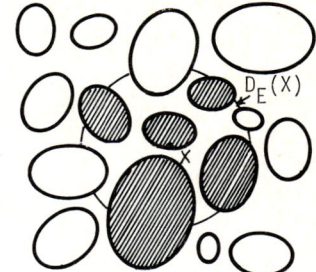

Fig. 5.8. *Fragmental functions*

$\mu_F(X)$ is a discrete estimator of $\mu_E(X)$ which is itself an extended estimator of the punctual function $\mu_P(X)$. The same holds true for $a_E(X)$ and $a_F(X)$.

5.3.4.4. Properties of the punctual, extended and fragmental functions : The critical content a_L of the lot L can be written :

$$a_L = \frac{\int_{(D_L)} a_P(X) \mu_P(X) dX}{\int_{(D_L)} \mu_P(X) dX} = \frac{\int_{(D_L)} a_E(X) \mu_E(X) dX}{\int_{(D_L)} \mu_E(X) dX} = \frac{\int_{(D_L)} a_F(X) \mu_F(X) dX}{\int_{(D_L)} \mu_F(X) dX}$$

The first equality is rigorous. The other two are only approximative but the smaller the ratio D_E/D_L, the better the approximation. For all practical purposes, however, the approximation is excellent.

5.3.5. Continuous model of a selection process :

According to this continuous model, each point X of the domain (D_L) occupied by the lot L is submitted to a given selecting process, according to a certain selection scheme, with a certain density $\Pi(X)$ of selection probability.

A selecting process is said to be <u>correct</u> (see definition in section 1.5.1) when the function $\Pi(X)$ is uniform throughout the domain (D_L) and equal to a constant Π_0. The density of selection probability is assumed to be nil outside (D_L). It is said to be incorrect when one of these conditions is not fulfilled.

5.3.6. Continuous model of the increment sampling process .

We saw in section 4.6. that the increment sampling process could be broken up into a sequence of four logical steps. We shall show in due course that under certain conditions the continuous model covers :

- the <u>point selection</u> step when applied to the punctual functions,
- the sequence: <u>point selection + increment delimitation</u>, when applied to the extended functions,
- the sequence: <u>point selection + increment delimitation + increment extraction</u>, when applied to the fragmental functions.

This property justifies the definition of the extended and fragmental functions that constitute the necessary bridge between the continuous model and the discrete reality. This point will be dealt with in the third part of this book.

5.4. DISCRETE MODEL

5.4.1. Discrete model of the lot L :

In the discrete perspective, the lot L is assimilated to the discrete set of N_{LU} unarranged units U_m (m = 1, 2, .. N_{LU}). These units U_m can be made up of :
- either single fragments or particles F_i (i = 1, 2, .. N_{LF})
- or groups of fragments G_n (n = 1, 2, .. N_{LG}).

The material contained in unit U_m is defined by two of the three parameters:

M_m (or M_i or M_n) : weight of active components in unit U_m (or fragment F_i or group

A_m (or A_i or A_n) : weight of critical component in unit U_m (or fragment F_i, group

a_m (or a_i or a_n) : critical content of unit U_m (or fragment F_i or group G_n).

and by the following obvious relationships :

$$a_m = \frac{A_m}{M_m} \qquad a_i = \frac{A_i}{M_i} \qquad a_n = \frac{A_n}{M_n}$$

The critical content of the lot is :

$$a_L = \frac{\sum_m A_m}{\sum_m M_m} = \frac{\sum_m a_m M_m}{\sum_m M_m} \qquad a_L = \frac{\sum_i A_i}{\sum_i M_i} = \frac{\sum_i a_i M_i}{\sum_i M_i} \qquad a_L = \frac{\sum_n A_n}{\sum_n M_n} = \frac{\sum_n a_n M_n}{\sum_n M_n}$$

m = 1, 2, .. N_{LU} \qquad i = 1, 2, .. N_{LF} \qquad n = 1, 2, .. N_{LG}

As far as the sampling of L by selection of a certain number of units U_m is concerned, the definition of U_m by means of two of the three parameters M_m, A_m and a_m is necessary and sufficient. We shall usually retain :
- a_m : critical content of U_m (quality parameter),
- M_m : weight of active components in U_m or more simply weight of U_m (weighting parameter).

These parameters correspond to the quality and weighting functions a(X) and µ(X of the continuous model.

5.4.2. Discrete model of a selection process :

According to this discrete model, each unit U_m of the lot L is submitted to a selecting process with a certain probability P_m of being selected. The selecting process is said to be "<u>correct</u>" when P_m is uniform and equal to a constant P_0 for all units belonging to L and nil for all units that do not belong to L. It is said to be incorrect when one of these conditions is not fulfilled.

5.5. OBJECTIVES PURSUED WHEN DEVELOPING SELECTION MODELS

The second part of this book and more specifically chapter 8 will be dedicated to the development of the continuous selection model. The fourth part and more specifically chapter 20 to the development of the discrete model. Our purpose, when developing these models, is to disclose the mathematical relationships that, of necessity, exist between three groups of variables and parameters :

- Properties of the material to be sampled : these are described by means of : the functions $a(X)$ and $\mu(X)$ within the domain (D_L) in the continuous model, the parameters a_m and M_m within the set N_{LU} in the discrete model.

They are characterized by means of more sophisticated functions (chapter 6) or parameters (chapter 19). These properties are usually to be regarded as more or less intangible data of the problem.

- Properties of the selection process : definition of the selection scheme and of :
·the density $\Pi(X)$ of selection probability in the continuous model,
·the selection probability P_m in the discrete model.

They can be estimated from the construction characteristics of the sampling method or device. The selection scheme and several characteristics of the sampling device can be freely chosen.

- Appreciation factors : these characterize various properties of the selection (or sampling) error SE and more specifically the following moments (definitions given in section 1.5.2.) :

mean $m(SE)$: first moment of the selection error, a measure of the selection accuracy,
variance $\sigma^2(SE)$: second moment about the mean of the selection error, a measure of the selection reproducibility,
mean square $r^2(SE)$: second moment of the selection error, a measure of the selected sample representativeness.

5.6. RESOLUTION OF SAMPLING PROBLEMS

When a group of relationships between these quantities have been obtained, two kinds of problems can be solved :

- Estimation of the moments of the sampling error in terms of the characteristics of the material to be sampled and of the sampling process,
- Choice of a sampling scheme and determination of the characteristics of the sampling devices in order to meet given accuracy and reproducibility requirements.

We shall have many opportunities to give practical examples of the resolution of these two problems. The seventh part of this book is dedicated to the practical resolution of sampling problems.

SECOND PART

CONTINUOUS MODEL OF THE INCREMENT SAMPLING PROCESS

The continuous selection model has been presented in section 5.3. According to its definition, the continuous model is applicable to any kind of material, irrespective of :

- its physical state : compact or particulate solids, liquids, gases, multi-phase systems, etc ..
- its origin : mineral, vegetable, animal, synthetic or materials such as town refuses, etc ..
- the critical component taken into consideration : mineralogical components, size fractions, moisture, solid phase of a pulp, etc ..
- the mathematical model retained to represent the lot and the selection process : three-, two-, one-dimensional geometrical models or one-dimensional time model (flowing streams).

We shall however restrict our demonstrations to a simple, concrete problem, characterized by the following properties :

1 - particulate solid, whether the interstitial fluid is air, water or any other liquid or gas,

2 - mineral origin, ores or concentrates and occasionally metallurgical products,

3 - unspecified critical component. Except in practical applications and examples, we shall not specify the critical content taken into consideration. It may be the proportion of a given mineral (whether valuable or not), that of a size fraction (whether coarse or fine), the moisture content or the proportion of solids in a pulp.

When several components are to be regarded as critical, which happens very often and for instance with a crushed porphyry copper ore :
- the copper content Cu %
- the molybdenum content Mo %
- the moisture content H_2O %
- the proportion of + 10 mm, etc ...

the problem must be independently solved for each of these. If we are estimating the weight of the multi-purpose sample to be taken (for example), the most exacting of all solutions must be retained.

4 - flowing stream of solids or pulp represented by a one-dimensional time model.

The second part of this book consists of a study of the continuous selection model. It is made of ten chapters :

Chapter 6 : Heterogeneity of a continuous set.
Chapter 7 : Reference selection schemes.
Chapter 8 : Development of the continuous selection model - Continuous selection error CE.
Chapter 9 : Breaking up of the continuous selection error CE.
Chapter 10 : Short-range quality fluctuation error QE_1.
Chapter 11 : Long-range non-periodic quality fluctuation error QE_2.
Chapter 12 : Periodic quality fluctuation error QE_3.
Chapter 13 : Weighting error WE.
Chapter 14 : Practical implementation of the continuous model - Variographic experiment.
Chapter 15 : Practical implementation of the continuous model - Error estimation.

CHAPTER 6

HETEROGENEITY OF A CONTINUOUS SET

6.1. INTRODUCTION

Consider a lot L of particulate material flowing past the sampling point from time $t = 0$ to time $t = T_L$. We know that the stream is completely described by three functions defined in section 5.3.2.3. :

$\mu(t)$: <u>rate of flow of active components</u>. It is the weight of active components flowing per time unit at time t,

$\alpha(t)$: <u>rate of flow of critical component</u>. It is the weight of critical component flowing per time unit at time t,

$a(t)$: <u>critical content</u>. It is the content of the slice of matter flowing from time t to time t + dt. By definition :

$$a(t) = \frac{\alpha(t)}{\mu(t)}$$

We shall only assume that the functions $\alpha(t)$ and $\mu(t)$ can be summed up and that their integrals extended to any time domain (T_L) are defined. This is intuitively supported by the fact that both $\alpha(t)$ and $\mu(t)$ represent a rate of flow and that their integrals represent a weight. We do not need to make any hypothesis on their continuity : in fact all three functions are liable to be discontinuous in a finite number of points of (T_L). We shall not specify whether we are dealing with punctual, extended or fragmental functions as our results are valid in any case.

The critical content a_L of the lot L is :

$$a_L = \frac{\int_{(T_L)} a(t) \mu(t) \, dt}{\int_{(T_L)} \mu(t) \, dt}$$

6.2. DEFINITION AND PROPERTIES OF A HOMOGENEOUS MATERIAL

A flowing material is said to be "<u>homogeneous</u>" as regards a certain critical component when the corresponding critical content is uniform and equal to a_0 throughout the domain (T_L). It is easy to show that $a_0 = a_L$. A flowing material is said to be "<u>heterogeneous</u>" when this condition is not fulfilled.

Any slice cut from a homogeneous stream has a critical content equal to a_L and a sample made of one or several such increments has a critical content equal to a_L : $a_S = a_0 = a_L$ with the consequence that : $SE \equiv 0$.

This important property can be stated as follows : <u>the sampling of a homogeneou</u>
<u>material is an exact process</u>.

Slices cut from a heterogeneous stream usually have different critical contents
If a sample is made of one or several such increments, its critical content a_S is
likely to be different from a_L :

$a_S \neq a_L$ with the consequence that SE $\neq$ 0.

<u>The sampling of a heterogeneous material is an error-generating process</u>. From
these properties we come to the conclusion that sampling errors are a direct conse
quence of heterogeneity, hence the necessity of analysing and quantifying the notion of heterogeneity in a work dedicated to the theory of sampling.

6.3. DESCRIPTION OF A HETEROGENEOUS MATERIAL

We shall call f(t) any non-specified function $\mu(t)$, $\alpha(t)$ or $a(t)$. The study of
a large number of non-uniform functions such as f(t) leads to the conclusion that
in the most general case, f(t) can be broken up into a sum of four terms :

$f(t) = f_0 + f_1(t) + f_2(t) + f_3(t)$ with :

f_0 : a constant describing the average properties of f(t) : <u>constant term</u>.

$f_1(t)$: a function describing the short-range non-periodic variations of f(t) :
        <u>short-range term</u>.

$f_2(t)$: a function describing the long-range non-periodic variations of f(t) :
        <u>long-range term</u>.

$f_3(t)$: a function describing the periodic variations of f(t) : <u>periodic term</u>.

Now, the functions f(t) and their components <u>describe</u> the time variations of
the flowing material but their mathematical expression is never known which makes
it impossible to express the moments of the sampling error in terms of f(t) and
its components. It remains therefore to <u>characterize</u> the time variations of the
flowing material, both in quality and in quantity, in such a way that the moments
of the sampling error can be related to the characteristic functions.

This problem has already been solved by the geostatisticians and more especiall
by G. Matheron who proposed the variogram that we define now.

6.4. DEFINITION OF THE VARIOGRAM $v_f(\theta)$

This and the following theoretical sections are completed by chapter 14 which
deals with all practical aspects of the problem, with the experimental determination of the variogram (variographic experiment) and the interpretation of the experimental results. For a given material, produced under routine conditions like
most products of the mineral industries (e.g. feed to a processing plant, to a
smelter, concentrates produced and shipped, etc ..) the quality and quantity va-

riations are usually under control. We may therefore assume that they are :

- <u>Stationary to the first order</u> : in other words, their first moment (mean) is time-stable or invariant under translation of the time axis, at least to some extent. Quality and quantity fluctuations are purposely regulated in such a way that their shifting mean over a certain length of time remains practically constant.

- <u>Stationary to the second order</u> : in other words, their mean, variance and mean square are time stable. The variance characterizes the fluctuations about the mean reflecting the handling and processing to which the material is submitted. Under routine conditions these fluctuations are likely to have time-stable properties.

As far as the periodic term is concerned, stationarity may be assumed only when the observation scale covers a length of time multiple of their period.

The stationary properties of f(t) can be characterized by means of various functions which are equivalent from a theoretical standpoint. We selected the semi-variogram that we are going to define now. We shall call :

f_L : the mean of f(t) throughout the domain (T_L). If M_L and A_L denote the weight of active components and the weight of critical component in the lot L respectively, then :

$$\mu_L = M_L/T_L \qquad \alpha_L = A_L/T_L \qquad a_L = A_L/M_L$$

θ : a time interval that may vary between 0 and T_L.

$\delta_f(t,\theta)$: the relative increase of f(t) between time $t - \theta/2$ and $t + \theta/2$:

$$\delta_f(t,\theta) = \frac{f(t + \theta/2) - f(t - \theta/2)}{f_L} \qquad \text{(dimensionless)}$$

$v_f(\theta)$: the relative semi-variogram, or more simply the <u>variogram</u> of f(t). It is defined as the half mean square of $\delta_f(t,\theta)$ throughout the domain (T_L).

$$v_f(\theta) = \frac{1}{2(T_L - \theta)} \int_{\theta/2}^{T_L - \theta/2} \delta_f^2(t,\theta) \, dt \qquad \text{(dimensionless)}$$

Thanks to the relative definition of $\delta_f(t,\theta)$, the variogram is also dimensionless. Furthermore, the variogram of kf(t) is identical with the variogram of f(t), a property which will be used later on.

6.5. GENERAL PROPERTIES OF THE VARIOGRAM $v_f(\theta)$

The functions $\mu(t)$, $\alpha(t)$ and $a(t)$ represented by the general notation f(t) are physical quantities with finite limits. The same holds true for the increase $\delta_f(t,\theta)$ and the variogram $v_f(\theta)$. The variogram can be experimentally determined and this point will be dealt with in chapter 14. The variogram of a stationary function is itself a stationary function : its general shape and the variographic parameters entering into its expression may to a certain extent, to be checked case by case, be regarded as invariant under translation of the time axis. This point has been constantly confirmed by experience with the consequence that a variogram determine on a certain day may be used to predict the variability of the same material obtained under similar conditions on another day.

The moments of the continuous selection error CE can be expressed in terms of the variographic parameters determined in a variographic experiment (chapter 8). The variogram function and the variographic experiment are therefore the key to the experimental estimation of the moments of the continuous selection error CE.

6.5.1. Breaking up of the variogram $v_f(\theta)$:

We have shown (Gy - 1975, section 10.3) that when $f(t)$ is a sum of four terms as assumed in section 6.3., its variogram is itself a sum of three terms, the variogram of the constant term f_0 being obviously zero.

$$v_f(\theta) = v_{f1}(\theta) + v_{f2}(\theta) + v_{f3}(\theta)$$

In this expression, $v_{f1}(\theta)$, $v_{f2}(\theta)$ and $v_{f3}(\theta)$ are the variograms of $f_1(t)$, $f_2(t)$ and $f_3(t)$ respectively.

6.5.2. Properties of $v_{f1}(\theta)$:

We shall define the following quantities :

v_{f1} : the relative mean square of $f_1(t)$ throughout the domain (T_L):

$$v_{f1} = \frac{1}{T_L \, f_L^2} \int_{(T_L)} f_1^2(t) \, dt \qquad \text{(dimensionless)}$$

$\rho_{f1}(\theta)$: the autocorrelation function of $f_1(t)$. The variogram $v_{f1}(\theta)$ can be written

$$v_{f1}(\theta) = \{1 - \rho_{f1}(\theta)\} \, v_{f1} \qquad \text{(dimensionless)}$$

The component $f_1(t)$ of $f(t)$ reflects the short-range variations of $f(t)$ resulting mainly from the particulate structure of the material investigated. For instance, all discontinuities observed in $f(t)$, if any, must be catalogued as typical short-range variations and are taken into account by $f_1(t)$. Furthermore, the continuous part of $f(t)$ is taken into account by the long-range term $f_2(t)$ with the consequence that for all practical purposes and for all values of θ larger than a certain value θ_{f1} called the range of $v_{f1}(\theta)$ (of the order of magnitude of a few second or a few tens of one second) we may safely assume that the autocorrelation function is zero. Then :

$$\theta \geq \theta_{f1} \qquad \rho_{f1}(\theta) = 0 \quad \text{hence} \quad v_{f1}(\theta) = v_{f1} = \text{constant}$$

The variogram of $f_1(t)$ is a constant : it is said to be "flat" in the useful domain of θ. This gives the component $f_1(t)$ the nature and the properties of a quasi-random function of time and the function $f(t)$ the nature of a quasi-stochastic function.

With particulate materials, the component $f_1(t)$ and its variogram $v_{f1}(\theta)$ are never identically nil whereas with continuous materials such as liquid streams, the variogram $v_{f1}(\theta)$ is usually identically nil.

6.5.3. Properties of $v_{f2}(\theta)$:

In the same way, we shall define the following quantities :

v_{f2} : the relative mean square of $f_2(t)$ throughout the domain (T_L) :

$$v_{f2} = \frac{1}{T_L \, f_L^2} \int_{(T_L)} f_2^2(t) \, dt \qquad \text{(dimensionless)}$$

$\rho_{f2}(\theta)$: the autocorrelation function of $f_2(t)$. The variogram $v_{f2}(\theta)$ can be written :

$$v_{f2}(\theta) = \{1 - \rho_{f2}(\theta)\} \, v_{f2} \qquad \text{(dimensionless)}$$

We know that $f_2(t)$ represents the continuous trends of $f(t)$, all discontinuities being taken into account by $f_1(t)$. In other words, this property means that :

$\theta = 0$: $\qquad \rho_{f2}(\theta) = 1 \qquad$ hence $\qquad v_{f2}(0) = 0$

Now, the shortest the interval θ the strongest the autocorrelation function $\rho_{f2}(\theta)$. This autocorrelation function is of the non-negative type as all cyclic trends are taken into account by $f_3(t)$ and we can surmise, which is constantly supported by experience, that for values of θ larger than a certain value θ_{f2} called the <u>range</u> of $v_{f2}(\theta)$ the autocorrelation function is nil :

$\theta \geq \theta_{f2}$: $\qquad \rho_{f2}(\theta) = 0 \qquad$ hence $\qquad v_{f2}(\theta) = v_{f2}$

Between $\theta = 0$ and $\theta = \theta_{f2}$, the variogram $v_{f2}(\theta)$ is an increasing function of θ.

6.5.4. Properties of $v_{f3}(\theta)$:

Any periodic function $f_3(t)$ can be broken up into a sum of a certain number of terms of the general form :

$$f_{31}(t) = f'_{31} \sin 2\pi t/T_{P1} + f''_{31} \sin 2\pi t/T_{P1}$$

$$f_{32}(t) = f'_{32} \sin 2\pi t/T_{P2} + f''_{32} \sin 2\pi t/T_{P2} \qquad \text{etc ...}$$

with f'_{31}, f''_{31}, f'_{32}, f''_{32} constants and T_{P1}, T_{P2} periods of the components of the phenomenon. By a simple translation along the time axis, $f_{31}(t)$ takes the simpler form :

$$f_{31}(t) = f_{31} \sin 2\pi t/T_{P1} \quad \text{which we shall retain.}$$

Thanks to the knowledge of the mathematical expression of $f_{31}(t)$ we can easily calculate $v_{f31}(\theta)$:

$$v_{f31}(\theta) = (1 - \cos 2\pi\theta/T_{P1}) \, v_{f31} \qquad \text{with} \qquad v_{f31} = \frac{f_{31}^2}{2 f_L^2}$$

The variogram of a simple periodic function is a periodic function with the same period.

K being an integer, we shall note the following properties:

$\theta = 2K \dfrac{T_{p1}}{2} \rightarrow v_{f31}(\theta) = 0$

$\theta = (2K+1) \dfrac{T_{p1}}{2} \rightarrow v_{f31}(\theta) = 2 v_{f31}$

The variogram oscillates indefinitely about its mean v_{f31} between 0 and $2 v_{f31}$.

As with any continuous function, we have:

$\theta = 0 \quad : v_{f31}(0) = 0$

For the sake of simplicity, we shall now assume that $f_3(t)$ is reduced to a simple term of the form:

$f_3(t) = f_3 \sin 2\pi t/T_p$ with a variogram of the form $v_{f3}(\theta) = (1 - \cos 2\pi\theta/T_p) v_{f3}$

6.5.5. Properties of $v_f(\theta)$:

$v_f(\theta) = v_{f1}(\theta) + v_{f2}(\theta) + v_{f3}(\theta)$

We shall call v_{f0} the largest quantity we accept to regard as negligible in comparison with v_{f1}. Then, we shall call "<u>threshold of the variogram $v_f(\theta)$</u>" the interval θ_{f0} defined by:

$v_{f2}(\theta_{f0}) + v_{f3}(\theta_{f0}) = v_{f0} \ll v_{f1}$

Hence, for values of θ such as $0 < \theta_{f1} < \theta \leq \theta_{f0}$ we have:

$v_f(\theta) = v_{f1}$

Now, for all values of θ such as $\theta \geq \theta_{f2}$ we have:

$v_{f1}(\theta) = v_{f1} \quad : \quad v_{f2}(\theta) = v_{f2} \quad : \quad v_{f3}(\theta) = (1 - \cos 2\pi\theta/T_p) v_{f3}$ and

$v_f(\theta) = v_{f1} + v_{f2} + (1 - \cos 2\pi\theta/T_p) v_{f3}$

If we put $v_f = v_{f1} + v_{f2} + v_{f3}$, then the variogram $v_f(\theta)$ oscillates indefinite about v_f.

Fig. 6.1. to 6.4. present an example of periodic variogram and of its component

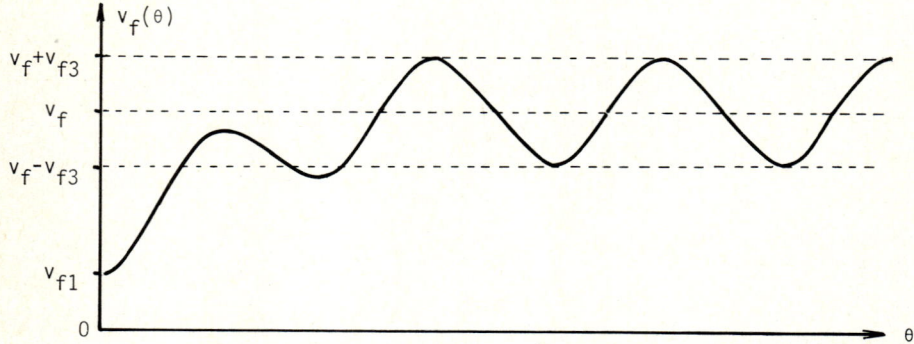

Fig. 6.1. Variogram $v_f(\theta)$ - General case - Example of a periodic variogram.

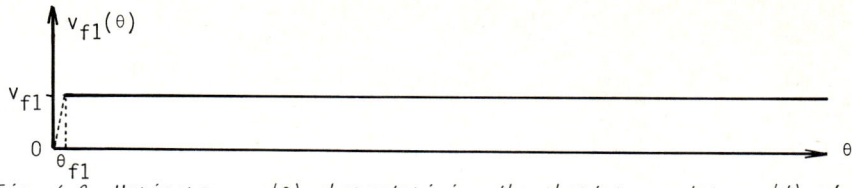
Fig. 6.2. Variogram $v_{f1}(\theta)$ characterizing the short-range term $a_1(t)$ of $a(t)$

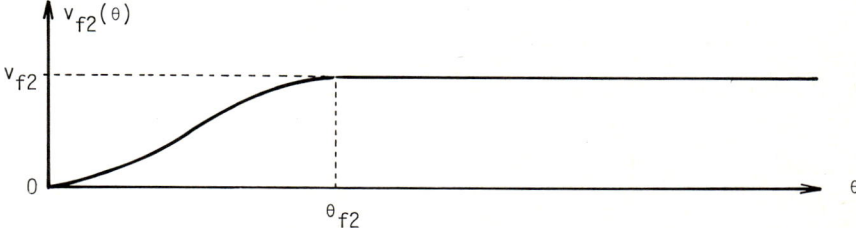
Fig. 6.3. Variogram $v_{f2}(\epsilon)$ characterizing the long-range term $a_2(t)$ of $a(t)$

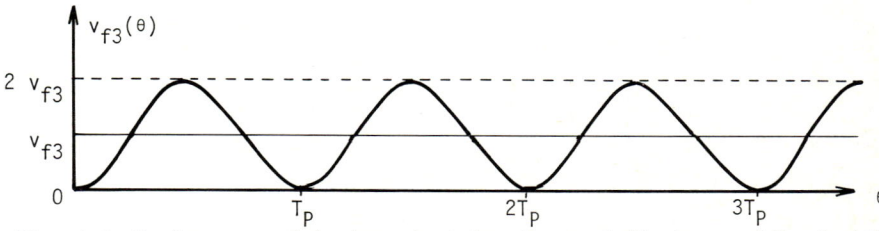
Fig. 6.4. Variogram $v_{f3}(\theta)$ characterizing the periodic term $a_3(t)$ of $a(t)$.

6.5.6. Particular non-periodic variograms :

Particular cases are derived from the general case when one or several components of the variogram are zero.

6.5.6.1. General non-periodic variogram : This case is defined by :
$$f_3(t) \equiv 0 \quad : \quad v_{f3}(\theta) \equiv 0 \quad : \quad f(t) = f_0 + f_1(t) + f_2(t)$$
$$v_f(\theta) = v_{f1}(\theta) + v_{f2}(\theta)$$

Such a variogram is represented in fig 6.5.

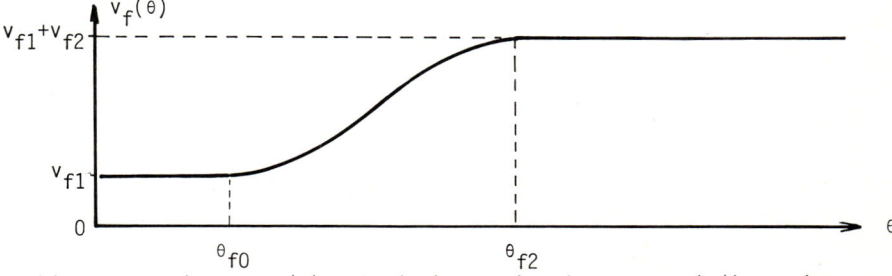
Fig.6.5. Variogram $v_f(\theta)$ - Typical example of a non-periodic variogram.

6.5.6.2. **Useful domain of a variogram** : We shall see in due course (chapter 8) that the estimation of the moments of the selection error requires a mathematical expression of the variogram valid between $\theta = 0$ and a certain value $\theta = \theta_{fu}$, these limits defining the useful domain of θ. For all practical purposes, θ_{fu} is the upper limit of the uniform interval between increments or the uniform strata length according to the selection scheme actually implemented (chapter 7). Now, as a very general rule (which must however be checked with every particular case) θ_{fu} always remains small in comparison with the range θ_{f2} beyond which $v_f(\theta)$ remai practically constant. Hence the possibility for a certain portion of the variogram to be represented by a function of θ that is not limited when $\theta \to \infty$.

To take a concrete example, if we don't consider using a uniform interval between systematic increments larger than 20 mn, we need a mathematical expression of the variogram valid between $\theta = 0$ and $\theta = 20$ mn.

6.5.6.3. **Flat variogram** : A variogram is said to be flat when it remains consta throughout its useful domain. The variogram $v_{f1}(\theta)$ characterizing the short-range fluctuation term $a_1(t)$ is flat. The variograms $v_{f2}(\theta)$ and $v_{f3}(\theta)$ are not. As a con sequence of this observation, the variogram $v_f(\theta)$ can be regarded as flat if and only if its components $v_{f2}(\theta)$ and $v_{f3}(\theta)$ are identically zero. Fig. 6.2. shows an example of a flat variogram. A flat variogram is therefore represented by :
$v_f(\theta) = v_{f1} =$ constant.

6.5.6.4. **Rectilinear variogram** : A variogram is said to be rectilinear when it can be represented by a straight line in its useful domain. Then :

$v_f(\theta) = v_{f1} + v'_{f2} \theta$ which means that : $v_{f1}(\theta) = v_{f1}$ and $v_{f2}(\theta) = v'_{f2} \theta$.

Fig. 6.6. represents a rectilinear variogram.

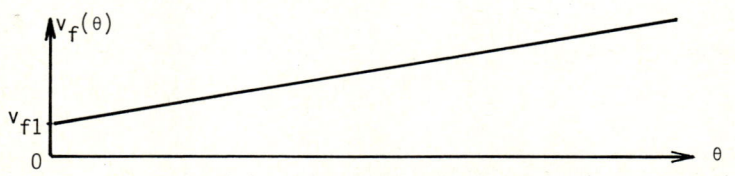

Fig. 6.6. Variogram $v_f(\theta)$ - Typical example of a rectilinear variogram.

The rectilinear approximation is very often acceptable for values of θ_{fu} as large as 30 or even 60 mn and may therefore be used in a number of practical cases

6.5.6.5. **Parabolic variogram** : A variogram is said to be parabolic when it can be represented by a parabola in its useful domain. Then :

$v_f(\theta) = v_{f1} + v'_{f2} \theta + v''_{f2} \theta^2$ which means : $v_{f1}(\theta) = v_{f1}$ and $v_{f2}(\theta) = v'_{f2}\theta + v''_{f2}\theta^2$

The parabola can be concave upwards (v''_{f2} positive) or downwards (v''_{f2} negative). According to our experience, the parabolic approximation is very often acceptable for values of θ_{fu} as large as several hours and practically always acceptable in the useful domain of the variogram. Fig. 6.7. represents a parabolic variogram.

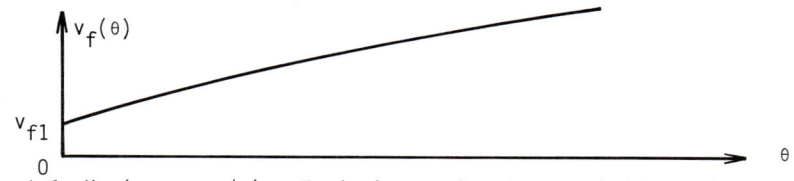

Fig. 6.7. Variogram $v_f(\theta)$ - Typical example of a parabolic variogram.

6.5.6.6. General expression of a non-periodic variogram : Both the flat and rectilinear variograms can be regarded as particular cases of a parabolic variogram. We shall therefore retain the parabolic variogram as the most general useful expression of a non-periodic variogram. When the value of θ_{fu} is small in comparison with the period T_p of an eventual periodic term, the parabolic approximation remains usually valid even in the presence of periodic fluctuations.

We shall therefore retain the general expression :

$v_f(\theta) = v_{f1} + v'_{f2}\theta + v''_{f2}\theta^2$ with v_{f1}, v'_{f2} and v''_{f2} the "variographic parameters".
A flat variogram is characterized by : $v'_{f2} = v''_{f2} = 0$
A rectilinear variogram by : $v''_{f2} = 0$

6.6. EXPERIMENTAL DETERMINATION OF THE VARIOGRAM - LOGICAL APPROACH

A "variographic experiment" is an experiment devised and implemented in order to determine the variograms of the functions $\mu(t)$, $\alpha(t)$ and $a(t)$. Such an experiment consists in :
- extracting from the flowing stream a certain number Q of increments I_q by means of a correct device (correct increment delimitation and extraction will be the object of chapters 17 and 18).
- estimating the weight M_q and critical content a_q of each increment I_q.
- calculating therefrom a series of estimates of the variograms $v_\mu(\theta)$, $v_\alpha(\theta)$ and $v_a(\theta)$.
- assessing the value of the variographic parameters $v_{\mu 1}$, $v'_{\mu 2}$, $v''_{\mu 2}$, $v_{\alpha 1}$, $v'_{\alpha 2}$, $v''_{\alpha 2}$, v_{a1}, v'_{a2}, v''_{a2} and of various coefficients of correlation which will be introduced in chapter 8. The complete variographic technique will be described for a non-specified function f(t). Examples of variographic experiments will be presented in chapter 14 together with practical advices on how to organize such an experiment.

6.6.1. Definition of the discrete variogram :

The variogram defined in section 6.4. is what we shall now call the "continuous variogram". From an experimental standpoint, the only thing we can do is to carry out a few punctual estimations of the "discrete variogram" that we are going to define now.

We shall assume that we know the true value of the function f(t) in a series of Q instants t_q defined as follows :

$t_q = (q - 1/2) h$ with q = 1, 2, ... Q.

In this expression, h is a uniform interval between consecutive punctual increments I_q. The number Q of increments is defined by the following inequalities :

$(Q - 1/2) h \leq T_L < (Q + 1/2) h$

which means that the Qth and last increment must fall within the limits of the domain (T_L) in order to belong to the lot. For the sake of simplicity, we shall assume that h is a submultiple of T_L and that T_L = Q h (Q being an integer).

We shall then calculate an estimate of f_L:
- with $f(t) \equiv \mu(t)$ or $\alpha(t)$:

$f_L = \frac{1}{Q} \sum_q f(t_q)$

- with $f(t) \equiv a(t)$

$f_L = \frac{\alpha_L}{\mu_L}$

We shall now define the "discrete variogram" $w_f(jh)$:

$w_f(jh) = \frac{1}{2(Q - j) f_L^2} \sum_q \{f(t_{q+j}) - f(t_q)\}^2$ with q = 1, 2, ..(Q-j) and j = 1, 2,

It is a discrete estimator of the continuous variogram $v_f(\theta)$. Such an estimator becomes poorly efficient when the number (Q-j) of squares involved in the calculation of the mean square becomes too small, say smaller than 20 or 30, with 20 a reasonable lower limit. J is therefore practically limited to $J \leq Q - 20$. It can b shown that when h tends towards zero, the product jh remaining constant and equal to h_0, the discrete variogram $w_f(jh)$ tends towards the continuous variogram $v_f(h_0)$ It is therefore always advisable to select a value of the interval h as small as economically reasonable. When h is small enough, it is possible to assimilate the variogram $v_f(h_0)$ to its estimator $w_f(jh)$.

Remark : Thanks to its relative definition, the variogram $w_f(jh)$ is not affecte when f(t) is multiplied by a constant factor. For this reason, the weights M_q and A_q being respectively proportional to the rates of flow $\mu(t_q)$ and $\alpha(t_q)$, the discrete variograms $w_M(jh)$ and $w_A(jh)$ are identical with the variograms $w_\mu(jh)$ and $w_\alpha(jh)$ and can be used directly as estimators of the variograms $v_\mu(\theta)$ and $v_\alpha(\theta)$.

6.6.2. Definition of the experimental variogram :

Actually, we never know the true values of $f(t_q)$ but experimental estimates resulting from various error-generating operations such as sampling, weighing, sample reduction and assaying carried out on the increments I_q. When we substitute the estimates $f'(t_q)$ for the true unknown values $f(t_q)$ in the expression of the discrete variogram $w_f(jh)$ we obtain the "experimental variogram" $w'_f(jh)$ which is a raw experimental estimate of $w_f(jh)$:

$$w'_f(jh) = \frac{1}{2(Q-j) f_L^2} \sum_q \{f'(t_{q+j}) - f'(t_q)\}^2 \text{ with } q = 1, 2, ..(Q-j) \text{ and } j = 1, 2,.. J.$$

Remark : The variogram of a discrete sequence of arranged data, such as the variogram of a zero-dimensional object is nothing else than the experimental variogram defined in this section.

6.6.3. Definition of the corrected variogram :

We shall call e_{fq} the relative estimation error of $f(t_q)$ with :

$$e_{fq} = \frac{f'(t_q) - f(t_q)}{f_L} \quad \text{(dimensionless)}$$

- when $f(t)$ is the rate of flow $\mu(t)$ of active components, e_{fq} is the resultant of the sampling and weighing errors,
- when $f(t)$ is the critical content $a(t)$, e_{fq} is the resultant of the sampling, sample reduction and assaying errors,
- when $f(t)$ is the rate of flow $\alpha(t)$ of critical component, calculated for each increment as the product $a(t) \mu(t)$, e_{fq} is the resultant of the sampling, weighing, sample reduction and assaying errors.

Since all variograms are defined as mean squares of the differences of two values of $f(t)$ or of its estimates, the procedure may (theoretically at least) be biased, the estimation bias does not interfere in the calculations. We shall call s_f^2 an experimental estimate of the true value of the variance of e_{fq} distribution. It is easy to show that :

$$w'_f(jh) = w_f(jh) + s_f^2$$

If we can obtain an estimate of s_f^2, we can define the "corrected variogram" :

$$w''_f(jh) = w'_f(jh) - s_f^2$$

This is the best available unbiased estimate of the discrete variogram $w_f(jh)$ and therefore the best available unbiased estimate of the continuous variogram $v_f(jh)$.

6.6.4. Definition of the model variogram :

We know how to obtain an estimate of the continuous variogram $v_f(\theta)$ for a series of points $\theta = jh$ with $j = 1, 2, .. J$ but we now need an analytical expression of $v_f(\theta)$. We shall call "model variogram" $v'_f(\theta)$ the mathematical expression that can be deduced from the punctual estimates $w''_f(jh)$ by means of any of the conventional interpolation methods. From a practical standpoint, however, the graphical method presented in section 6.7. and illustrated in chapter 14 is the most convenient. The model variogram $v'_f(\theta)$ is the best available representation of the variogram $v_f(\theta)$.

6.6.5. Recapitulation :

We have defined five variograms of a same function $f(t)$:

$v_f(\theta)$: continuous variogram
$w_f(jh)$: discrete variogram - estimator of $v_f(\theta)$ for $\theta = jh$
$w'_f(jh)$: experimental variogram - raw estimate of $w_f(jh)$
$w''_f(jh)$: corrected variogram - unbiased experimental estimate of $w_f(jh)$
$v'_f(\theta)$: model variogram - continuous representation of $w''_f(jh)$ - model of $v_f(\theta)$

Thanks to the definition of the relative increase $\delta_f(t,\theta)$, all variograms are dimensionless, irrespective of the unit used to express $f(t)$ provided that the same unit is used everywhere. As far as the time unit is concerned, we definitely recommend the use of the "<u>decimal minute</u>" in all calculations, even if for the sake of convenience we speak of 2 seconds or 3 hours. A 30 second interval will be expressed by 0.5 mn , a 3 mn 45 s interval will be expressed by 3.75 mn and a 3 hour interval by 180 mn.

6.7. INTERPRETATION OF THE RESULTS OF A VARIOGRAPHIC EXPERIMENT

This section deals with the logical background to this interpretation whilst chapter 14 deals with the practical aspects of the variographic experiment and with numerical examples. We shall assume here that :

- A variographic experiment has been carefully devised and carried out and that several series of Q increments I_q have been extracted from the lot at an interval h which is uniform within a given series and different from one series to the next

- The Q increments of each series have been carefully weighed, prepared and assayed. Each increment I_q is therefore fully described by three estimates :
M_q : weight of active components in I_q
a_q : critical content of the increment I_q
A_q : weight of critical component in I_q calculated from $A_q = a_q M_q$

- Each increment has been diverted from the flowing stream during a constant time T_I that can be calculated from the geometrical and mechanical features of the sampling device (e.g. for a straight-path cross-stream cutter with a width W and a velocity V, the diversion time is $T_I = W/V$). If necessary we can then calculate the following estimates :

$\mu(t_q) = M_q/T_I$ and $\alpha(t_q) = A_q/T_I = a_q M_q/T_I$

This is usually unnecessary since we know (section 6.6.1. in fine) that the variograms of $\mu(t)$ and $\alpha(t)$ are identical with those of M_q and A_q. We can therefore directly calculate the variograms of M_q and A_q.

- The estimation variances s_M^2 and s_A^2 have been determined. For all practical purposes, we can safely admit that $s_M^2 = 0$, $s_A^2 = s_a^2$ (assaying variance).

- A certain number $J = Q - 20$ of points of the corrected variograms $w_M''(jh)$, $w_A''(jh)$ and $w_a''(jh)$ have been computed and plotted on a graph against $\theta = jh$.

- The curves of the model variograms $v_M'(\theta)$, $v_A'(\theta)$ and $v_a'(\theta)$ have been drawn between the experimental points.

Like the continuous variograms, the model variograms $v_f'(\theta)$ can be regarded as sums of three terms :

$$v_f'(\theta) = v_{f1}'(\theta) + v_{f2}'(\theta) + v_{f3}'(\theta)$$

Each of these three terms must be determined separately as illustrated in sections 14.3. and 14.4.

6.7.1. Estimation of v_{f1}' : see also section 14.3.2 :

We know that when θ tends towards zero, the terms $v_{f2}(\theta)$ and $v_{f3}(\theta)$ also tend towards zero whilst $v_{f1}(\theta)$ remains constant and equal to v_{f1}. We shall assume that the threshold θ_{f0} is smaller than the half-period $T_p/2$ of an eventual periodic term. Then, for all values of θ smaller than $T_p/2$, the variograms $v_{f2}(\theta)$, $v_{f3}(\theta)$ and therefore $v_f(\theta)$ are increasing functions of θ. All values $w''(jh)$ of the corrected variogram obtained for $jh \leq \theta_{f0}$ can be regarded as estimators of the variographic parameter v_{f1}. The best estimator is that which has been obtained with the largest number of degrees of freedom, i.e. $w_f''(h)$ calculated from $Q-1$ differences. We shall therefore retain :

$$v_{f1}' = w_f''(h)$$

as the best available unbiased estimator of v_{f1}.

6.7.2. Estimation of $v_{f2+3}'(\theta) = v_{f2}'(\theta) + v_{f3}'(\theta)$: see also section 14.3.3.

We can now calculate and eventually tabulate the $J = Q - 20$ estimates of :

$w_f''(jh)$	column 1
$w_{f1}''(jh) = w_f''(h) = v_{f1}'$	column 2 = first line of 1
$w_{f2+3}''(jh) = w_{f2}''(jh) + w_{f3}''(jh) = w_f''(jh) - w_f''(h)$	column 3 = col.1 - col.2

The latter is an estimate of $v_{f2+3}(\theta)$ for $\theta = jh$. By definition of the threshold, (section 6.5.4.) $w_f''(jh)$ and $w_f''(h)$ are two independent estimates of the same parameter v_{f1} as long as jh remains smaller than or equal to the threshold θ_{f0}. Their difference shown in column 3 is a random variable liable to take negative values. Now, when jh is larger than the threshold, this difference which is also by definition the sum of two variograms, is necessarily a non-negative quantity. This property provides a good objective, experimental definition of the threshold θ_{f0} : The threshold of a variogram is the largest value of jh for which the difference shown in column 3 is negative.

6.7.3. Checking the existence of a periodic term : see also section 14.3.4.1. :

This can usually be deduced from the examination of the graph of $w_f''(jh)$. The existence of a periodic term is graphically obvious whenever :

- the interval h is small as compared with the period T_p of the phenomenon (say 1/5 or less),
- the sum $v_{f1}' + s_f^2$ is small as compared with the variographic parameter v_{f3}.

In some instances, however, the existence of a periodic term is doubtful and must be checked by means of the test of the differential consisting in :

- computing for all values of j the difference $\Delta w_{fj}'' = w_f''(j+1)h - w_f''(jh)$. This difference is an estimator of the differential of $v_f(\theta)$ for $\theta = (2j+1)h/2$.
- plotting $\Delta w_{fj}''/h$ against $\theta = (2j+1)h/2$ and drawing a line between the experimental points.
- admitting, if $\Delta w_{fj}''/h$ is a never decreasing function, that, at the scale of our experiment the periodic term is identically nil,
- admitting, if $\Delta w_{fj}''/h$ is alternately positive and negative with a marked periodic character that the variogram contains a periodic term.

By amplifying the periodic character of the function, the test of the differential is more sensible than a simple examination of the graph of the corrected variogram. We shall see an example of this amplifying effect in section 14.3.4.1.

6.7.4. Analysis of a non-periodic variogram : see also section 14.4. :

Such a variogram is characterized by :

$$v_{f3}(\theta) \equiv 0 \rightarrow v_f(\theta) = v_{f1}(\theta) + v_{f2}(\theta)$$

Knowing the estimate v_{f1}' of $v_{f1}(\theta)$ (section 6.7.1.) we can easily calculate :

$$w_{f2}''(jh) = w_f''(jh) - w_f''(h)$$ already tabulated in section 6.7.2. (column 3),

retaining only the positive values corresponding to values of jh larger than the threshold. There remains to plot $w_{f2}''(jh)$ against $\theta = jh$. The experiment does not provide any information on the behaviour of $v_f(\theta)$ for values of θ smaller than the threshold, but we know the point $v_{f2}(0) = 0$. It will practically never be necessary to resort to algebraical interpolation methods. In all examples in our possession, covering a great variety of materials, it has been possible to solve the problem in a simple graphical way.

As a general rule, graphical interpolation is easy if a straight line can be drawn between experimental points. We have previously mentioned that our representation was required only in the useful domain of the variogram which, with systematic and stratified random sampling, is usually limited to 30 mn, very seldom more than that.

According to this general rule we shall successively try :

- to plot $w''_{f2}(jh)$ against jh : if the points form a straight line, with minor random deviations, the variogram is rectilinear and can be represented by :

$$v'_{f2}(\theta) = v'_{f2}\theta \rightarrow v'_f(\theta) = v'_{f1} + v'_{f2}\theta$$

The gradient v'_{f2} of the straight line is easily estimated. If the curve shows a slight bend up- or downwards, we can try :

- to plot $w''_{f2}(jh)/jh$ against jh : if the points form a straight line, the variogram is parabolic and can be represented by :

$$v'_{f2}(\theta) = v'_{f2}\theta + v''_{f2}\theta^2 \rightarrow v'_f(\theta) = v'_{f1} + v'_{f2}\theta + v''_{f2}\theta^2$$

The gradient v''_{f2} of the straight line and the ordinate v'_{f2} for $jh = 0$ can be easily estimated. We shall present an example of parabolic representation in section 14.3.4.5.

6.7.5. Analysis of a periodic variogram : see also section 14.3 :

Let's come back to the point of the demonstration (section 6.7.2.) where we have obtained :

$$w''_{f2+3}(jh) = w''_{f2}(jh) + w''_{f3}(jh) = w''_f(jh) - w''_f(h) = w''_f(jh) - v'_{f1}$$

6.7.5.1. Logical approach : for the sake of simplicity, we shall assume that the term $f_3(t)$ is sinusoidal with a variogram :

$$v_{f3}(\theta) = (1 - \cos 2\pi\theta/T_p) v_{f3}$$

We must now estimate the parameters v_{f3} and T_p. We shall use the property :

For $\theta = 2K\dfrac{T_p}{2}$ (K integer) : $v_{f3}(\theta) = 0 \rightarrow v_f(\theta) = v_{f1} + v_{f2}(\theta)$

For $\theta = (2K + 1)\dfrac{T_p}{2}$: $v_{f3}(\theta) = 2v_{f3} \rightarrow v_f(\theta) = v_{f1} + v_{f2}(\theta) + 2v_{f3}$

The variogram $v_f(\theta)$ oscillates between two curves that can be superposed by a translation of ordinates equal to $2v_{f3}$. Now, the experiment may fail to give a true picture of this phenomenon for two main reasons :

- the discrete values of jh do not usually coincide exactly with $2K\dfrac{T_p}{2}$ or $(2K+1)\dfrac{T_p}{2}$
- the estimates $w''_f(jh)$ may differ from the true values $v_f(jh)$.

Fig 6.8. shows the example of a periodic variogram actually observed. The function $f(t)$ was the Zn content of the feed to a flotation plant. The periodic nature of $v_f(\theta)$ is doubtless and will be confirmed (section 14.3) by the test of the differential. With a corrected variogram such as this, our first step will consist in drawing two simple curves Γ_1 and Γ_2, superposable by translation parallel to the ordinates axis and enveloping the broken line Γ linking up the experimental points $w''_f(jh)$.

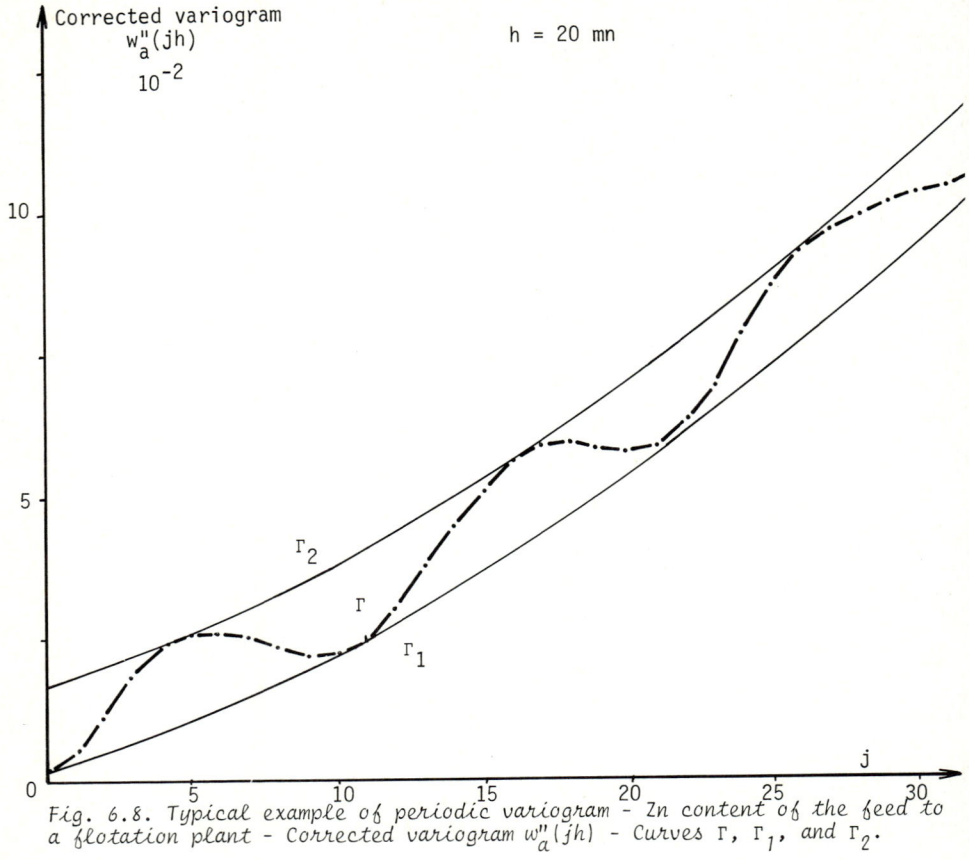

Fig. 6.8. Typical example of periodic variogram - Zn content of the feed to a flotation plant - Corrected variogram $w''_a(jh)$ - Curves Γ, Γ_1, and Γ_2.

On this graph,

- Γ_1 is a rough graphical representation of $v'_{f1} + v'_{f2}(\theta)$. It should be practically tangent to Γ for all values of $\theta = KT_p$ (K integer).

- Γ_2 is a rough graphical representation of $v'_{f1} + v'_{f2}(\theta) + 2v'_{f3}$ (v'_{f3} being an experimental estimate of v_{f3}). It should be practically tangent to Γ for all value of $\theta = (2K + 1)T_p/2$.

We shall describe two methods for estimating T_p and v_{f3} : a simple and fast one and a more accurate one.

<u>6.7.5.2. Simple graphical method for estimating</u> T_p and v_{f3} : see section 14.3.4

- the abscissa of the first contact of Γ with Γ_2 is an estimate of $T_p/2$.
- the abscissa of the first (non-zero) contact of Γ with Γ_1 is an estimate of T_p.
- the abscissa of the second contact of Γ with Γ_2 is an estimate of $3T_p/2$, .. etc.

The second of these three estimates is usually the best. It directly gives T_p. The distance between Γ_1 and Γ_2 is an estimate of $2v_{f3}$.

6.7.5.3. Accurate method for estimating T_p and v_{f3} : see section 14.3.4.3. :

From the first method, we retain that the first estimate of T_p falls between two experimental points :

$Kh \leq T_p \leq (K+1)h$ (K integer)

In order to eliminate the periodic term $v_{f3}(\theta)$ we shall use the following property of the sinusoid :

$$\frac{1}{T_p}\int_0^{T_p} \cos \frac{2\pi}{T_p}(\theta + \theta') d\theta = 0 \quad : \text{ irrespective of } \theta'$$

In the same way, assuming K to be an even integer and $(K+1)h$ to be equal to T_p :

$$\frac{1}{K+1}\sum_{j'} \cos \frac{2\pi}{T_p}(j + j')h = 0 \quad : \text{ irrespective of } j'.$$

In this expression, j' is an integer taking all possible values from $-K/2$ to $+K/2$. Still with K even but with $(K+1)h$ slightly different from T_p, this equality is approximately satisfied, but for a periodic residue that may be neglected. Then :

$$v_{f3}\{(j+j')h\} = \{1 - \cos\frac{2\pi}{T_p}(j+j')h\} v_{f3} \rightarrow \frac{1}{K+1}\sum_{j'} v_{f3}\{(j+j')h\} = v_{f3}$$

The method will therefore consist in calculating the successive values of the shifting mean $W_f(jh)$ of K or K+1 consecutive values of $w_f''(jh)$. Assuming K to be even, we shall calculate :

$$W_f(jh) = \frac{1}{K+1}\sum_{j'} w_f''\{(j+j')h\} \quad \text{with } j' \text{ integer and } j' = -K/2, \ldots 0 \ldots, +K/2.$$

Then, assuming $(K+1)h$ to be equal to T_p, we have :

$$W_f(jh) = \frac{1}{K+1}\sum_{j'} w_{f2}''\{(j+j')h\} + v_{f1} + v_{f3}$$

From a practical standpoint, we shall plot $W_f(jh)$ against jh and distinguish two cases :

- a periodic modulation remains noticeable on the curve linking up the points $W_f(jh)$. This happens when K is small and when $(K+1)h$ is too different from T_p. The best solution is to eliminate these residual fluctuations by a graphical method. The non-periodic curve that can be drawn between the points of $W_f(jh)$ is then treated in the following way :

- no periodic modulation remains noticeable. Here again, two cases may arise :
a) $W_f(jh)$ is practically rectilinear in the interval $0 < jh \leq 2T_p$ (this can be approximative). Then, we obtain a good estimate $W_{f2}(jh)$ of $v_{f2}(jh)$ with :

$$W_{f2}(jh) = \frac{1}{K+1}\sum_{j'} w_{f2}''\{(j+j')h\}$$

This is also an estimate of $w_{f2}''(jh)$.
We can therefore write :

$W_f(jh) = W_{f2}(jh) + v_{f1} + v_{f3}$

$w_f''(jh) = w_{f2}''(jh) + v_{f1} + (1 - \cos \frac{2\pi}{T_p} jh) v_{f3}$

A simple substraction gives (assimilating both estimates of $v_{f2}(jh)$:

$W_f(jh) - w_f''(jh) = v_{f3} \cos \frac{2\pi}{T_p} jh =$ estimate of $v_{f3} - v_{f3}(jh) = w_{f3}'(jh)$

By plotting the estimates of $w_{f3}'(jh)$ against jh we obtain points that should fit a sine curve from which it will be easy to estimate T_p and v_{f3}. A near perfect example of sinusoidal variogram $w_{f3}'(jh)$ will be presented in section 14.3.4.3. (fig. 14.7).

b) $W_{f2}(jh)$ cannot be regarded as rectilinear. Then, the best thing to do is to revert to the simple graphical method described in section 6.7.5.2.

6.7.5.4. Expression of $v_{f3}'(\theta)$: see also section 14.3.4.4. :

The model variogram $v_{f3}'(\theta)$ is immediately deduced from the estimates T_p' and v_{f3}' of T_p and v_{f3} :

$v_{f3}'(\theta) = (1 - \cos \frac{2\pi\theta}{T_p'}) v_{f3}'$

6.7.5.5. Expression of $v_{f2}'(\theta)$: see also section 14.3.4.5. :

We obtain a good estimate of $v_{f2}(jh)$ by calculating :

$w_{f2}''(jh) = W_f(jh) - (v_{f1}' + v_{f3}')$

using the values of $W_f(jh)$, v_{f1}' and v_{f3}' estimated in the preceding sections. We plot $w_{f2}''(jh)$ against jh and deduce the expression of $v_{f2}'(\theta)$ as already shown in section 6.7.4. The branch $0 < \theta < Kh/2$ of $v_{f2}'(\theta)$ must be interpolated between $v_{f2}'(0) = 0$ and $w_{f2}''(Kh/2) = W_f(Kh/2) - (v_{f1}' + v_{f3}')$. The best way to do it is to follow the natural shape of the curve down to zero.

6.7.5.6. Expression of $v_f'(\theta)$: see also section 14.3.4.6. :

We know the expression of the three components of $v_f'(\theta)$ and we just have to write :

$v_f'(\theta) = v_{f1}' + v_{f2}'(\theta) + v_{f3}'(\theta)$

It is always advisable to calculate the values of $v_f'(jh)$ and to compare them to the corrected experimental values $w_f''(jh)$. The differences should be small.

6.7.6. Objective definition of the components of f(t) :

We now know how to estimate the three components of the variogram of $f(t)$, and we know that these components are the variograms of $f_1(t)$, $f_2(t)$ and $f_3(t)$ respectively. We are therefore in a position to give now an objective definition of the components of $f(t)$.

- $f_1(t)$ is a function with a zero mean which is characterized by a variogram
$v'_{f1}(\theta) = v'_{f1} =$ constant.
- $f_2(t)$ is a function with a zero mean which is characterized by a variogram
$v'_{f2}(\theta) = v'_{f2}\theta + v''_{f2}\theta^2$
- $f_3(t)$ is a function with a zero mean (assuming T_L to be a multiple of T_p) which is characterized by a variogram such as :
$v'_{f3}(\theta) = (1 - 2\pi\theta/T_p) v'_{f3}$
or a sum of terms of the same form such as :
$v'_{f31}(\theta) = (1 - 2\pi\theta/T_{p1}) v'_{f31}$, $v'_{f32}(\theta) = (1 - 2\pi\theta/T_{p2}) v'_{f32}$ etc...

6.7.7. Variographic parameters :

The variogram $v'_f(\theta)$ is completely determined by the following group of parameters :

v'_{f1} : characterizing the short-range term $f_1(t)$ (dimensionless)
v'_{f2} : characterizing the long-range term $f_2(t)$ (dimension t^{-1})
v''_{f2} : characterizing the long-range term $f_2(t)$ (dimension t^{-2})
v'_{f3} : characterizing the periodic term $f_3(t)$ (dimensionless)
T_p : characterizing the periodic term $f_3(t)$ (dimension t^{+1})

These are the <u>variographic parameters</u>. We shall see in chapters 8 to 13 how to express the moments of the various errors arising from the implementation of the continuous selection model by means of these parameters. Assuming the variogram to be non-periodic, a variographic experiment provides nine variographic parameters :

v'_{M1} , v'_{M2} , v''_{M2} characterizing the rate of flow function $\mu(t)$ (active components),
v'_{A1} , v'_{A2} , v''_{A2} characterizing the rate of flow function $\alpha(t)$ (critical component)
v'_{a1} , v'_{a2} , v''_{a2} characterizing the critical content function $a(t)$.

To these nine parameters, we must add the coefficient of correlation $\rho(A_q, M_q)$ of the distribution of :

- M_q : weight of active components in the increment I_q ,
- A_q : weight of critical component in the increment I_q.

6.8. PARTICULAR CASE OF ZERO-DIMENSIONAL LOTS

These lots have been defined in section 5.3.2.4. The variogram characterizing a sequence of units is nothing else than the discrete variogram defined in section 6.6.1. with h = 1 (dimensionless). The corresponding raw experimental variogram, corrected variogram and model variogram are accordingly defined. When the chronological sequence is random the variogram is flat. A variographic experiment can therefore be used in order to check the randomness of a sequence of data.

6.9. CONCLUSIONS

We have had the opportunity of studying the variograms of hundreds of mineral commodities such as the feed to mineral processing plants, smelters, blending systems, cement kilns, ores and concentrates being shipped or unloaded, etc ... usually in relation to the estimation of sampling errors or to the appraisal of sampling facilities, we can summarize our experience as follows :

1) The variogram is an indispensable tool every time that the variability of a material is to be characterized,

2) The variability, characterized by means of one or several variograms, may differ considerably from one case to the next, both qualitatively and quantitatively,

3) In a given material, different components (different metals for instance) may be represented by very different variograms,

4) For a given material, produced under routine conditions (feed to a processing plant for example), the variability expressed by one or several variograms may be regarded as a relatively time-stable characteristic. This point should however be checked whenever possible,

5) Periodic phenomena are much more frequent than is usually imagined. According to their causes, their periods may vary from a few seconds to several days. These may be very dangerous and generate very large sampling errors (chapter 13). The variogram detects and analyses the periodic fluctuations much more efficiently than any other mathematical tool,

6) Hand computation of variograms is long, tedious and liable to introduce calculation errors. The help of a computer or at least of a serious desk calculator is strongly advisable,

7) As the moments of the sampling errors can be expressed in terms of variographic parameters, the variographic experiment that will be described in great detail in chapter 14 appears to be the key to the experimental estimation of the sampling errors and to the evaluation of sampling facilities, hence the importance of this chapter, not only from a theoretical but especially from a practical standpoint.

8) The reader should keep in mind, however, that the interpretation of the results of a variographic experiment is nearly always delicate and should be carried out by somebody well acquainted with variograms.

CHAPTER 7

REFERENCE SELECTION SCHEMES

7.1. INTRODUCTION

As previously mentioned, we shall restrict our study to the one-dimensional problem of the flowing streams of particulate materials of mineral origin. By definition, the lot L is made of the material flowing past the sampling point between time $t = 0$ and time $t = T_L$. In relation to the point selection step of the increment sampling process, the lot L is completely defined as the continuous set of all instants belonging to the time domain :

$$(T_L) \equiv \{ t \mid 0 < t \leq T_L \}$$

7.1.1. Punctual sample S :

The punctual sample S is obtained by reunion of Q punctual increments I_q selected from (T_L) in a series of instants t_q and is completely defined as the set of these Q instants t_q.

The purpose of the present chapter is to describe three reference point selection schemes, namely :

- Systematic selection with random positioning or, more shortly, systematic selection,
- Random stratified selection or, more shortly, stratified selection,
- Random selection,

and to study a certain number of their general properties. Founded as they are on a random selection of one kind or another, these three selection schemes are probabilistic. In the next chapters we shall study the properties of the continuous selection error and of its components when implementing each of these three selection schemes.

7.1.2. Definition of a "random selection" :

A selection is said to be <u>random</u> when it is carried out with a uniform probability (discrete perspective) or probability density (continuous perspective). In the present case, the selection of one and only one instant t_o within a certain time domain (T_o) is said to be "<u>random</u>" if the density of selection probability $\Pi(t)$ is uniform throughout (T_o) and equal to a constant Π_o. It is assumed to be nil outside (T_o). By hypothesis, the probability of selecting one and only one instant t_o is a certainty, i.e. a probability equal to unity. Hence :

$$\int_{(T_o)} \pi(t_o)dt_o = \pi_o \int_{(T_o)} dt = \pi_o T_o = 1 \rightarrow \pi_o = 1/T_o$$

7.2. SYSTEMATIC SELECTION WITH RANDOM POSITIONING

This will be more simply referred to as "<u>systematic selection</u>" (subscript sy) but the reader must remember that a systematic scheme may be regarded as probabilistic and the results of the following chapters may be applied if and only if the system is positioned at random on the time axis.

7.2.1. Definition : see fig. 7.1.

A systematic selection with random positioning can be schematized in the following way :

- A time interval T_{sy} is adopted according to the degree of representativeness to be achieved by the sample. T_{sy} is always small as compared with T_L.

- An instant t_1 is selected "at random" within the domain (T_1) defined as the set
$$(T_1) \equiv \{ t \mid 0 < t \leq T_{sy} \}$$

- A series of instants t_q are linked to t_1 according to a systematic pattern with a uniform interval between consecutive increments : by definition :

$$t_q = t_1 + (q - 1) T_{sy} \qquad (7.1.)$$

The instant t_q belongs to the domain (T_q) of extent T_{sy} defined as the set :
$$(T_q) \equiv \{ t \mid (q - 1) T_{sy} < t \leq q T_{sy} \} \qquad (7.2.)$$

- The number Q of instants t_q actually falling within the domain (T_L) occupied by the lot L on the time axis is defined by the following inequalities :

$$t_1 + (Q - 1) T_{sy} \leq T_L < t_1 + Q T_{sy} \qquad (7.3.)$$

Q is usually a random variable. This problem will be dealt with in section 7.2.2.

- These Q instants t_q are the seats of the punctual increments I_q and the set of these increments I_q (with q = 1, 2, ..Q) is the punctual sample S retained to represent the lot L.

7.2.2. Properties of the number Q of increments :

For given values of T_L and T_{sy}, the number Q of increments actually falling within the limits of the domain $(T_L) \equiv \{ t \mid 0 < t \leq T_L \}$ is a random variable depending on the value of the random variable t_1 as illustrated in fig. 7.2. and 7

- <u>Euclidean division of T_L by T_{sy}</u> - <u>Definition of Q'</u> :
By definition of the Euclidean division, we can write :

$$T_L = Q' T_{sy} + T'_{sy} \text{ with Q' integer and } 0 \leq T'_{sy} < T_{sy} \qquad (7.4.)$$

Q' is called the <u>Euclidean quotient</u> and T'_{sy} the <u>Euclidean remainder</u> of the division of T_L by T_{sy}.

- Definition of Q" number of domains (T_q) having at least one point in common with (T_L) : the domain ($T_{Q"}$) containing the point T_L is defined as follows :

$$(Q" - 1) T_{sy} < T_L \leq Q" T_{sy} \qquad (7.5.)$$

Two possibilities may arise :
  a) $T'_{sy} = 0$: then, $T_L = Q' T_{sy}$ and : $Q" - 1 < Q' \leq Q"$
With both Q' and Q" integer the only solution is $Q" = Q'$.
  b) $0 < T'_{sy} < T_{sy}$: then, $Q" - 1 < Q' + T'_{sy}/T_{sy} \leq Q"$
With $T'_{sy}/T_{sy} < 1$ the only solution is $Q" = Q' + 1$

Whereas Q is a random variable, Q' and Q" are perfectly defined as soon as T_L and T_{sy} are known.

- Definition of Q number of increments I_q actually falling within (T_L) :
According to (7.3.) and (7.4.) we can write :

$$t_1 + (Q - 1) T_{sy} \leq Q' T_{sy} + T'_{sy} < t_1 + Q T_{sy} \quad \text{or}$$

$$(Q - 1) T_{sy} \leq Q' T_{sy} + T'_{sy} - t_1 < Q T_{sy} \qquad (7.6.)$$

Here again, two possibilities may arise :

  a) $T'_{sy} = 0 \rightarrow Q" = Q'$: Then, irrespective of the random value of t_1 we have : $Q = Q' = Q"$. The number Q is defined.

  b) $T'_{sy} \neq 0 \rightarrow Q" = Q' + 1$: The domain ($T_{Q"}$) overlaps the domain (T_L). Let's call :
(T'_1) : the domain defined by : $(T'_1) \equiv \{ t \mid 0 < t \leq T'_{sy} \}$
($T"_1$) : the domain defined by : $(T"_1) \equiv \{ t \mid T'_{sy} < t \leq T_{sy} \}$
$(T_1) \equiv (T'_1) + (T"_1)$

According to the value of the random variable t_1, two cases must be taken into consideration :

  ○ t_1 belongs to (T'_1) : the instant $t_{Q"}$ selected in the overlapping domain ($T_{Q"}$) falls within the domain (T_L) and therefore belongs to the sample S. Then :
$Q = Q" = Q' + 1$ with   Prob $\{ Q = Q' + 1 \} = T'_1/T_1 = T'_{sy}/T_{sy}$
This case is illustrated in fig. 7.2.

  ○ t_1 belongs to ($T"_1$) : the instant $t_{Q"}$ selected in the overlapping domain ($T_{Q"}$) falls outside the domain (T_L) and therefore does not belong to the sample S. Then :
$Q = Q" - 1 = Q'$ with   Prob $\{ Q = Q' \} = T"_1/T_1 = (T_{sy} - T'_{sy})/T_{sy}$
This case is illustrated in fig. 7.3.

We shall note that :

Prob $\{ Q = Q' + 1 \}$ + Prob $\{ Q = Q' \} = 1$

The sample contains either $Q' + 1$ or Q' increments.

Table 7.1. summarizes the properties of the number Q as a function of T'_{sy} and t_1.

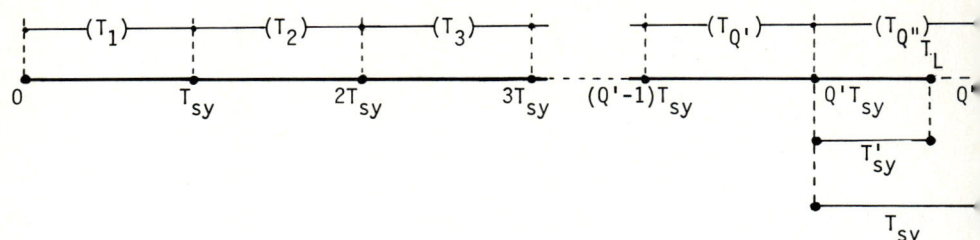

Fig. 7.1. Euclidean division of T_L by T_{sy} - Definition of Q', Q'' and T'_{sy}.

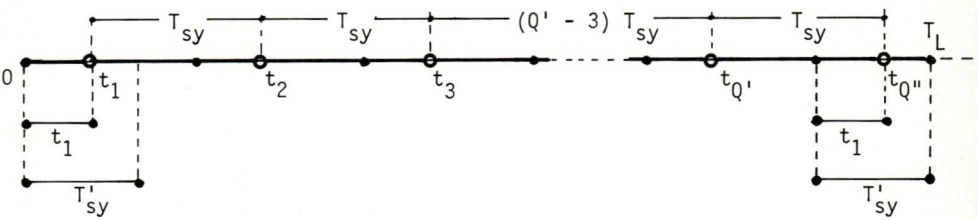

Fig. 7.2. Distribution of the instants t_q when t_1 is smaller than T'_{sy}.

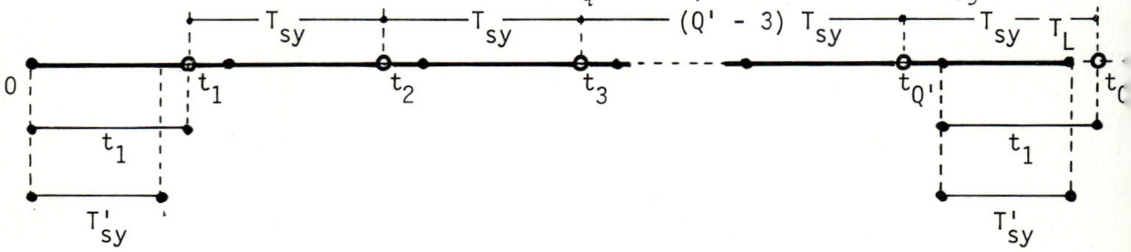

Fig. 7.3. Distribution of the instants t_q when t_1 is larger than T'_{sy}.

Table 7.1. Number Q of punctual increments actually falling within (T_L).

	$T'_{sy} = 0$	$T'_{sy} \neq 0$
$0 < t_1 \leq T'_{sy}$	$Q = Q'$	$Q = Q' + 1$ Prob$(Q=Q'+1) = T'_{sy}/T_{sy}$
$T'_{sy} < t_1 \leq T_{sy}$	Prob$(Q=Q') = 1$	$Q = Q'$ Prob$(Q=Q') = (T_{sy}-T'_{sy})/T_{sy}$
Nature of Q	Non-random	Random

In the most general case, Q is a random variable. This is a cause of sampling bias, which justifies a rather lengthy development.

Remark : The reader should not forget that it is the random selection of t_1 in the domain (T_1) that makes the systematic selection a probabilistic process. We shall however admit that the selection of t_1 is random whenever there is no correlation between t_1 and the properties of the stream to be sampled. From a practical standpoint this hypothesis can always be accepted as the selection of t_1 is left to external circumstances usually uncorrelated with the properties of the stream.

7.3. RANDOM STRATIFIED SELECTION

This will be more simply referred to as "stratified selection" (subscript st).

7.3.1. Definition : see fig. 7.4. :

A random stratified selection can be schematized in the following way :

- A length of time T_{st} is adopted according to the degree of representativeness to be achieved by the sample S. T_{st} is always small as compared with T_L.

- The time axis is "stratified" or in other words broken up into adjacent "strata" (T_q) of equal extent T_{st}, in such a way that :
$(T_q) \equiv \{ t \mid (q - 1) T_{st} < t \leq q T_{st} \}$

- We shall call Q" the number of strata (T_q) having at least one point in common with (T_L). The last stratum $(T_{Q"})$ may be overlapping.

- One and only one instant t_q is selected at random within the stratum (T_q) with $q = 1, 2, \ldots Q"$:
$(q - 1) T_{st} < t_q \leq q T_{st}$

- The Q instants t_q actually falling within (T_L) are retained as the seats of the punctual increments I_q and the set of these increments I_q (with $q = 1, 2, \ldots Q$) is the punctual sample S retained to represent the lot L.

7.3.2. Properties of the number Q of increments :

For given values of T_L and T_{st}, the number Q of increments actually falling within the limits of the domain (T_L) is a random variable depending on the random value of $t_{Q"}$ as illustrated in fig. 7.5. and 7.6.

- <u>Euclidean division of T_L by T_{st}</u> - <u>Definition of Q'</u> :
By following the same demonstration as in section 7.2.2. we shall write :

$T_L = Q' T_{st} + T'_{st}$ with Q' integer and $0 \leq T'_{st} < T_{st}$

Q' is the Euclidean quotient and T'_{st} the Euclidean remainder of the division of T_L by T_{st}.

- <u>Definition of Q"</u> : Q" is the number of strata having at least one point in common with (T_L). The demonstration is the same as in section 7.2.2. We shall re-

tain that :
  a) $T'_{st} = 0$: then, $Q" = Q'$
  b) $T'_{st} \neq 0$: then, $Q" = Q' + 1$

Whereas Q is a random variable, Q' and Q" are perfectly defined as soon as T_L and T_{st} are known.

- <u>Definition of Q</u>, number of increments actually falling within (T_L) : we shall call t'_q the abscissa of I_q in the stratum (T_q). By definition :

$$t'_q = t_q - (q - 1) T_{st} \quad \text{with} \quad 0 < t'_q \leq T_{st}$$

Two possibilities may arise :

  a) $T'_{st} = 0$: then $Q" = Q'$: the lot L is made of a whole number of strata. As by definition one and only one increment is selected from each stratum, then :
$Q = Q' = Q"$

  b) $T'_{st} \neq 0$: then $Q" = Q' + 1$: the last stratum $(T_{Q"})$ is overlapping. Let's cal $(T'_{Q"})$: the domain defined by $(T'_{Q"}) \equiv \{ t \mid (Q" - 1) T_{st} < t \leq T_L \}$
$(T"_{Q"})$: the domain defined by $(T"_{Q"}) \equiv \{ t \mid T_L < t \leq Q" T_{st} \}$

$(T_{Q"}) \equiv (T'_{Q"}) + (T"_{Q"})$

According to the value of $t'_{Q"}$, two cases must be taken into consideration :

  ○ $t'_{Q"}$ belongs to $(T'_{Q"})$: the instant $t_{Q"}$ selected in the overlapping stratum falls within the domain (T_L) and therefore belongs to the sample S. Then :

$Q = Q" = Q' + 1$ with $\text{Prob}(Q = Q' + 1) = T'_{Q"}/T_{Q"} = T'_{st}/T_{st}$

This case is illustrated in fig 7.5.

  ○ $t'_{Q"}$ belongs to $(T"_{Q"})$: the instant $t_{Q"}$ selected in the overlapping stratum falls outside the domain (T_L) and therefore does not belong to the sample S. Then :

$Q = Q" - 1 = Q'$ with $\text{Prob}(Q = Q') = T"_{Q"}/T_{Q"} = (T_{st} - T'_{st})/T_{st}$

This case is illustrated in fig. 7.6.

We shall note that :

$\text{Prob}(Q = Q' + 1) + \text{Prob}(Q = Q') = 1$

The sample contains either Q' + 1 or Q' increments.

Table 7.2. summarizes the properties of the number Q as a function of T'_{st} and t_C

In the most general case characterized by $T'_{st} \neq 0$ the number Q of increments is a random variable. This is one of the causes of sampling bias, which justifies a rather lengthy development.

<u>Remarks</u> : - in the stratified selection scheme, the random selection of points cannot be left to external, independent circumstances as suggested for the selec-

tion of t_1 in the systematic scheme. The best way to obtain a sequence of instants corresponding to a random stratified selection is to use a random stratified timer (British and U.S. patents see Gy - 1966 and 1968).

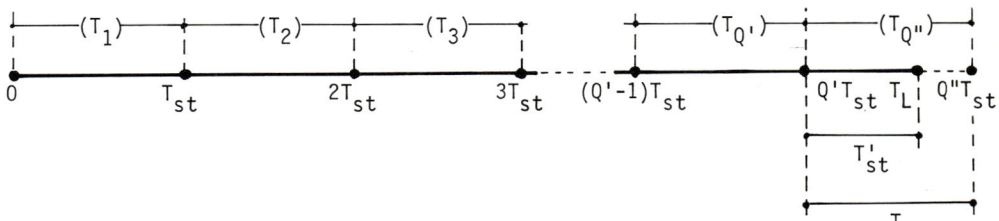

Fig. 7.4. *Euclidean division of T_L by T_{st} - Definition of Q', Q'' and T'_{st}.*

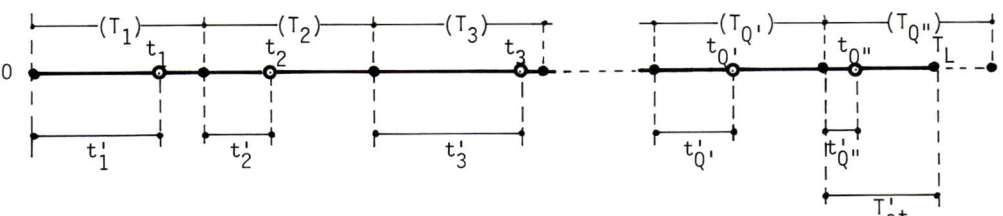

Fig. 7.5. *Distribution of the instants t_q when $t'_{Q''}$ is smaller than T'_{st}.*

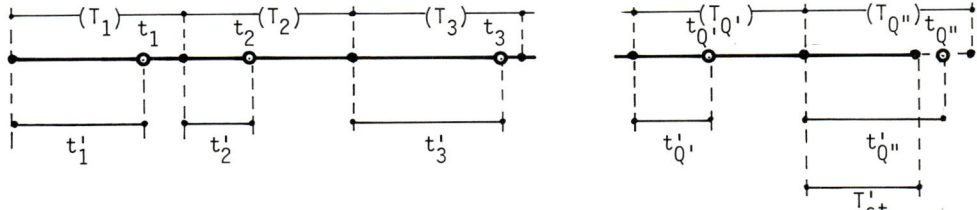

Fig. 7.6. *Distribution of the instants t_q when $t'_{Q''}$ is larger than T'_{st}.*

Table 7.2. *Number Q of punctual increments actually falling within (T_L).*

	$T'_{st} = 0$	$T'_{st} \neq 0$
$0 < t'_{Q''} \leq T'_{st}$	$Q = Q'$	$Q = Q' + 1$ $Prob(Q=Q'+1) = T'_{st}/T_{st}$
$T'_{st} < t'_{Q''} \leq T_{st}$	$Prob(Q = Q') = 1$	$Q = Q'$ $Prob(Q=Q') = (T_{st}-T'_{st})/T_{st}$
Nature of Q	Non-random	Random

Remarks (continuation) : - Random stratified selection schemes where strata are either of unequal size or of equal size but with several increments selected from each stratum do not fall within the province of this section. In both cases, each stratum is to be regarded as a separate lot and treated as such.

7.4. RANDOM SELECTION

The random selection is characterized by the subscript ra.

7.4.1. Definition :

A random selection can be schematized in the following way :

- A number Q_{ra} of increments is adopted according to the degree of representativeness to be achieved by the sample S.

- Q_{ra} instants t_q are selected at random within the domain (T_L) occupied by L.

- These Q_{ra} instants t_q are retained as the seats of the punctual increments I_q and the set of these increments, with q = 1, 2, .. Q_{ra}, is the punctual sample S retained to represent the lot L.

7.4.2. Number Q of increments :

According to the definition of the random selection scheme, the number Q of increments is a non-random number with :

$Q = Q_{ra}$

7.5. FULL DESCRIPTION OF A SELECTION SCHEME

A selection scheme is fully described by :

- <u>Its type</u> : systematic, stratified or random,
- <u>Its "free parameter"</u>. We call free parameter the parameter which can be chosen according to our will in order to meet given reproducibility or representativeness standards. The free parameters are :

　○ Systematic selection : the uniform interval T_{sy} between consecutive increments
　○ Stratified selection : the uniform strata extent T_{st} ,
　○ Random selection : The number Q_{ra} of increments.

7.6. COMPARISON OF THE THREE SELECTION SCHEMES

Fig. 7.7. shows on an example how the punctual increments can be distributed throughout the domain (T_L) by application of the three reference selection schemes. In this example we have retained :

- $T_{sy} = T_L/7$
- $T_{st} = T_L/7$
- $Q_{ra} = 7$

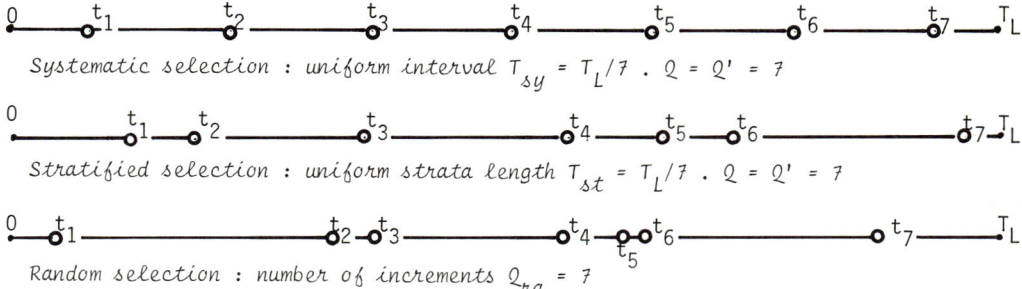

Fig. 7.7. Distribution of the punctual increments throughout (T_L) by application of the three reference selection schemes.

7.7. FIELD OF APPLICATION OF THE THREE REFERENCE SELECTION SCHEMES

<u>7.7.1. Systematic selection</u> : it is by far the most common of all selection schemes. It is easy to carry out and is nearly always the most reproducible. There is however a very important exception, the sampling of periodic functions, where a systematic selection is liable to generate unacceptable errors. This point will be dealt with in great detail in chapter 12.

<u>7.7.2. Stratified selection</u> : it is not yet widely known to the users, except in the cement industry which was the first to recognize its interest. A stratified selection scheme should be implemented whenever the critical content or the rate of flow of the stream to be sampled is liable to fluctuate according to a more or less periodic pattern. The study of hundreds of variograms shows that such periodic fluctuations are much more frequent than is usually imagined. A stratified selection can be easily carried out by means of special timers. As it is always almost as reproducible as the systematic scheme, and as it suppresses the important error liable to take place when sampling a periodic function according to a systematic scheme, the stratified scheme always constitutes the safest of all solutions.

<u>7.7.3. Random selection</u> : it is never implemented for the sampling of flowing streams but is often used when sampling zero-dimensional objects. It is usually the less reproducible of the three selection schemes. It has however to be investigated here, for a number of authors on sampling as well as Standards Organizations currently apply to systematic selection formulas valid only with random selection. The result of such a confusion is to unnecessarily increase the sample weight and the cost of sample reduction when a given reproducibility standard is to be achieved.

CHAPTER 8

DEVELOPMENT OF THE CONTINUOUS SELECTION MODEL
CONTINUOUS SELECTION ERROR CE

8.1. INTRODUCTION

The "continuous selection model" has been defined in section 5.3. It has been devised in order to study the moments of the "continuous selection error" CE incurred when representing the lot L by a sample S made of a set of increments I_q.

When applied to the punctual functions, this model covers the point selection step of the increment sampling process.

When applied to the extended functions, it covers the sequence: point selection + increment delimitation, of the increment sampling process (correct delimitation).

When applied to the fragmental functions, it covers the sequence: point selection + increment delimitation + increment extraction, of the increment sampling process (correct delimitation and extraction).

The conditions of correct delimitation and extraction will be studied in great detail in chapters 17 and 18 respectively. In the present chapter, we shall process the model without specifying the nature of the functions $\mu(t)$, $\alpha(t)$ and $a(t)$ taken into consideration, remembering that :

- This model is rigorous with the punctual functions and approximate with the extended and fragmental functions. The approximation is however excellent whenever the extent T_E of the extension domain remains very small in comparison with the extent T_L of the lot. This condition is always respected in practice.

- A variographic experiment provides fragmental increments and the variographic parameters that can be estimated from the results of such an experiment directly apply to fragmental functions on condition that both increment delimitation and increment extraction be correctly carried out. In this case, the extension domain T_E coincides with the domain T_I covered by the increments extracted from the lot.

We shall use the same notations as in the preceding chapters.

8.2. DEFINITION OF THE MODEL

8.2.1. Definition of the lot L : the lot L is made of the material flowing past the sampling point between time $t = 0$ and time $t = T_L$. It covers the time domain (T_L) of extent T_L. We shall retain the following notations :

M_L : weight of active components in L (or more simply weight of L when no confusion is possible).
A_L : weight of critical component in L,
a_L : critical content of L. By definition : $a_L = \dfrac{A_L}{M_L}$

8.2.2. Definition of the material to be sampled :

The material flowing at instant t can be <u>described</u> by means of the functions :
$\mu(t)$: rate of flow of active components at instant t,
$\alpha(t)$: rate of flow of critical component at instant t,
$a(t)$: critical content at instant t. By definition : $a(t) = \dfrac{\alpha(t)}{\mu(t)}$

These functions can be <u>characterized</u> by their variograms (defined in chapter 6)
$v_\mu(\theta)$: identical with the variogram $v_M(\theta)$ of the increment weights M_q
$v_\alpha(\theta)$: identical with the variogram $v_A(\theta)$ of the weights A_q
$v_a(\theta)$: variogram of the critical contents a_q.

We shall assume that we have been able to explicit the model variograms $v'_M(\theta)$, $v'_A(\theta)$ and $v'_a(\theta)$, and to estimate all variographic parameters.

8.2.3. Definition of the selecting process :

This process is defined when we have fixed :
- the <u>type</u> of selection scheme : systematic (subscript sy), stratified (subscript st) or random (subscript ra).
- the <u>free parameter</u> of the selection scheme : interval T_{sy} , strata extent T_{st} or number of increments Q_{ra}.

The selecting process fixes Q instants t_q (with q = 1, 2, .. Q) at which increments are extracted from the stream.

8.2.4. Definition of the increment delimitation process :

The delimitation is defined by means of a single parameter T_I which is the extent of the increments. T_I is assumed to be uniform from one increment to the next. We shall call :

I_q : the increment extracted from the stream at instant t_q ,
M_q : the weight of active components in I_q (or more simply the weight of I_q),
A_q : the weight of critical component in I_q ,
a_q : the critical content of I_q . By definition : $a_q = \dfrac{A_q}{M_q}$

8.2.5. Definition of the sample S :

We shall call :
S : the sample obtained by reunion of the Q increments I_q ,
M_S : the weight of active components in S : $M_S = \sum_q M_q$
A_S : the weight of critical component in S : $A_S = \sum_q A_q$
a_S : the critical content of S : $a_S = A_S/M_S = \sum_q A_q / \sum_q M_q = \sum_q a_q M_q / \sum_q M_q$

8.2.6. Definition of the continuous selection error CE :

It is defined as the relative difference :

$$CE = \frac{a_S - a_L}{a_L} \quad \text{with} \quad a_L = \frac{\int_0^{T_L} a(t)\,\mu(t)\,dt}{\int_0^{T_L} \mu(t)\,dt}$$

The purpose of this chapter is to study the properties of the random error CE and more specifically :
- its distribution law,
- its mean $m(CE)$ which is a measure of the selection accuracy,
- its variance $\sigma^2(CE)$ which is a measure of the selection reproducibility,
- its mean square $r^2(CE)$ which is a measure of the selection representativeness :

$$r^2(CE) = m(CE^2) = m^2(CE) + \sigma^2(CE)$$

These moments cannot be directly estimated. We must begin with a study of the distributions of M_S (sample weight), A_S (weight of critical component in the sample) and with a study of the coefficient of correlation $\rho(A_S, M_S)$.

Remark : we shall assume in this chapter that the periodic terms $\mu_3(t)$, $\alpha_3(t)$ and $a_3(t)$ are identically nil. The very important problem of periodic fluctuations will be dealt with in chapter 12.

8.3. DISTRIBUTION OF THE WEIGHT M_S OF ACTIVE COMPONENTS IN THE SAMPLE S

8.3.1. Distribution law of M_S : By definition :

$$M_S = \sum_q M_q \quad \text{with} \quad q = 1, 2, \ldots Q \quad \text{and} \quad M_q = T_I\,\mu(t_q)$$

The population of all possible values of M_q is, save for a constant factor T_I, identical with the population of all possible values of $\mu(t)$ which is one of the data of the problem and does not belong to any specific type. As for the distribution law of M_S we must distinguish the various selection schemes.

8.3.1.1. Systematic or stratified selection schemes :

We shall retain the notations of chapter 7, T_S denoting either T_{sy} or T_{st}. We know from sections 7.2.2 and 7.3.2. that in the most general case ($T_S' \neq 0$) the number Q of increments is a random variable with two possible values Q' or Q' + 1. We shall call :

ω : the set of all possible samples S. This set can be split up into two sub-sets :
ω_1 : the sub-set of all possible samples S_1 made of Q' + 1 increments with :
   Prob (Q = Q' + 1) = T_S' / T_S
ω_2 : the sub-set of all possible samples S_2 made of Q' increments with :
   Prob (Q = Q') = $(T_S - T_S') / T_S = 1 - T_S' / T_S$

Each of these two sub-sets is homogeneous with respect to the number of increments making up the sample.

In each sub-set, according to the Central Limit Theorem of Mathematical statistics, the distribution of M_S tends to become normal as the number of increments increases. We shall admit that normality is practically achieved when it is larger than 30, which is somewhat arbitrary but acceptable from a strictly practical poin of view, irrespective of the distribution law of the increment weights. When the number is small (say smaller than 30) the distribution may deviate from normalit and cannot be specified.

In the complete set of samples, the distribution of M_S results from the merging of two distributions with different means (which is intuitive but will nevertheles be shown in section 8.3.2.). Two cases may arise :

- Q' is "large enough" : this notion depends on the absolute value of Q' but al so on the relative standard deviation $u(M_q)$ of the increment weight M_q. Though remaining bimodal, the distribution may present the appearances of a unimodal distribution. The distribution is practically normal within each sub-set (because Q' is "large") and the means of M_S in the two sub-sets are not different enough to be distinguished from each other (because they are respectively proportional to Q' + and Q' and $u(M_q)$ is relatively large). For all practical purposes we may assimilate the distribution of M_S to a single, unimodal, normal distribution.

- Q' is not "large enough" : the distribution is definitely bimodal and furthermore the distribution within each of the two sub-sets cannot be specified.

8.3.1.2. Random selection scheme :

The number Q_{ra} is specified and all possible samples are made of the same numbe of increments. The distribution of M_S is therefore unimodal and tends to become normal as the number Q_{ra} increases. We shall admit that normality is achieved when Q_{ra} is larger than 30.

8.3.1.3. Incorrect selection schemes :

We have assumed so far that the density of selection probability was uniform throughout (T_L) and this is true for the three reference selection schemes presented in chapter 7. If this condition is not fulfilled, the distribution of M_S still tends to become normal as the number Q of increments increases.

8.3.2. Mean of M_S :

By definition : $m(M_S) = \sum_q m(M_q)$ with q = 1, 2, .. Q .

Here again we must distinguish the three selection schemes :

8.3.2.1. Systematic selection scheme :

The punctual increment I_q falls within the domain (T_q) with :

$(T_q) \equiv \{ t \mid (q - 1) T_{sy} < t \leq q T_{sy} \}$

The instant t_1 is selected in the domain (T_1) with a density of selection probability $\Pi(t_1)$ assumed to be uniform throughout (T_1) and nil outside.

$$\Pi(t_1) = \Pi_o = 1 / T_{sy}$$

According to the definition of the systematic selection scheme, the instant t_q falls within the domain (T_q) with the same density of selection probability :

$$\Pi(t_q) = \Pi_o = 1 / T_{sy}$$

We must however distinguish the overlapping domain $(T_{Q''})$ when $T'_{sy} \neq 0$. The instant $t_{Q''}$ is retained as the seat of a punctual increment if and only if it falls within (T_L). Let's split the domain $(T_{Q''})$ into two sub-domains :

$(T'_{Q''}) \equiv \{ t \mid (Q'' - 1) T_{sy} < t \leq T_L \}$: extent T'_{sy}

$(T''_{Q''}) \equiv \{ t \mid T_L < t \leq Q'' T_{sy} \}$: extent $T_{sy} - T'_{sy}$

Then :

If $t_{Q''}$ belongs to $(T'_{Q''})$: $\Pi(t_{Q''}) = 1 / T_{sy}$

If $t_{Q''}$ belongs to $(T''_{Q''})$: $\Pi(t_{Q''}) = 0$

Now, $m_{sy}(M_q)$ denoting the mean of M_q :

$$m_{sy}(M_q) = T_I \, m_{sy}\{\mu(t_q)\} = T_I \frac{\int_{(T_q)} \mu(t) \, \Pi(t) \, dt}{\int_{(T_q)} \Pi(t) \, dt}$$

We must now distinguish :

- $q = 1, 2, \ldots Q'$:

$$m_{sy}(M_q) = \frac{T_I}{T_{sy}} \int_{(T_q)} \mu(t) \, dt$$

- $q = Q'' = Q' + 1$:

$$m_{sy}(M_q) = \frac{T_I}{T'_{sy}} \int_{(T'_{Q''})} \mu(t) \, dt$$

The mean $m_{sy}(M_S) = \sum_q m_{sy}(M_q)$ with $q = 1, 2, \ldots Q$ can therefore be written :

- <u>In the sub-set ω_1</u> : with $q = 1, 2, \ldots Q'$:

$$m_{sy}(M_{S1}) = \frac{T_I}{T_{sy}} \sum_q \int_{(T_q)} \mu(t) \, dt + \frac{T_I}{T'_{sy}} \int_{(T'_{Q''})} \mu(t) \, dt$$

- <u>In the sub-set ω_2</u> : with $q = 1, 2, \ldots Q'$:

$$m_{sy}(M_{S2}) = \frac{T_I}{T_{sy}} \sum_q \int_{(T_q)} \mu(t) \, dt$$

- <u>In the set ω</u> : with $q = 1, 2, \ldots Q'$

$$m_{sy}(M_S) = m_{sy}(M_{S1}) \, \text{Prob (S in } \omega_1) + m_{sy}(M_{S2}) \, \text{Prob (S in } \omega_2)$$

$$m_{sy}(M_S) = \frac{T'_{sy}}{T_{sy}} m_{sy}(M_{S1}) + \frac{T_{sy} - T'_{sy}}{T_{sy}} m_{sy}(M_{S2}) = \frac{T_I}{T_{sy}} \{\sum_q \int_{(T_q)} \mu(t) \, dt + \int_{(T'_{Q''})} \mu(t) \, dt$$

$$m_{sy}(M_S) = \frac{T_I}{T_{sy}} \int_{(T_L)} \mu(t) \, dt = \frac{T_I}{T_{sy}} M_L = \tau_{sy} M_L \quad \text{with} \quad \tau_{sy} = \frac{T_I}{T_{sy}}$$

The ratio T_I / T_{sy} is what we shall call the "<u>time sampling ratio</u>" τ_{sy}, proportion of the time during which the stream is diverted.

8.3.2.2. Stratified selection scheme :

The demonstration and the results are the same as with the systematic scheme and we shall directly write :

$$m_{st}(M_S) = \frac{T_I}{T_{st}} M_L = \tau_{st} M_L \quad \text{with} \quad \tau_{st} = \frac{T_I}{T_{st}}$$

8.3.2.3. Random selection scheme :

Each of the Q_{ra} instants t_q is selected from (T_L) with a uniform density of selection probability $\Pi(t) = \Pi_o = 1 / T_L$. Then :

$$m_{ra}(M_q) = T_I \, m_{ra}\{\mu(t_q)\} = T_I \frac{\int_{(T_L)} \mu(t) \, \Pi(t) \, dt}{\int_{(T_L)} \Pi(t) \, dt} = \frac{T_I}{T_L} \int_{(T_L)} \mu(t) \, dt = \frac{T_I}{T_L} M_L$$

$$m_{ra}(M_S) = Q_{ra} \, m_{ra}(M_q) = \frac{Q_{ra} T_I}{T_L} M_L = \tau_{ra} M_L \quad \text{with} \quad \tau_{ra} = \frac{Q_{ra} T_I}{T_L}$$

Here again, τ_{ra} is the "time sampling ratio".

8.3.2.4. Estimation by sampling of the weight M_L of a lot L :

The time sampling ratio τ can be calculated with great accuracy from the mechanical features of a sampling device. If the sampling operation is correctly carried out (a notion that will be precised in chapters 17 and 18) it readily follows that $M'_L = \frac{M_S}{\tau}$ is an unbiased estimator of the true weight M_L of the lot L. As the sample weight M_S can be easily measured and with great accuracy by means of scales, this is often a fairly accurate way of estimating the weight of a flowing lot of material.

Furthermore, in a processing plant for instance, at a full scale or at pilot scale, the sampling of all branches of the flow-sheet by means of sampling devices achieving exactly the same sampling ratio, provides a set of samples whose weight is proportional to the tonnage flowing through the corresponding branch. This makes it possible to calculate with great accuracy and at a very acceptable cost the various metallurgical balances characterizing the processing efficiency (e.g. metal recoveries).

8.3.3. Relative variance of the sample weight M_S :

8.3.3.1. General formulas :

According to Matheron's results (references in appendix), the variance of M_S can be expressed in terms of the variogram $v'_M(\theta)$ of the function $\mu(t)$. Let's introduce the auxiliary functions :

$$X_M(\theta) = \frac{1}{\theta} \int_0^\theta v'_M(\theta') \, d\theta'$$

$$\Phi_M(\theta) = \frac{2}{\theta^2} \int_0^\theta \theta' \, X_M(\theta') \, d\theta'$$

For the three reference selection schemes, the relative variance $u^2(M_S) = \dfrac{\sigma^2(M_S)}{m^2(M_S)}$ can be expressed in the following way :

- <u>Systematic scheme</u> with interval T_{sy} :

$$u^2_{sy}(M_S) = \frac{T_{sy}}{T_L} \{2 \, X_M(T_{sy}/2) - \Phi_M(T_{sy})\}$$

- <u>Stratified scheme</u> with strata extent T_{st} :

$$u^2_{st}(M_S) = \frac{T_{st}}{T_L} \Phi_M(T_{st})$$

- <u>Random scheme</u> with Q_{ra} increments :

$$u^2_{ra}(M_S) = \frac{1}{Q_{ra}} \Phi_M(T_L)$$

- <u>Classical statistics</u> (subscript cl) : Let's assume that instead of being extracted from a continuous stream, all possible increments are naturally separated and form a zero-dimensional lot. In this population, the increment weight has a mean $m(M_q)$ and a relative variance $u^2(M_q)$. A sample S made of Q_{ra} increments extracted at random from the population will be characterized by a weight M_S with :
  - a mean : $m_{cl}(M_S) = Q_{ra} \, m(M_q) = m_{ra}(M_S)$
  - a variance : $u^2_{cl}(M_S) = u^2(M_q) / Q_{ra} = u^2_{ra}(M_S)$

The latter equalities express the fact that it is equivalent to extract Q_{ra} increments by both methods. As the variographic experiment provides a good estimate of $u^2(M_q)$ but a usually bad expression of the variogram for values of θ as large as T_L, it is always much more accurate to calculate $u^2_{cl}(M_S)$ than $u^2_{ra}(M_S)$.

<u>Remark</u> : The above formula assumes that the Q_{ra} increments are extracted from an infinite population. When the number N_{LU} of units making up the population is not very large as compared with Q_{ra}, the right expression of $u^2_{cl}(M_S)$ is :

$$u^2_{cl}(M_S) = \left[\frac{1}{Q_{ra}} - \frac{1}{N_{LU}} \right] \frac{N_{LU}}{N_{LU} - 1} u^2(M_q)$$

8.3.3.2. Practical formulas - Parabolic variogram :

We shall assume that the variogram $v'_M(\theta)$ is parabolic, rectilinear or flat and can be represented by the general expression :

$$v'_M(\theta) = v'_{M1} + v'_{M2}\,\theta + v''_{M2}\,\theta^2$$

at least in the useful domain of the variogram covering the usual values of T_{sy} and T_{st}. Thanks to the fact that $u^2_{ra}(M_S)$ is equal to $u^2_{cl}(M_S)$ we do not need a mathematical expression of the variogram valid up to $\theta = T_L$. Then :

$$X_M(\theta) = v'_{M1} + \frac{v'_{M2}}{2}\,\theta + \frac{v''_{M2}}{3}\,\theta^2$$

$$\Phi_M(\theta) = v'_{M1} + \frac{v'_{M2}}{3}\,\theta + \frac{v''_{M2}}{6}\,\theta^2$$

- <u>Systematic scheme</u> with interval T_{sy} :

$$u^2_{sy}(M_S) = \frac{1}{T_L}\,(v'_{M1}\,T_{sy} + \frac{v'_{M2}}{6}\,T^2_{sy})$$

- <u>Stratified scheme</u> with strata extent T_{st} :

$$u^2_{st}(M_S) = \frac{1}{T_L}\,(v'_{M1}\,T_{st} + \frac{v'_{M2}}{3}\,T^2_{st} + \frac{v''_{M2}}{6}\,T^3_{st})$$

- <u>Random scheme</u> with Q_{ra} increments :

$$u^2_{ra}(M_S) = \frac{1}{Q_{ra}}\,(v'_{M1} + \frac{v'_{M2}}{3}\,T_L + \frac{v''_{M2}}{6}\,T^2_L)$$

This expression of $u^2_{ra}(M_S)$ is valid only on condition that the parabolic expression of the variogram covers the whole domain $0 < \theta \leq T_L$. As already mentioned, it is always preferable to use the formulas derived from classical statistics

$$u^2_{ra}(M_S) = u^2_{cl}(M_S) = \frac{1}{Q_{ra}}\,u^2(M_q)$$

8.3.3.3. Practical formulas - Rectilinear variogram :

This is a particular case of a parabolic variogram with : $v''_{M2} = 0$. Then :

- <u>Systematic scheme</u> with interval T_{sy} :

$$u^2_{sy}(M_S) = \frac{1}{T_L}\,(v'_{M1}\,T_{sy} + \frac{v'_{M2}}{6}\,T^2_{sy})$$

- <u>Stratified scheme</u> with strata extent T_{st} :

$$u^2_{st}(M_S) = \frac{1}{T_L}\,(v'_{M1}\,T_{st} + \frac{v'_{M2}}{3}\,T^2_{st})$$

- <u>Random scheme</u> with Q_{ra} increments :

$$u^2_{ra}(M_S) = \frac{1}{Q_{ra}}\,(v'_{M1} + \frac{v'_{M2}}{3}\,T_L)$$

8.3.3.4. Practical formulas - Flat variogram :

This is another particular case of a parabolic variogram with : $v'_{M2} = v''_{M2} = 0$.

- Systematic scheme with interval T_{sy} : $\quad u^2_{sy}(M_S) = \dfrac{v'_{M1} T_{sy}}{T_L} = \dfrac{v'_{M1}}{Q}$

- Stratified scheme with strata extent T_{st} : $\quad u^2_{st}(M_S) = \dfrac{v'_{M1} T_{st}}{T_L} = \dfrac{v'_{M1}}{Q}$

- Random scheme with Q_{ra} increments : $\quad u^2_{ra}(M_S) = \dfrac{v'_{M1}}{Q_{ra}}$

8.3.3.5. Relationship between $u^2(M_q)$ and the variographic parameters :

Whenever the parabolic variogram is valid up to T_L we deduce :

$$u^2(M_q) = v'_{M1} + \frac{v'_{M2}}{3} T_L + \frac{v''_{M2}}{6} T_L^2$$

8.4. DISTRIBUTION OF THE WEIGHT A_S OF CRITICAL COMPONENT IN THE SAMPLE S

8.4.1. Distribution law of A_S :

All conclusions of section 8.3.1. remain valid here and more particularly the fact that from a practical standpoint it is nearly always reasonable to assume that A_S follows a normal distribution. The only exception concerns the materials with a very small critical content such as the ores of precious minerals and metals and the tailings of some processing plants. This also happens when the critical component is an impurity penalized at a very low level, a frequent case in the food industries. When the critical content is very low, the distribution of A_S tends to follow a log-normal type rather than a normal one.

8.4.2. Mean of A_S :

From section 8.3.2. we immediately deduce :

$$m_{sy}(A_S) = \frac{T_I}{T_{sy}} A_L = \frac{T_I}{T_{sy}} a_L M_L = \tau_{sy} a_L M_L = a_L m_{sy}(M_S)$$

$$m_{st}(A_S) = \frac{T_I}{T_{st}} A_L = \frac{T_I}{T_{st}} a_L M_L = \tau_{st} a_L M_L = a_L m_{st}(M_S)$$

$$m_{ra}(A_S) = \frac{Q_{ra} T_I}{T_L} A_L = \frac{Q_{ra} T_I}{T_L} a_L M_L = \tau_{ra} a_L M_L = a_L m_{ra}(M_S)$$

Irrespective of the selection scheme, we shall retain :

$$m(A_S) = \tau A_L = \tau a_L M_L = a_L m(M_S)$$

The absence of subscripts meaning that the result is valid for the three selection schemes, τ being the time sampling ratio, proportion of the time during which the stream is diverted.

8.4.3. Relative variance of A_S :

From section 8.3.3. we deduce :

8.4.3.1. Practical formulas - Parabolic variogram :

- <u>Systematic scheme</u> with interval T_{sy} :

$$u^2_{sy}(A_S) = \frac{1}{T_L} (v'_{A1} T_{sy} + \frac{v'_{A2}}{6} T^2_{sy})$$

- <u>Stratified scheme</u> with strata extent T_{st} :

$$u^2_{st}(A_S) = \frac{1}{T_L} (v'_{A1} T_{st} + \frac{v'_{A2}}{3} T^2_{st} + \frac{v''_{A2}}{6} T^3_{st})$$

- <u>Random scheme</u> with Q_{ra} increments :

$$u^2_{ra}(A_S) = \frac{1}{Q_{ra}} (v'_{A1} + \frac{v'_{A2}}{3} T_L + \frac{v''_{A2}}{6} T^2_L)$$

- <u>Classical statistics</u> :

$$u^2_{cl}(A_S) = \frac{1}{Q_{ra}} u^2(A_q)$$

8.4.3.2. Practical formulas - Rectilinear and flat variograms :

As already pointed out, these are particular cases of a parabolic variogram wit

- Rectilinear variogram : $v''_{A2} = 0$
- Flat variogram : $v'_{A2} = v''_{A2} = 0$

8.5. CORRELATION BETWEEN THE DISTRIBUTIONS OF M_S AND A_S

The coefficient of correlation between the distributions of M_S and A_S will be needed when studying the distribution of the critical content a_S of the sample S.

8.5.1. First hypothesis
: the characteristics of two increments I_q and $I_{q'}$ belonging to the sample S may be regarded as independent, irrespective of q and q'. Then it can be easily shown that :

$$\rho(M_S, A_S) = \rho(M_q, A_q)$$

The coefficient of correlation $\rho(M_q, A_q)$ can be easily calculated from the results of a variographic experiment.

8.5.2. Second hypothesis
: the characteristics of the increments I_q and $I_{q'}$ can not be regarded as independent. Then there is no simple relationship between $\rho(M_q,$ and $\rho(M_S, A_S)$ and there is no way of estimating the latter.

8.5.3. Validity of the first hypothesis
: we can define two simple cases where this hypothesis may be accepted :

- One at least of the variograms $v'_M(\theta)$ and $v'_A(\theta)$ is flat : then, irrespective of the selection scheme, the independence hypothesis can be accepted. The true coefficient of correlation is zero. The coefficient of correlation that can be calculated from the results of a variographic experiment is then very small. When the rate of flow $\mu(t)$ is regulated, the variogram $v'_M(\theta)$ is likely to be flat which is seldom true with the variogram $v'_A(\theta)$, except when the material to be sampled has just been reclaimed from a blending pile.

- None of the variograms $v'_M(\theta)$ and $v'_A(\theta)$ is flat but the selection scheme is random : the independence hypothesis can be accepted though the coefficient of correlation is not zero.

But there are cases where the independence hypothesis can hardly be accepted :

- None of the variograms $v'_M(\theta)$ and $v'_A(\theta)$ is flat and the selection scheme is either systematic or stratified. Then, the smaller the ratios v'_{M2}/v'_{M1} and v'_{A2}/v'_{A1}, the better the independence approximation.

- None of the variograms $v'_M(\theta)$ and $v'_A(\theta)$ is flat and one of them at least is periodic : independence can be achieved only if the selection scheme is random.

Now, in most modern sampling facilities the rate of flow is regulated prior to any sampling stage. We shall admit in the next sections that this condition is fulfilled and that we can estimate $\rho(M_S, A_S) = \rho(M_q, A_q)$ from the results of the variographic experiment.

8.6. DISTRIBUTION OF THE CRITICAL CONTENT a_S OF THE SAMPLE S - INTRODUCTION

8.6.1. Review of the relevant factors :

Six factors have been found to be relevant to the problem :

Material to be sampled :
1) $a(t)$: the "quality function ",
2) $\mu(t)$: the "weighting function",
3) $\rho\{a(t), \mu(t)\}$: the coefficient of correlation between the distributions of the quality and weighting functions. We shall write more simply $\rho_{a\mu}$

Selection scheme :
4) Type of the selection scheme, free parameter T_{sy}, T_{st} or Q_{ra},
5) $\Pi(t)$: density of selection probability,
6) T'_S : Remainder of the Euclidean division of T_L by T_{sy} or T_{st} with the systematic or stratified selection schemes.

Factors No. 1) to 3) are data of the problem. We must accept them as they are but for a few exceptions that will be discussed in due course.

Factors No. 4) to 6) can be controlled, theoretically at least, even though this control is not so simple in practice as it looks on paper.

8.6.2. State of the material to be sampled :

We shall retain four possible states of the material to be sampled :

State No. 1 : the descriptive functions $a(t)$ and $\mu(t)$ are correlated. This is the most general case, characterized by $\rho_{a\mu} \neq 0$.

State No. 2 : the descriptive functions are uncorrelated. This state is characterized by $\rho_{a\mu} = 0$.

State No. 3 : the quality function $a(t)$ is uniform throughout (T_L) : the flowing material is homogeneous and characterized by $a(t) \equiv a_L$. This is a particular case of state No. 2.

State No. 4 : the weighting function $\mu(t)$ is uniform throughout (T_L) : the rate of flow is constant and characterized by $\mu(t) \equiv \mu_L$ = constant. This is a particular case of state No. 2.

8.6.3. Selecting conditions :

We shall retain three possibilities :

Condition No. 1 : the selection is "absolutely correct", which means that the two following conditions are simultaneously fulfilled :
- the selection is "correct" : the density of selection probability is uniform throughout (T_L) : $\pi(t) \equiv \pi_0$ = constant.
- the remainder T'_S is zero : $T'_S = 0$ or $T_L = Q' T_S$ (Q' integer).

Condition No. 2 : the selection is "simply correct" and characterized by :
$\pi(t) \equiv \pi_0$ = constant. T'_S may be non-zero.

Condition No. 3 : the selection is "incorrect" and characterized by :
$\pi(t) \neq \pi_0$. The density of selection probability is no longer uniform throughout (T_L). In order to separate the sources of bias we shall assume here that $T'_S = 0$.

8.6.4. Cases studied in the following sections :

We shall retain the following "cases". These practically cover all possible situations :

Case No. 1 (section 8.7.) : State No. 1 (correlated functions), Condition No. 1 (absolutely correct selection).

Case No. 2 (section 8.8.) : State No. 2 (uncorrelated functions), Condition No. (absolutely correct selection).

Case No. 3 (section 8.9.) : State No. 3 (homogeneous material), Condition No. 1 2 or 3 (the selecting conditions are irrelevant).

Case No. 4 (section 8.10.): State No. 4 (uniform weighting), Condition No. 1 (absolutely correct selection).

Case No. 5 (section 8.11.): State No. 4 (uniform weighting), Condition No. 2 (simply correct selection).

Case No. 6 (section 8.12.) : State No. 4 (uniform weighting), Condition No. 3 (incorrect selection).

8.6.5. Sources of selection bias :

From a practical standpoint, it is of the utmost importance to know whether or no a selecting operation is liable to be biased. As regards the continuous selection model, we have been able to disclose three sources of bias :
- the "correlation bias" $B_C(a_S)$ resulting from the existence of a correlation between the descriptive functions $a(t)$ and $\mu(t)$.
- the "residual bias" $B_R(a_S)$ resulting from the fact that T_S (i.e. T_{sy} or T_{st}) is not a sub-multiple of T_L or in other words that the remainder T'_S is not nil.
- the "incorrectness bias" $B_I(a_S)$ resulting from the non-uniformity of the density of selection probability throughout T_L.

In fact two other sources of error may eventually result in a bias :
- non-uniformity of the extent T_I of the increments,
- non-uniformity of the interval T_{sy} or of the strata extent T_{st}.

These may indeed slightly vary from one increment to the next and if these variations are correlated with the function $a(t)$ they may result in a bias. From a practical standpoint, however, we shall observe that :
- the possible fluctuations of T_I, T_{sy} or T_{st} are small and practically negligible,
- These fluctuations are unlikely to be correlated with $a(t)$.

In the next sections, we shall therefore assume that T_I, T_{sy} or T_{st} are rigorously constant.

We shall call $B_T(a_S)$ the "total selection bias" associated to the continuous model. As the three souces of bias reviewed overleaf act independently from one another, we can write :

$$B_T(a_S) = B_C(a_S) + B_R(a_S) + B_I(a_S)$$

Our six cases have been combined in order to isolate the three sources of bias :
- the correlation bias $B_C(a_S)$ is isolated in case No. 1 (section 8.7.),
- the residual bias $B_R(a_S)$ is isolated in case No. 5 (section 8.11.),
- the incorrectness bias $B_I(a_S)$ is isolated in case No. 6 (section 8.12.).

It is usually easy enough to cancel the incorrectness bias $B_I(a_S)$. The cancellation of the residual bias $B_R(a_S)$ is theoretically easy but actually very difficult. The correlation bias being tied to the state of the material to be sampled cannot be cancelled. We must take it as it is.

8.7. DISTRIBUTION OF THE CRITICAL CONTENT a_S OF THE SAMPLE S - CASE No. 1

This case is characterized by the three hypotheses :

a) $\rho_{a\mu} \neq 0$: the descriptive functions are correlated,
b) $\Pi(t) \equiv \Pi_0$ = constant : the density of selection probability is uniform,
c) $T_S' = 0 : T_{sy}$ or T_{st} is a sub-multiple of T_L.

8.7.1. Distribution law of a_S :

The critical content a_S of the sample S is defined as follows : $a_S = \dfrac{A_S}{M_S}$.

It is the quotient of two random variables and it has been shown (Geary, 1930 and Bastien, 1960) that such a quotient never follows a simple law. When, however, both the numerator and the denominator have normal distributions and when furthermore the relative standard deviation of the denominator remains small in comparison with unity (e.g. 0.03) the quotient has a practically normal distribution.

We have already mentioned (sections 8.3.1. and 8.4.1.) that when the number Q of increments is "large enough" (e.g. larger than 30) A_S and M_S can be regarded as practically normal. Now, the condition $u(M_S) \leq 0.03$ or practically $u^2(M_S) \leq 10^-$ can be checked by using the results of section 8.3.3. If we want it to be fulfilled we just have to solve the relevant equation for either T_{sy}, T_{st} or Q_{ra} and implement a selection scheme ensuring that T_{sy} or T_{st} are smaller or that Q_{ra} is larger than the calculated value. We know what to do in order that both A_S and M_S should be normal and that the relative standard deviation of M_S should be small enough. We shall therefore assume from now on that these conditions are fulfilled and that a_S is a random variable with a normal (or practically normal) distribution.

8.7.2. Moments of a_S - Introduction :

8.7.2.1. Notations : it is convenient to use the following auxiliary variables

ε : relative deviation of A_S from its mean $m(A_S)$

η : relative deviation of M_S from its mean $m(M_S)$

ζ : difference $\zeta = \varepsilon - \eta$.

By definition :

$\varepsilon = \dfrac{A_S - m(A_S)}{m(A_S)}$ or $A_S = (1 + \varepsilon)\, m(A_S)$

$\eta = \dfrac{M_S - m(M_S)}{m(M_S)}$ or $M_S = (1 + \eta)\, m(M_S)$

$a_S = \dfrac{A_S}{M_S} = \dfrac{(1 + \varepsilon)\, m(A_S)}{(1 + \eta)\, m(M_S)} = \dfrac{1 + \varepsilon}{1 + \eta}\, a_L$

CE : "__continuous selection error__" or relative deviation of a_S from the critical content a_L of the lot L. By definition :

$$CE = \frac{a_S - a_L}{a_L} = \frac{\varepsilon - \eta}{1 + \eta} = \frac{\zeta}{1 + \eta}$$

ε, η, ζ and CE are random variables.

8.7.2.2. Moments of ε, η, ζ and CE :

By definition :

$m(\varepsilon) = 0 \qquad m(\eta) = 0 \qquad m(\zeta) = 0$

$\sigma^2(\varepsilon) = u^2(A_S) \qquad \sigma^2(\eta) = u^2(M_S) \qquad \rho(\varepsilon, \eta) = \rho(A_S, M_S)$

$\sigma^2(\zeta) = \sigma^2(\varepsilon) + \sigma^2(\eta) - 2\rho(\varepsilon, \eta) \sigma(\varepsilon) \sigma(\eta)$

The moments of CE are related to those of a_S by the following relationships :

$m(a_S) = \{1 + m(CE)\} a_L \qquad \text{or} \qquad B_T(a_S) = m(CE)$

$\sigma^2(a_S) = a_L^2 \sigma^2(CE) \qquad \text{or} \qquad u^2(a_S) = \sigma^2(CE)$

Since in this case the selection is assumed to be absolutely correct, the total bias $B_T(a_S)$ is reduced to the correlation bias $B_C(a_S)$. Then in case No. 1 :

$m(CE) = B_C(a_S)$

8.7.2.3. Development of $\frac{1}{1 + \eta}$:

The difficulty obviously arises from the fact that the denominator $1 + \eta$ of CE is a random variable. This difficulty can be overcome by observing that when $|\eta| < 1$ we may write the quotient $1/(1 + \eta)$ under the form of an absolutely convergent development :

$$\frac{1}{1 + \eta} = 1 - \eta + \eta^2 - \eta^3 + \eta^4 - \eta^5 + \text{etc} \ldots$$

When in addition $|\eta|$ is relatively small, each term of the development can be regarded as negligible in comparison with the preceding term. Let's observe that when the condition of normality is fulfilled, i.e. when $u(M_S) \leq 0.03$ or $3.33\, u(M_S)$ smaller than 0.1, then Prob $(|\eta| < 0.1) = 0.99914$ which can be regarded as a practical certainty. The development is certainly convergent and each term is negligible in comparison with the preceding term.

The development of $1/(1 + \eta)$ can be limited to different orders, according to the degree of precision sought in the estimation of the moments of a_S.

8.7.2.4. First approximation : (subscript 1) : We shall simply write :

$\frac{1}{1 + \eta} = 1 \qquad \text{or} \qquad (CE)_1 = \zeta$

$(CE)_1$ is the first estimator of CE. For all practical purposes, this first approximation is sufficient, except when dealing with low-grade materials such as ores of precious metals or minerals. This exception justifies the second approximation, especially for the selection bias.

8.7.2.5. Second approximation (subscript 2) : we shall write :

$$\frac{1}{1 + \eta} = 1 - \eta \quad \text{or} \quad (CE)_2 = \zeta\,(1 - \eta)$$

$(CE)_2$ is the second estimator of CE. It would be pointless to look for a higher approximation.

8.7.3. Moments of a_S - First approximation :

The first approximation is defined by : $(CE)_1 = \zeta$

8.7.3.1. Mean of a_S : we shall call $(a_S)_1$ the first approximation of a_S. It readily follows that :

$$m(CE)_1 = m(\zeta) = 0 \quad \text{or} \quad m(a_S)_1 = a_L \quad \text{or} \quad B_C(a_S)_1 = 0$$

In first approximation the correlation bias is zero : the mean of the critical content a_S of the sample S is equal to the critical content a_L of the lot L.

8.7.3.2. Variance of a_S : in the same way, we can write :

$$u^2(a_S)_1 = \sigma^2(CE)_1 = \sigma^2(\zeta) = u^2(A_S) + u^2(M_S) - 2\,\rho(A_S, M_S)\,u(A_S)\,u(M_S)$$

All the terms of this equality can be estimated from the results of a variographic experiment. We shall show in chapter 15 how this expression can be used in practice.

8.7.4. Moments of a_S - Second approximation :

The second approximation is defined by : $(CE)_2 = \zeta\,(1 - \eta)$

8.7.4.1. Mean of a_S : we shall call $(a_S)_2$ the second approximation of a_S. We easily calculate :

$$m(CE)_2 = m(\zeta) - m(\zeta\eta) = -m(\zeta\eta) = m(\eta^2) - m(\varepsilon\eta)$$

$$m(\eta^2) = u^2(M_S) \quad \text{and} \quad m(\varepsilon\eta) = m(\varepsilon)\,m(\eta) + \rho(\varepsilon,\eta)\,\sigma(\varepsilon)\,\sigma(\eta) = \rho(A_S, M_S)\,u(A_S)\,u(M_S)$$

$$m(a_S)_2 = \{1 + m(CE)_2\}\,a_L \quad \text{or} \quad B_C(a_S)_2 = m(CE)_2 = u^2(M_S) - \rho(A_S, M_S)\,u(A_S)\,u(M_S)$$

This bias is usually non-zero and this is a very important point. Even though the selection scheme is absolutely correct (the best we can do) it is biased. This bias results from the nature of the random variable a_S which is the quotient of two random variables that in the present case are correlated to each other. It can be estimated from the results of a variographic experiment. Fortunately, with the already mentioned exception of very low-grade materials such as the ores of precious metals and minerals, the bias $B_C(a_S)_2$ is always negligible. This will be shown on examples in chapter 15.

8.7.4.2. <u>Variance of</u> a_S : from a practical point of view the first approximation $u^2(a_S)_1$ is sufficient. Nevertheless, the method which has been developed in the preceding sections can be applied without any difficulty to the calculation of higher approximations of the first two moments or for the calculation of higher moments (this has been carried out in Gy, 1975).

8.8. DISTRIBUTION OF THE CRITICAL CONTENT a_S OF THE SAMPLE S - CASE No. 2

This case is characterized by three hypotheses :
a) $\rho_{a\mu} = 0$: the descriptive functions are un-correlated,
b) $\Pi(t) \equiv \Pi_0$ = constant : the density of selection probability is uniform,
c) $T_S' = 0$: T_{sy} or T_{st} is a sub-multiple of T_L.

8.8.1. <u>Mean of</u> a_S :

The absence of correlation between $a(t)$ and $\mu(t)$ involves that there is no correlation between the critical content a_q and the weight M_q of the increment I_q nor between the critical content a_S and the weight M_S of the sample S. From this property we may deduce :

$$m(A_S) = m(a_S) \, m(M_S)$$

But we already know (section 8.4.2.) that :

$$m(A_S) = a_L \, m(M_S) \quad \text{hence} \quad m(a_S) = a_L \quad \text{or} \quad B_C(a_S) \equiv 0$$

The only difference between case No. 1 and case No. 2 is the hypothesis that the coefficient of correlation $\rho_{a\mu}$ is nil. This justifies the name of "<u>correlation bias</u>" given to $B_C(a_S)$. This bias cancels in the same time as the coefficient of correlation between $a(t)$ and $\mu(t)$.

8.8.2. <u>Variance of</u> a_S :

The absence of correlation between $a(t)$ and $\mu(t)$ leads to the equality :

$$u^2(A_S) = u^2(a_S M_S) = u^2(a_S) \{1 + u^2(M_S)\} + u^2(M_S)$$
$$u^2(a_S) = \frac{u^2(A_S) - u^2(M_S)}{1 + u^2(M_S)} = u^2(A_S) - u^2(M_S)$$

The latter equality is justified by the fact that $u^2(M_S)$ is always very small and negligible as compared with unity.

8.9. DISTRIBUTION OF THE CRITICAL CONTENT a_S OF THE SAMPLE S - CASE No. 3

This case is characterized by the unique hypothesis :

$a(t) \equiv a_L$ = constant : the material to be sampled is homogeneous or in other words uniform throughout the domain (T_L). Since $a(t)$ is reduced to a constant a_L it is obviously un-correlated with $\mu(t)$. Case No. 3 is therefore a particular case of case No. 2.

Any increment I_q extracted from the stream has a critical content $a_q = a_L$.

Any sample S made of Q increments such as I_q has a critical content $a_S = a_L$. Then, irrespective of the selecting conditions, the selection is <u>exact</u>.

$a_S \equiv a_L$ hence $m(a_S) \equiv a_L$ or $B_C(a_S) \equiv 0$ and $\sigma^2(a_S) \equiv 0$ or $u^2(a_S) \equiv$

8.10. DISTRIBUTION OF THE CRITICAL CONTENT a_S OF THE SAMPLE S - CASE No. 4

This case is characterized by three hypotheses :

a) $\mu(t) \equiv \mu_L$ = constant : the rate of flow is uniform throughout (T_L),
b) $\Pi(t) \equiv \Pi_o$ = constant : the density of selection probability is uniform,
c) $T'_S = 0 : T_{sy}$ or T_{st} is a sub-multiple of T_L.

Since $\mu(t)$ is uniform throughout (T_L) it is obviously un-correlated with $a(t)$. Case No. 4 is therefore a particular case of case No. 2, the results of which remain valid in this section. It can be easily shown that :

$\mu_L = M_L/T_L$

8.10.1. General formulas :

From section 8.8. we know that :

$m(a_S) \equiv a_L$ or $B_C(a_S) \equiv 0$

$u^2(a_S) = u^2(A_S) - u^2(M_S)$

The third hypothesis entails that $Q = Q'$: the number of increments is fixed.

$M_q = T_I \mu_L = \dfrac{T_I}{T_L} M_L$ = constant. hence: $M_S = \dfrac{Q T_I}{T_L} M_L = \tau M_L$ = constant.

From which we deduce :

$u^2(M_S) = 0$ and $u^2(a_S) = u^2(A_S)$

But since $\alpha(t) = a(t) \mu(t) = a(t) \mu_L$, then $v'_\alpha(\theta) = v'_a(\theta)$.

The variance $u^2(a_S)$ can be expressed by means of the variogram $v'_a(\theta)$.

8.10.2. Practical formulas :

We shall assume that the variogram $v'_a(\theta)$ is parabolic, rectilinear or flat :

$v'_a(\theta) = v'_{a1} + v'_{a2}\theta + v''_{a2}\theta^2$ where v'_{a2} and v''_{a2} may eventually be zero.

The variance $u^2(a_S)$ can be written :

- <u>Systematic scheme</u> with interval T_{sy} :

$u^2_{sy}(a_S) = \dfrac{1}{T_L} (v'_{a1} T_{sy} + \dfrac{v'_{a2}}{6} T^2_{sy})$

- <u>Stratified scheme</u> with strata extent T_{st} :

$u^2_{st}(a_S) = \dfrac{1}{T_L} (v'_{a1} T_{st} + \dfrac{v'_{a2}}{3} T^2_{st} + \dfrac{v''_{a2}}{6} T^3_{st})$

- **Random scheme** with Q_{ra} increments :

$$u_{ra}^2(a_S) = \frac{1}{Q_{ra}}(v'_{a1} + \frac{v'_{a2}}{3} T_L + \frac{v''_{a2}}{6} T_L^2)$$

8.11. DISTRIBUTION OS THE CRITICAL CONTENT a_S OF THE SAMPLE S - CASE No. 5

This case is characterized by the three following hypotheses :
a) $\mu(t) \equiv \mu_L$ = constant : the rate of flow is uniform throughout (T_L),
b) $\Pi(t) \equiv \Pi_0$ = constant : the density of selection probability is uniform,
c) $T'_s \neq 0$: T_{sy} or T_{st} is not a submultiple of T_L.

The first hypothesis ensures that the correlation bias is cancelled. The second that the incorrectness bias is cancelled too. The third that the residual bias, if any, is not cancelled. This case isolates the residual bias $B_R(a_S)$.

We showed (GY 1975 section 13.8.2) that the residual bias $B_R(a_S)$ can be expressed as follows (systematic scheme) :

$$B_R(a_S) = \frac{(a_{L2} - a_{L1})}{a_L} \cdot \frac{T'_{sy} (T_{sy} - T'_{sy})}{T_{sy} T_L}$$

where a_{L1} and a_{L2} are the critical contents of the complementary sub-lots L_1 and L_2 occupying the domains (T_{L1}) and (T_{L2}) defined in fig. 8.1.

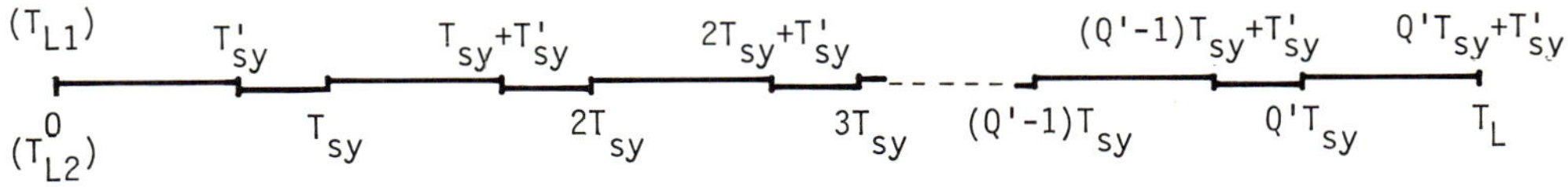

```
(T_L1)     T'sy        Tsy+T'sy      2Tsy+T'sy         (Q'-1)Tsy+T'sy   Q'Tsy+T'sy
  ├─────────┼─────────────┼─────────────┼──── ── ── ────┼─────────────┼─────────┤
  0        Tsy          2Tsy          3Tsy           (Q'-1)Tsy       Q'Tsy     TL
(T_L2)
```

Fig. 8.1. Definition of the sub-domains (T_{L1}) and (T_{L2}) of (T_L).

This bias obviously cancels when one at least of the three following conditions is fulfilled :

- $a_{L1} = a_{L2} = a_L$: as the function $a(t)$ is one of the data of the problem, and as it is usually not uniform, this condition is usually not fulfilled.

- $T'_{sy} = 0$: this solution is excluded from the present case by hypothesis c).

- $T'_{sy} = T_{sy}$: this solution is excluded from the present case by the definition of the Euclidean division according to which T'_{sy} is necessarily smaller than T_{sy}.

From a practical standpoint, this bias cannot be easily estimated. When Q' is large, it is likely to be small and probably negligible, except when $a(t)$ contains a periodic term and when T_{sy} is a multiple of the period T_p of the phenomenon. This problem, very important on account of the sampling errors it is liable to generate, will be studied in chapter 12. From this section we shall only retain that it is always advisable to select T_{sy} and T_{st} among the sub-multiples of T_L.

8.12. DISTRIBUTION OF THE CRITICAL CONTENT a_S OF THE SAMPLE S - CASE No. 6

This case is characterized by the three following hypotheses :

a) $\mu(t) \equiv \mu_L$ = constant : the rate of flow is uniform throughout (T_L),
b) $\Pi(t) \neq \Pi_0$: the density of selection probability is not uniform throughout (T_L)
c) $T'_S = 0$: T_{sy} or T_{st} is a sub-multiple of T_L.

The first hypothesis ensures that the correlation bias is cancelled. The second that the incorrectness bias is not cancelled. The third that the residual bias is cancelled. This case isolates the incorrectness bias $B_I(a_S)$.

This case had to be mentioned but it is purely academic since, at least as far as the sampling of flowing streams is concerned, it is always possible and easy to implement a correct selection scheme.

It is not at the level of the point selection that sampling is likely to deviate from correctness but at the level of increment delimitation and increment extraction as we shall see in the third part of this book.

CHAPTER 9

BREAKING UP OF THE CONTINUOUS SELECTION ERROR CE

9.1. INTRODUCTION

So far we have regarded the continuous selection error as a whole. We have developed a general model and proposed general formulas for estimating its mean and variance based on the knowledge of variographic parameters calculated from the results of a variographic experiment. Our purpose is now to deepen our analysis of the problem and to show that the continuous selection error CE is in fact the resultant of several errors independent of one another. The final aim of this analysis is a better understanding of the sampling error generating mechanisms.

Retaining the notations used in the preceding chapters with which the reader is now familiar, this analysis will be based on the investigation of :

- the parts played by the quality function $a(t)$ and the weighting function $\mu(t)$ used for the description of the material to be sampled,
- the parts played by the components of the quality function $a(t)$.

We shall retain in the next sections that the function $a(t)$ is :

$a(t) = a_0 + a_1(t) + a_2(t) + a_3(t)$ with :

a_0 : <u>a constant</u> describing the <u>average properties</u> of $a(t)$. It is convenient to use :

$$a_0 = \frac{1}{T_L} \int_0^{T_L} a(t)\, dt$$

$a_1(t)$: <u>short-range non-periodic term</u> describing the <u>local phenomena</u> tied to the particulate structure of the material to be sampled.

$a_2(t)$: <u>long-range non-periodic term</u> describing the <u>continuous tendencies</u> of the quality fluctuations.

$a_3(t)$: <u>periodic term</u> describing the <u>cyclic phenomena</u> that may be involved.

We shall assume in this chapter that neither $a(t)$ nor $\mu(t)$ is uniform throughout the domain (T_L) and that the selection is absolutely correct, i.e. that the density of selection probability $\Pi(t)$ is uniform throughout (T_L) and equal to Π_0 and that T'_S is nil:

$\Pi(t) \equiv \Pi_0 = $ constant
$T_L = Q'\, T_{sy}$ or $T_L = Q'\, T_{st}$ with Q' integer.

9.2. DEFINITION OF THE WEIGHTING ERROR WE AND OF THE QUALITY FLUCTUATION ERROR QE

The first step of our analysis will consist in separating the effects of the fluctuations of the functions $a(t)$ and $\mu(t)$. We shall define :

L' : an imaginary "unweighted lot" that would be obtained if the weighting function $\mu(t)$ should be uniform throughout (T_L) and equal to $\mu_L = M_L/T_L$ whereas the function $a(t)$ should remain unaltered.

$a_{L'}$: the critical content of L'. By definition :

$$a_{L'} = \frac{1}{T_L} \int_0^{T_L} a(t)\, dt = a_o$$

We shall now assume that the lots L and L' are submitted to the same selection and define :

S' : an imaginary "unweighted sample" representing the unweighted lot L'.

$a_{S'}$: the critical content of S'. By definition :

$$a_{S'} = \frac{1}{Q} \sum_q a(t_q) \quad \text{with} \quad q = 1, 2, \ldots Q.$$

The difference between L and L' on the one hand, between S and S' on the other reflects the influence of the weighting function $\mu(t)$. The moments of a_S have been expressed in section 8.7. and those of $a_{S'}$ in section 8.10. (cases No. 1 and 4 respectively). We shall now call or recall :

CE : the <u>continuous selection error</u> whose moments have been expressed in section 8.

$$CE = \frac{a_S - a_L}{a_L} \qquad m(CE) = B_C(a_S) \qquad \sigma^2(CE) = u^2(a_S)$$

QE : the "<u>quality fluctuation error</u>", part of the continuous selection error that is due to the sole fluctuations of the quality function $a(t)$. By definition :

$$QE = \frac{a_{S'} - a_{L'}}{a_{L'}}$$

Its moments have been expressed in section 8.10. :

$$m(QE) = B_C(a_{S'}) \qquad \sigma^2(QE) = u^2(a_{S'})$$

WE : the "<u>weighting error</u>", part of the continuous selection error due to the amplifying effect of the weighting function $\mu(t)$ on the fluctuations of the quality function $a(t)$. By definition, WE is the error with the following moments :

$$m(WE) = m(CE) - m(QE) = B_C(a_S) - B_C(a_{S'})$$
$$\sigma^2(WE) = \sigma^2(CE) - \sigma^2(QE) = u^2(a_S) - u^2(a_{S'})$$

Thanks to this definition, CE may be regarded as the sum of two independent erro

$$CE = QE + WE \qquad m(CE) = m(QE) + m(WE) \qquad \sigma^2(CE) = \sigma^2(QE) + \sigma^2(WE)$$

The quality fluctuation error QE will be analysed in the next section, its moments will be expressed in section 9.4. and its components will be studied in chapters 10 to 12. The weighting error WE will be studied in chapter 13.

9.3. ANALYSIS OF THE QUALITY FLUCTUATION ERROR QE

As recalled in section 9.1. the function $a(t)$ can be broken up into a sum of four terms :

$$a(t) = a_0 + a_1(t) + a_2(t) + a_3(t)$$

We shall define (with $j = 1, 2, 3$) :

$$a_{L'j} = \frac{1}{T_L} \int_{(T_L)} a_j(t) \, dt \quad \text{and} \quad a_{S'j} = \frac{1}{Q} \sum_q a_j(t_q) \quad \text{with } q = 1, 2, \ldots Q.$$

$$QE_j = \frac{a_{S'j} - a_{L'j}}{a_{L'}}$$

We can write :

$$a_{L'} = a_0 + a_{L'1} + a_{L'2} + a_{L'3}$$
$$a_{S'} = a_0 + a_{S'1} + a_{S'2} + a_{S'3}$$

and by substracting :

$$QE = QE_1 + QE_2 + QE_3$$

As the components of $a(t)$ take independent phenomena into account, we may regard QE as the sum of three independent errors :

QE_1 : "short-range quality fluctuation error" (studied in chapter 10),
QE_2 : "long-range quality fluctuation error" (studied in chapter 11),
QE_3 : "periodic quality fluctuation error" (studied in chapter 12).

9.4. MOMENTS OF THE QUALITY FLUCTUATION ERROR QE

We shall assume in this section that the periodic component $a_3(t)$ of $a(t)$ is nil.

9.4.1. Mean of QE :

We shall recall the results obtained in section 8.10 with the notations retained in the present chapter : irrespective of the selection scheme, we know that :

$$m(QE) = B_C(a_{S'}) \equiv 0$$

The sampling of the isolated function $a(t)$ is unbiased.

9.4.2. Variance of QE :

We shall assume that the variogram of $a(t)$ can be represented by the general parabolic formula of which the rectilinear and flat variograms are particular cases :

$$v'_a(\theta) = v'_{a1} + v'_{a2}\theta + v''_{a2}\theta^2 \quad \text{with } v'_{a2} \text{ and } v''_{a2} \text{ possibly zero.}$$

- **Systematic scheme** with interval T_{sy} :

$$\sigma_{sy}^2(QE) = u_{sy}^2(a_{S'}) = \frac{1}{T_L}(v'_{a1} T_{sy} + \frac{v'_{a2}}{6} T_{sy}^2)$$

- **Stratified scheme** with strata extent T_{st} :

$$\sigma_{st}^2(QE) = u_{st}^2(a_{S'}) = \frac{1}{T_L}(v'_{a1} T_{st} + \frac{v'_{a2}}{3} T_{st}^2 + \frac{v''_{a2}}{6} T_{st}^3)$$

- **Random scheme** with Q_{ra} increments :

$$\sigma_{ra}^2(QE) = u_{ra}^2(a_{S'}) = \frac{1}{Q_{ra}}(v'_{a1} + \frac{v'_{a2}}{3} T_L + \frac{v''_{a2}}{6} T_L^2)$$

9.5. RECAPITULATION

The continuous selection error CE can be regarded as the sum of four independent errors :

$$CE = QE_1 + QE_2 + QE_3 + WE$$

$$m(CE) = m(QE_1) + m(QE_2) + m(QE_3) + m(WE)$$

$$\sigma^2(CE) = \sigma^2(QE_1) + \sigma^2(QE_2) + \sigma^2(QE_3) + \sigma^2(WE)$$

The properties of these errors will be studied in chapters 10 to 13.

CHAPTER 10

SHORT-RANGE QUALITY FLUCTUATION ERROR QE_1

10.1. DEFINITION

The <u>short-range non-periodic quality fluctuation term</u> $a_1(t)$ has been objectively defined (section 6.7.6.) as the component of $a(t)$ that is characterized by the constant term v'_{a1} of the model variogram $v'_a(\theta)$. More subjectively, we know that $a_1(t)$ takes into account all that $a(t)$ contains which is tied to the particulate structure of the material to be sampled : discontinuities, local disorder, segregation, etc...

As we did in chapter 9, we shall assume here that the selection is absolutely correct.

By definition, the <u>short-range quality fluctuation error</u> QE_1 is the component of QE containing the variographic parameter v'_{a1}. It may be interesting to note that the quality fluctuation error QE is reduced to QE_1 when the variogram is flat, i.e. characterized by a constant v'_{a1}.

10.2. MEAN OF QE_1

According to its definition, it is easy to show that :
$$m(QE_1) = 0$$
Irrespective of the selection scheme, the sampling of the component $a_1(t)$ of $a(t)$ is unbiased.

10.3. VARIANCE OF QE_1

This variance can be easily deduced from the general formulas recalled in section 9.3. by simply putting : $v'_{a2} = v''_{a2} = 0$:

- <u>Systematic selection</u> with interval T_{sy} :

$$\sigma^2_{sy}(QE_1) = \frac{T_{sy}}{T_L} v'_{a1} = \frac{v'_{a1}}{Q}$$

- <u>Stratified selection</u> with strata extent T_{st} :

$$\sigma^2_{st}(QE_1) = \frac{T_{st}}{T_L} v'_{a1} = \frac{v'_{a1}}{Q}$$

- <u>Random selection</u> with Q_{ra} increments :

$$\sigma^2_{ra}(QE_1) = \frac{1}{Q_{ra}} v'_{a1} = \frac{v'_{a1}}{Q}$$

Irrespective of the selection scheme the variance of QE_1 can be written :

$$\sigma^2(QE_1) = \frac{v'_{a1}}{Q} \quad \text{with Q number of increments.}$$

As far as the short-range quality fluctuation error is concerned, the three selection schemes are strictly equivalent.

10.4. CANCELLING AND MINIMIZING OF $\sigma^2(QE_1)$

This variance cancels if and only if $v'_{a1} = 0$. Experience shows and this point will be confirmed in chapter 21 dealing with the analysis of QE_1 in the discrete perspective, that with particulate materials v'_{a1} is never zero. With continuous materials such as compact solids or liquid streams, on the contrary, v'_{a1} is nearly always nil or at least negligible. We shall therefore conclude that within the scope of this book the short-range quality fluctuation error is never zero.

A bed blending theory developed by the author on the basis of the present theory of sampling shows that even bed blending does not reduce v'_{a1}. The effect of blending is to cancel v'_{a2} and v''_{a2} or at least to reduce these parameters in a large proportion. Mixing of the whole lot would reduce v'_{a1} but is uneconomical as soon as large tonnages are concerned. We must therefore consider v'_{a1} as irreducible. Then, the only way of minimizing $\sigma^2(QE_1)$ is to increase the number Q of increments by retaining the smallest interval T_{sy} or extent T_{st} or the largest number Q_{ra} economically acceptable.

As the cost of sample reduction increases with the number Q of increments, a compromise must be found somewhere between a fairly reproducible but costly sampling and a cheap but poorly reproducible one. This compromise is not easy to define in objective terms as it is not always possible to figure out the economical asset of reproducible sampling nor the risks involved when departing from a given reproducibility standard.

10.5. FURTHER ANALYSIS OF THE SHORT-RANGE QUALITY FLUCTUATION ERROR QE_1

It is at the level of the function $a_1(t)$ and of the error QE_1 that the continuous model and the discrete model can be linked up to each other. The analysis of the logical content of QE_1 will be carried out in chapter 21 on the basis of the results of the discrete model developed in chapter 20 and of the analysis of the notion of heterogeneity of a discrete set performed in chapter 19.

CHAPTER 11

LONG-RANGE QUALITY FLUCTUATION ERROR QE_2

11.1. DEFINITION

The <u>long-range non-periodic quality fluctuation term</u> $a_2(t)$ has been objectively defined (section 6.7.6.) as the component of $a(t)$ that is characterized by the non-constant non-periodic term of the model variogram $v'_a(\theta)$. When the parabolic representation is valid, $a_2(t)$ is the fraction of $a(t)$ characterized by the variogram :

$$v'_{a2}(\theta) = v'_{a2}\theta + v''_{a2}\theta^2$$

More subjectively, we know that $a_2(t)$ takes into account all that remains of the original continuity of the material to be sampled, as discontinuities, local disorder and segregation have been disposed of by the term $a_1(t)$, and as all kinds of periodic phenomena will be taken into account by the term $a_3(t)$. In other words $a_2(t)$ describes the long-range non-periodic tendencies of $a(t)$.

When nothing remains of the original continuity nearly always observed in mineral deposits, then $a_2(t) \equiv 0$ which entails $v'_{a2} = v''_{a2} = 0$: the variogram $v'_a(\theta)$ is flat.

As we did in chapter 9, we shall assume in this chapter that the selection is absolutely correct.

By definition, the <u>long-range quality fluctuation error</u> QE_2 is the component of QE containing the variographic parameters v'_{a2} and v''_{a2}.

11.2. MEAN OF QE_2

According to its definition, it is easy to show that the mean $m(QE_2)$ is zero if and only if T_{sy} or T_{st} (systematic and stratified schemes) is a sub-multiple of T_L. When this condition is not fulfilled but when Q is large (say larger than 30) the resulting bias is likely to be negligible. For all practical purposes, we shall admit that :

$m(QE_2) = 0$ irrespective of the selection scheme.

11.3. VARIANCE OF QE_2

This variance can be easily deduced from the general formulas recalled in section 9.3 by simply putting $v'_{a1} = 0$. It readily follows :

- <u>Systematic selection</u> with interval T_{sy} :

$$\sigma^2_{sy}(QE_2) = \frac{1}{T_L} \left(\frac{v'_{a2}}{6} T^2_{sy} \right)$$

- <u>Stratified selection</u> with extent T_{st} :

$$\sigma^2_{st}(QE_2) = \frac{1}{T_L} \left(\frac{v'_{a2}}{3} T^2_{st} + \frac{v''_{a2}}{6} T^3_{st} \right)$$

- <u>Random selection</u> with Q_{ra} increments :

$$\sigma^2_{ra}(QE_2) = \frac{1}{Q_{ra}} \left(\frac{v'_{a2}}{3} T_L + \frac{v''_{a2}}{6} T^2_L \right)$$

11.4. COMPARISON OF THE THREE SELECTION SCHEMES

As regards the variance $\sigma^2(QE_2)$ it can be easily shown that :

$$\frac{\sigma^2_{st}(QE_2)}{\sigma^2_{sy}(QE_2)} = 2 + \frac{v''_{a2}}{v'_{a2}} \cdot \frac{T_L}{Q} \qquad \text{and} \qquad \frac{\sigma^2_{ra}(QE_2)}{\sigma^2_{st}(QE_2)} = Q$$

The factor $v''_{a2} T_L / v'_{a2} Q$ is often positive, sometimes negative but usually small as compared with 2. From a practical standpoint, we shall retain that the systematic variance is about twice smaller than the stratified variance which is itself Q times smaller than the random variance. With Q larger than 30 the random selection scheme is practically disqualified. In other words, for a given reproducibility, the systematic selection is cheaper than the stratified selection which is itself much cheaper than the random selection.

11.5. CANCELLING AND MINIMIZING OF $\sigma^2(QE_2)$

By definition, $\sigma^2(QE_2)$ cancels if and only if the variographic parameters v'_{a2} and v''_{a2} are nil or in other words if the variogram is flat. As already mentioned, bed blending transforms any variogram into a flat variogram. The reader should however remember that most reclaiming systems are likely to introduce a non-negligible periodic term $a_3(t)$ liable to generate a periodic quality fluctuation error if sampling according to a systematic scheme with an interval T_{sy} multiple of the period of the reclaiming device (e.g. bucket wheel). As this period is very well known, the interval T_{sy} can be chosen accordingly. This point will be dealt with in chapter 12. Now we shall observe that bed blending is a very expensive technique and that, as far as we know, no blending system has ever been resorted to for the sole purpose of suppressing the error QE_2.

If we regard v'_{a2} and v''_{a2} as intangible data of the problem, there is no way of cancelling QE_2 and the only way of reducing it is to take as large a number Q of increments as economically acceptable. The notions of acceptable representativeness and of acceptable cost are analysed in chapter 29.

CHAPTER 12

PERIODIC QUALITY FLUCTUATION ERROR QE_3

12.1. DEFINITION

The periodic quality fluctuation term $a_3(t)$ has been objectively defined (section 6.7.6.) as the component of $a(t)$ that is characterized by the periodic term of the variogram $v'_a(\theta)$. This periodic term can be written :

$$v'_{a3}(\theta) = v'_{a31}(\theta) + v'_{a32}(\theta) + \ldots$$

each term of the sum having the simple general form :

$$v'_{a31}(\theta) = (1 - \cos 2\pi\theta/T_{P1}) \, v'_{a31}$$
$$v'_{a32}(\theta) = (1 - \cos 2\pi\theta/T_{P2}) \, v'_{a32} \qquad \text{etc...}$$

and representing one of the components $a_{31}(t)$, $a_{32}(t)$, etc.. of $a(t)$ with :

$$a_{31}(t) = a_{31} \sin 2\pi t/T_{P1} + a'_{31} \cos 2\pi t/T_{P1} \quad \text{and} \quad v'_{a31} = (a_{31}^2 + a_{31}^{'2})/2 \, a_L^2$$
$$a_{32}(t) = a_{32} \sin 2\pi t/T_{P2} + a'_{32} \cos 2\pi t/T_{P2} \quad \text{and} \quad v'_{a32} = (a_{32}^2 + a_{32}^{'2})/2 \, a_L^2 \quad \text{etc..}$$

As already mentioned in section 6.7.6., we should have no illusion : as soon as $v'_{a3}(\theta)$ is made of more than two simple terms, its analysis becomes very difficult and the error estimation inextricable. For this reason we shall assume in this chapter that the periodic phenomenon is simply sinusoidal :

$$a_3(t) = a_3 \sin 2\pi t/T_p + a'_3 \cos 2\pi t/T_p$$

and is characterized by a simple variogram :

$$v'_{a3}(\theta) = (1 - \cos 2\pi\theta/T_p) \, v'_{a3} \qquad \text{with} \qquad v'_{a3} = (a_3^2 + a_3^{'2})/2 \, a_L^2$$

Should we know the respective values of a_3 and a'_3 or in other words the respective positions on the time axis of the origin of the flow and that of the sinusoid, the knowledge of the instant t_q of the increments would enable us to calculate exactly the selection error QE_3, due to the fact that the mathematical expression of $a_3(t)$ would be exactly known. As we never do, we shall assume that the origin of the flow is positioned at random within the domain (T_p) occupied by one cycle of the sine curve. This amounts to assuming that the flow is un-correlated with the periodic phenomenon. Then, the origin of the flow being irrelevant, we shall study the simple term :

$$a_3(t) = a_3 \sin 2\pi t/T_p \qquad v'_{a3}(\theta) = (1 - \cos 2\pi\theta/T_p) \, v'_{a3} \qquad \text{with} \qquad v'_{a3} = \frac{a_3^2}{2a_L^2}$$

We shall use the following notations :

a''_{L3} : mean of $\sin 2\pi t/T_p$ throughout the domain (T_L) occupied by the lot L.

$$a''_{L3} = \frac{T_p}{2\pi T_L} (1 - \cos 2\pi T_L/T_p)$$

a''_{S3} : mean of the sample extracted from the function $\sin 2\pi t/T_p$:

$$a''_{S3} = \frac{1}{Q} \sum_q \sin 2\pi t_q/T_p$$

QE_3 : the periodic quality fluctuation error incurred when sampling $a_3(t)$

$$QE_3 = \frac{a_3}{a_L} (a''_{S3} - a''_{L3})$$

We shall study the moments of a''_{S3} and deduce those of QE_3 in the following way

$$m(QE_3) = \frac{a_3}{a_L} \{m(a''_{S3}) - a''_{L3}\} \qquad \text{and} \qquad \sigma^2(QE_3) = \frac{a_3^2}{a_L^2} \sigma^2(a''_{S3})$$

Remarks : The conditions of first order stationarity are not satisfied with a function such as $a_3(t)$ which forbids the use of the variogram to estimate the moments of QE_3. Now, the knowledge of the analytical expression of $a_3(t)$ makes it possible to estimate directly the moments of QE_3.

12.2. OCCURRENCE OF PERIODIC FLUCTUATIONS

The study of a large number of a(t) functions provides a great variety of periodic fluctuations presenting a more or less regular pattern of more or less equal minima alternating with more or less equal maxima. This is confirmed by the study of a number of variograms. The periodic phenomenon does not need to be strictly sinusoidal for its effects to be detrimental to sampling.

Periodic fluctuations are occasionally observed at the scale of mineral deposits but must be regarded as exceptional. The periodic phenomena actually observed after the ore has been extracted may have different origins :

1) Human activity and more specifically industrial activity is always arranged according to cyclic patterns with periods such as shift , day, week, etc.. Such patterns may generate either abrupt or more or less progressive quality or/and quantity changes occurring at constant interval but these intervals are usually large enough not to be dangerous for sampling.

2) Automated processing flow-sheets very often work in a cyclic way, a certain characteristic of the flowing material being allowed to vary between a high and a low mark. During the first part of the cycle this characteristic is liable (for instance) to increase until it reaches the high mark and triggers some regulating device. Then, the increase is progressively slowed down and stopped and a decrea-

sing tendency is likely to be observed until the critical characteristic reaches the low mark where the process is reversed again, etc... Regulation aims at maintaining constant a certain characteristic and actually achieves a cyclic fluctuation between known limits. In such a case, the parameter a_3 and its variographic counterpart v'_{a3} can be easily estimated, a_3 being the half of the difference between the high and the low mark. The period is unknown but experience shows that it varies little. As a very general rule, the periods of such phenomena may be of the order of a few minutes to one hour. More often than not, it falls within the dangerous range of usual intervals between systematic increments.

3) Certain pieces of equipment to be found in most mineral processing plants are built in such a way that they progressively accumulate certain size or density fractions of the material until a natural limit is attained. Then, the process is reversed and this accumulation decreases until a low point is reached and the process is reversed again. One of the best examples of such a behaviour is provided by closed grinding circuits. With mechanical classifiers, the period of the phenomenon is of the order of 10 to 20 minutes, with hydrocyclones, the period is much shorter and less dangerous. Another good example is given by centrifugal pumps extracting a pulp from a sump. In both cases careful studies of the outgoing material (overflow of the classifier or pumped material) have shown cyclic fluctuations of the size analysis with correlated cyclic fluctuations of the mineral composition at a point of the flow-sheet where sampling is likely to take place.

These are but a few examples presented in order to show that periodic fluctuations actually occur in everyday practice. We shall see in the next sections that they may become dangerous, which justifies their being seriously taken into consideration in a complete theory of sampling.

12.3. MOMENTS OF QE_3

The moments of QE_3 have been calculated in a previous work (Gy, 1975) for the three reference selection schemes. It would be pointless to reproduce the demonstrations in this book since the formulas expressing the moments are so complex that no one would ever think of using them in a practical work. We must however summarize our conclusions.

Among the relevant factors, the properties of the Euclidean division of :

T_L by T_{sy} or T_{st} ,
T_L by T_P
T_{sy} or T_{st} by T_P

have been found of major importance when implementing a systematic or a stratified selection scheme.

We shall use the following notations :

Q_{Ls} : quotient of the Euclidean division of T_L by T_s (i.e. T_{sy} or T_{st}),
T_{Ls} : remainder of the same division. By definition :

$$T_L = Q_{Ls} T_s + T_{Ls} \quad \text{with} \quad 0 \leq T_{Ls} < T_s$$

Q_{LP} : quotient of the Euclidean division of T_L by T_P,
T_{LP} : the remainder of the same division. By definition :

$$T_L = Q_{LP} T_P + T_{LP} \quad \text{with} \quad 0 \leq T_{LP} < T_P$$

Q_{sP} : quotient of the Euclidean division of T_s (i.e. T_{sy} or T_{st}) by T_P,
T_{sP} : remainder of the same division. By definition :

$$T_s = Q_{sP} T_P + T_{sP} \quad \text{with} \quad 0 \leq T_{sP} < T_P$$

12.3.1. Most general case :

This case is characterized by the following hypotheses :

$$T_{Ls} \neq 0 \qquad T_{LP} \neq 0 \qquad T_{sP} \neq 0$$

In this case, the selection is biased, due to the remainder T_{Ls} being non-zero as already pointed out in chapter 7, except with the random selection scheme. The variances of the three selection schemes can be ordered in the following way :

T_{sP} near $T_P/2$: $\sigma^2_{sy} < \sigma^2_{st} < \sigma^2_{ra}$

T_{sP} near 0 or T_P : $\sigma^2_{st} < \sigma^2_{ra} < \sigma^2_{sy}$

12.3.2. First particular case :

This case is characterized by the following hypotheses :

$$T_{Ls} = 0 \qquad T_{LP} \neq 0 \qquad T_{sP} \neq 0$$

T_L, extent of the lot L, is a multiple of either T_{sy} or T_{st}. Then, irrespective of the selection scheme, the selection is unbiased. The variances can be ordered in the same way as above.

12.3.3. Second particular case :

This case is characterized by the following hypotheses :

$$T_{Ls} = 0 \qquad T_{LP} = 0 \quad \text{but} \quad T_{sP} \neq 0$$

T_L is a multiple of both T_s and T_P but, and this is very important, T_s <u>is not a multiple of</u> T_P. Then, it can be shown that when implementing a systematic selection scheme the selection is exact (mean and variance of QE_3 identically zero) and we can wonder whether it is possible to take a practical advantage of this property. For a given value of T_L multiple of the period T_P (we can to some extent adjust T_L in order to respect this condition) the variance becomes zero for all values of T_{sy} belonging to the series : $T_{sy0} = T_L / Q_{Ls}$ with Q_{Ls} integer.

Let's for instance assume that :

T_L = 480 mn and T_P = 80 mn.

The values of T_{sy0} of the order of 5 mn are :

480 ÷ 95 = 5.052 mn
480 ÷ 96 = 5.000 mn
480 ÷ 97 = 4.948 mn etc...

For each of these values of T_{sy} the variance $\sigma^2_{sy}(QE_3)$ is zero but we must expect that between two consecutive values of T_{sy0}, this variances reaches at least one (in fact three) local maximum. This means that about mid-way between two consecutive zero minima, or very near 5.026 and 4.974 mn in the present example, the variance will reach a local maximum. We can see that a very small error in the setting of the timer (± 1.5 s or 0.5 % relative) can be responsible for a maximum variance whilst we were expecting a zero minimum.

Furthermore, the period T_P is very seldom known with such a precision that we can be certain that it is actually a sub-multiple of T_L. Thus, this very attractive mathematical solution is practically worthless.

12.3.4. Third particular case :

This case is characterized by the single hypothesis :

$T_{sP} = 0$:

a) <u>Systematic selection</u> with interval T_{sy} : T_{sy} is a multiple of the period T_P of the phenomenon. As shown in fig. 12.1. all increments are extracted from the same point of the sine curve. Q systematic increments provide exactly the same amount of information on the curve $a_3(t)$ as a single increment.

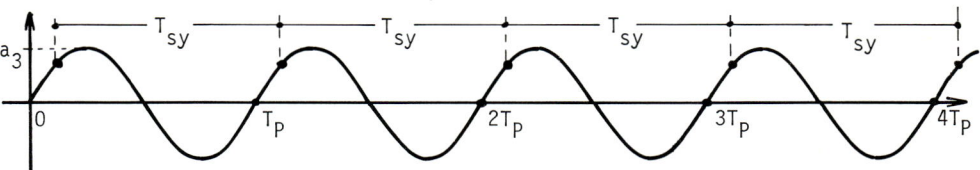

Fig. 12.1. Systematic sampling of a sine curve with $T_{sy} = T_P$.

In this case, it has been shown that the variance $\sigma^2_{sy}(QE_3)$ reaches a "<u>maximum maximorum</u>" :

$\sigma^2_{sy}(QE_3)_{max} = v'_{a3}$

b) <u>Stratified selection</u> with strata extent T_{st} : under the same conditions it can be shown that $\sigma^2_{st}(QE_3)$ reaches a local maximum with :

$\sigma^2_{st}(QE_3)_{max} = v'_{a3} / Q$ with Q number of increments.

12.4. COMPARISON OF THE THREE SELECTION SCHEMES

The most important conclusion is that the systematic selection which so far seemed to be the best of all selection schemes is liable to introduce a very impotant error when implemented to sample periodic functions. This is not an academic but a practical point : on several occasions, we met periodic fluctuations charac terized by variographic parameters v'_{a3} of the order of 0.01. The maximum maximorum of the standard deviation in the case of a systematic sampling is therefore 0.1 or 10 % relative. This means that the risk incurred when applying a systematic scheme is a relative error of ± 20 %, a risk which is far from being negligible.

This risk is however tempered by several factors :

a) the period T_p is seldom strictly constant (exception : certain reclaiming systems used in bed blending facilities). When it is constant, it can be measured with a good precision. Assuming the period T_p to be known, T_{sy} can be selected among the odd multiples of the half-period. This does not minimizes the variance but eliminates the risk represented by the maximum maximorum.

b) the interval T_{sy} is seldom strictly constant,

c) the probability of T_{sy} being an exact multiple of T_p seems to be very remote. One may nearly always expect some sort of a stroboscopic effect thanks to which the danger cancels in the long run.

We must nevertheless remain conscious of the risk and be very careful. Our practical conclusions can be summarized as follows :

1) T_p and v'_{a3} are unknown : no variographic experiment has been carried out but we have reasons to suspect the existence of periodic fluctuations. There is a risk but we cannot precise its importance. Then, the safest solution is always to implement a stratified selection scheme.

2) T_p and v'_{a3} are known : for instance as the result of a variographic experiment, but T_p may be known by direct analysis of the phenomenon (e.g. reclaiming systems of bed blending facilities). Two cases may arise :

a) the maximum variance $\sigma^2_{sy}(QE_3)_{max} = v'_{a3}$ is acceptable : the corresponding confidence interval is $\pm 2\sqrt{v'_{a3}}$. Then, we can safely implement a systematic scheme and forget about periodic fluctuations. But it is always safe, nevertheless, to carry out a stratified selection.

b) the maximum variance $\sigma^2_{sy}(QE_3)_{max} = v'_{a3}$ is definitely unacceptable : two solutions are available. The first and most advisable is to implement a stratified selection scheme. The second is to implement a systematic scheme, chosing T_{sy} among the odd multiples of the half-period. When T_p is known with great accuracy and when using a very accurate timer, we can thus minimize the variance.

As a very general rule, when dealing with possibly periodic functions, a random stratified selection scheme is always the safest of all solutions. Systematic selection can be a little better or a lot worse (in terms of selection variance) according as we are lucky or not. The choice between stratified and systematic selection can reduce itself to a very simple question : are we in a position to gamble ?

In no case is the random selection scheme better than the stratified scheme.

12.5. EXAMPLE

We have calculated the three variances $\sigma^2_{sy}(QE_3)$, $\sigma^2_{st}(QE_3)$ and $\sigma^2_{ra}(QE_3)$ in a practical example characterized by T_p = 13 mn and v'_{a3} = 0.0001. A sample was to be taken every four hours which means T_L = 240 mn. The three variances have been plotted against T_{sy}, T_{st} or Q_{ra} and are represented in fig 12.2. for values of T_{sy} and T_{st} ranging from 35 to 58 mn and covering several periods. This graph is typical. It shows the peaks representing the maximum maximorum of $\sigma^2_{sy}(QE_3)$ whenever T_{sy} is a multiple of the period T_p = 13 mn.

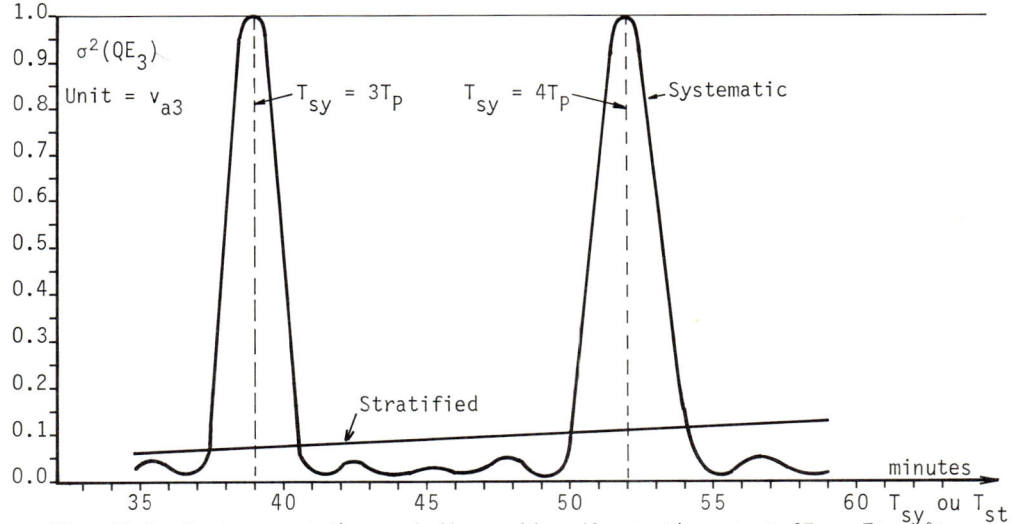

Fig. 12.2. *Variance of the periodic quality fluctuation error* QE_3. *Example.*

12.6. GENERAL CONCLUSION CONCERNING THE CHOICE OF A SELECTION SCHEME

When chosing a selection scheme, we should never forget that we don't sample separately the various components of a(t) but their sum. In the same way our problem is not to minimize $\sigma^2(QE_1)$, $\sigma^2(QE_2)$ or $\sigma^2(QE_3)$ but their sum $\sigma^2(QE)$.

We already know that $\sigma^2(QE_1)$ does not depend on the selection scheme. As far as the variance $\sigma^2(QE_2)$ is concerned, the systematic scheme is about twice better than the stratified scheme. Now, we have just seen the dangers of systematic selection when sampling periodic functions and concluded that stratified sampling was always the safest solution.

As a general conclusion, we shall state that the most reproducible selection scheme will be, assuming a constant sampling and sample reduction cost :

- either the systematic selection scheme when it has been proved that there were no periodic quality fluctuations or that they were harmless,

- or the stratified selection scheme when periodic fluctuations are likely to take place and to introduce inacceptable errors with a systematic selection.

- in no case is the random scheme better than the other two, which rules it out from a practical standpoint.

Since periodic fluctuations are much more frequent than is usually imagined, we would definitely advise the generalization of the stratified selection. It is the safest of all solutions. It can be easily implemented by means of special timers (see section 7.3.2. remark in fine).

CHAPTER 13

WEIGHTING ERROR WE

13.1. DEFINITION

The <u>weighting error WE</u> has been objectively defined (section 9.2.) as the error whose moments can be calculated by difference between those of the continuous selection error CE (section 8.7.) and those of the quality fluctuation error QE (section 8.10.). More subjectively, the weighting error WE represents the part of the continuous selection error CE which is due to the amplifying effect of the weighting function $\mu(t)$ on the fluctuations of the quality function $a(t)$.

We shall assume in this chapter that the selection scheme is absolutely correct.

13.2. MEAN OF THE WEIGHTING ERROR WE

By definition :

$m(WE) = m(CE) - m(QE)$

From sections 8.7. and 8.10. we know that :

$m(CE) = B_C(a_S)$ and $m(QE) = B_C(a_{S'}) = 0$

Then :

$m(WE) = B_C(a_S) = u^2(M_S) - \rho(A_S, M_S)\, u(A_S)\, u(M_S)$

The "<u>weighting bias</u>" $m(WE)$ coincides with the "<u>correlation bias</u>" $B_C(a_S)$. It can be estimated from the results of a variographic experiment. It is usually non-zero and we already know (section 8.8.) that it cancels when the coefficient of correlation $\rho_{a\mu}$ between the functions $a(t)$ and $\mu(t)$ tends toward zero. This may happen independently, none of the functions $a(t)$ and $\mu(t)$ being uniform throughout (T_L), but also when one of these functions is uniform (homogeneity or uniform weighting). The properties of the functions $a(t)$ and $\mu(t)$ are data of the problem on which we have practically no means of action.

When the selection is absolutely or even simply correct, the weighting bias is always negligible, the coefficient of bias $b(WE) = |m(WE)| / \sigma(WE)$ being very small as compared with unity.

13.3. VARIANCE OF THE WEIGHTING ERROR WE

13.3.1. Expression of the variance :

By definition (sections 8.7. and 8.10.) :

$$\sigma^2(WE) = \sigma^2(CE) - \sigma^2(QE) = u^2(a_S) - u^2(a_{S'})$$

This definition obviously requires that the variance $u^2(a_S)$ be larger than $u^2(a_{S'})$ which is usually observed. When it is not, both variances are of the same order of magnitude which simply means that they are estimates of the same variance with the result that the weighting variance may be regarded as nil.

The variances $u^2(a_S)$ and $u^2(a_{S'})$ are expressed for each of the three selection schemes by formulas respectively given in sections 8.7.5.2. and 8.10.2. These variances can be estimated from the results of a variographic experiment.

13.3.2. Weighting fluctuation factor :

It is convenient to express the relative importance of the nuisance resulting from the weighting error by means of the "weighting fluctuation factor" $\Gamma(WE)$ defined as follows :

$$\Gamma(WE) = \frac{\sigma^2(WE)}{\sigma^2(QE)} = \frac{u^2(a_S)}{u^2(a_{S'})} - 1 \quad \text{which entails :} \quad \sigma^2(CE) = \{1 + \Gamma(WE)\}\, \sigma^2(QE)$$

From a practical standpoint, we shall therefore admit that the weighting error

- acceptable : when $\Gamma(WE) \leq 1$
- negligible : when $\Gamma(WE) \leq 0.1$

13.3.3. Weighting variance and fluctuations of the rate of flow :

The weighting variance depends on the fluctuations of the rate of flow $\mu(t)$ on the one hand but also on the coefficient of correlation between $a(t)$ and $\mu(t)$. As a general rule but with possible exceptions when this correlation is low, the weighting variance can be regarded as :

- acceptable : when the fluctuations of the rate of flow characterized by the confidence interval $\pm\, 2u(M_q)$ do not exceed $\pm\, 20\,\%$.
- negligible : when $\pm\, 2u(M_q)$ do not exceed $\pm\, 10\,\%$.

13.3.4. Comparison of the three selection schemes :

The latter conclusion is valid, irrespective of the selection scheme (systematic or stratified) with a tendency for the random scheme to generate smaller weighting variances than the other two schemes.

13.4. CONSTANT TONNAGE SAMPLING SYSTEMS

Some manufacturers of sampling equipment, worried by the possible importance of the weighting error, have attempted to reduce or suppress the effects of the fluctuations of the rate of flow by implementing a systematic selection scheme based on a constant tonnage rather than on a constant time interval. In plain words, this means that an increment is extracted from the stream , say every 200

tons instead of every 10 mn. In such devices, the sampling cutter is set in action
by electrical impulses generated by integrating scales. But is this an improvement ?

13.4.1. The problem to be solved :

The problem arises from the fact that non-uniform tonnages of material are represented by non-uniform non-proportional increment weights. We already know that the weighting error is suppressed when the weighting function is uniform throughout (T_L) but it can be easily shown that it is suppressed in the same way if the increment weight is proportional to the tonnage it is supposed to represent. In a constant tonnage systematic sampling system, the weighting error will therefore be cancelled if and only if the increment weight is also constant. We shall see now how this condition can be fulfilled. Let's call :

W : the constant width of a straight path sampler cutter opening,
$V(t_q)$: the cutter velocity when extracting the increment I_q at instant t_q,
M_q : the weight of increment I_q : it readily follows that :

$$M_q = W \frac{\mu(t_q)}{V(t_q)}$$

Since W is a constant, the increment weight will be a constant too if and only if the cutter velocity $V(t_q)$ can be set proportionally to the rate of flow $\mu(t_q)$. Now what do we observe in practice ? Constant tonnage sampling systems belong to two categories :

- those using a uniform speed cutter,
- those using a proportional speed cutter.

13.4.2. Systems using a uniform speed cutter :

When the cutter speed is uniform from one increment to the next, the increment weight is proportional to the rate of flow $\mu(t_q)$. This means that a constant tonnage will be represented by a variable increment weight. The problem has been displaced, it has not been solved. An error has been replaced by another error. One can even fear that the resulting weighting error be larger than with the constant time interval sampling systems, since in this latter case there is always a correlation (a shy step towards proportionality) between the increment weight and the tonnage flowing during the period $t_q \pm T_{sy} / 2$ it is supposed to represent (section 13.6.2. below).

Such systems reduce in no way the weighting error and can even be suspected of increasing it. They are uselessly expensive. Though mentioned or even recommended by various national and international standards, we can see no reason to adopt them.

13.4.3. Systems using a proportional speed cutter :

From a theoretical standpoint, these systems are very satisfactory since they are supposed to deliver increments of uniform weight representing uniform tonnages of material. Practically, however, several difficulties arise when attempting at :

- measuring the instantaneous rate of flow with a sufficient accuracy,
- setting a cutter speed which is actually proportional to the measured rate of flow,
- realizing such a system at an acceptable cost.

These systems would be more efficient than the constant time systems if they could achieve fluctuations of the ratio $\mu(t)/V(t)$ smaller than the fluctuations of the rate of flow $\mu(t)$. This can undoubtedly be technically achieved but at a cost that does not seem to be justified. We therefore consider that if a financial effort is to be made in order to reduce the weighting error, it will be cheaper and more efficient to regulate the rate of flow.

13.5. ERROR RESULTING FROM THE NON-UNIFORMITY OF THE CUTTER SPEED FROM ONE INCREMENT TO THE NEXT

Let's assume that the rate of flow $\mu(t)$ is uniform throughout (T_L) but that, due for instance to fluctuations of the electric current, the velocity $V(t_q)$ of the cutter is not uniform from one increment to the next. We can express the increment weight M_q as follows (with W cutter width) :

$$M_q = W \frac{\mu(t_q) \quad \text{(assumed to be uniform)}}{V(t_q) \quad \text{(assumed to be non-uniform)}}$$

The resulting error is obviously of the same nature as the weighting error. With electrically driven samplers, this error is always negligible and is anyway accounted for by the weighting error WE. With pneumatic, hydraulic or hand samplers, however, the speed variations are much larger and may result in unacceptable errors. This is one of the reasons for which electrically driven samplers are the only reliable samplers.

13.6. PARTICULAR RATE OF FLOW FUNCTIONS

The functions $\mu(t)$ very often correspond to relatively small fluctuations about an average value that may be regarded as stationary. Such functions are characterized by flat variograms. We must however take into consideration two particular types of rate of flow functions :

- crenellated functions,
- long-wave functions

This is the subject of the following sections.

13.6.1. Crenellated rate of flow functions :

A function μ(t) is said to be "crenellated" when it remains practically constant for a certain lapse of time, then drops more or less quickly to zero for a while and eventually regains its former nearly constant value, etc... Fig. 13.1. illustrates such variations that can be observed at the discharge of a weight regulating feeder disposed at the bottom of a surge bin when the average outgoing rate of flow is slightly higher than the incoming rate, which is a practical necessity in order to prevent filling up and overflowing of the bin.

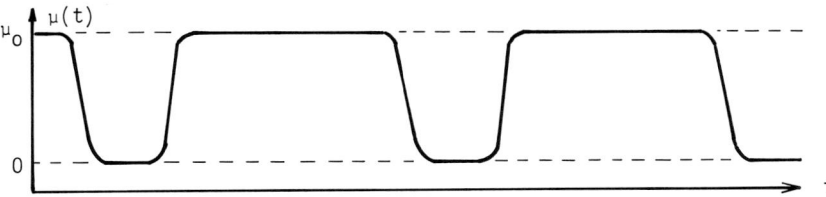

Fig. 13.1. Crenellated variations of the rate of flow

From a theoretical standpoint, this can result in weighting errors larger than expected. The solution is simple and cheap : the bin must be equipped with two level detectors, one near the top, one near the bottom, working in the following way :

1) When the level in the bin is between the detectors, the constant weight feeder extracting the material from the bin and the sampler timer are switched on,

2) When the level reaches the lower detector, extractor and timer are simultaneously switched off. The bin begins to fill up again.

3) When the level reaches the higher detector, extractor and timer are simultaneously switched on again.

This solution is theoretically excellent and easy to implement. Now, the rate of flow delivered by some types of extractors varies according to the load or in other words according to the level of material in the bin. The above solution remains applicable but both detectors should be placed near to one another and near to the top of the bin.

13.6.2. Long-wave variations :

Fig 13.2. illustrates such variations that can be frequently observed at the discharge of certain types of feeders or of certain types of reclaiming systems. This case is characterized by the fact that the rate of flow can practically be regarded as a rectilinear function of time throughout most domains $t_q \pm T_{sy}/2$. The increment weight is practically proportional to $\mu(t_q)$ as well as the tonnage M_{Lq} flowing during this interval.

$M_q = T_I \, \mu(t_q)$ and $M_{Lq} = T_{sy} \, \mu(t_q)$ then $M_q = M_{Lq} \, T_I \, / \, T_{sy}$

with the factor T_I / T_{sy} practically constant. We already know that in such conditions, the weighting error is practically cancelled.

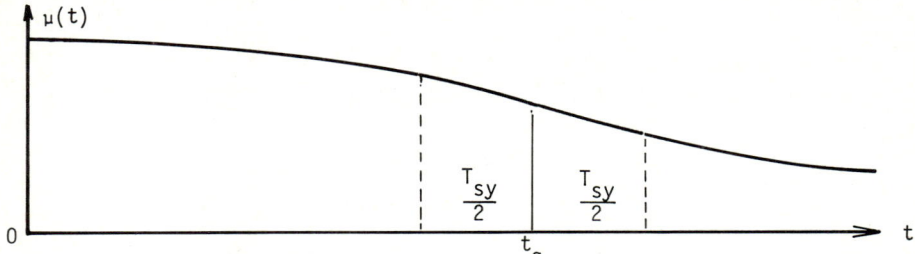

Fig. 13.2. Long-wave variations of the rate of flow.

13.6.3. Variogram of the rate of flow functions :

When the variogram $v'_M(\theta)$ has been determined under the conditions that will be actually met with in the sampling plant, the formulas given in chapter 8 remain valid irrespective of the type of rate of flow function, including the two particular cases reviewed in the preceding sections.

When the rate of flow is regulated, either in weight or in volume, the variogram is usually flat :

$$v'_M(\theta) = v'_{M1} = \text{constant}$$

The constant term v'_{M1} is likely to be smaller and even much smaller with weight regulation than with volume regulation.

With long-wave fluctuations the variogram is usually rectilinear :

$$v'_M(\theta) = v'_{M1} + v'_{M2} \theta$$

Experience shows that the weighting variance depends much more on the term v'_{M1} than on the term v'_{M2} but this dependence cannot be expressed in a simple mathematical way.

Now, periodic fluctuations of the rate of flow are not unusual, as well with solids as with pulps. The periods actually observed may vary from a few seconds to 20 to 30 mn. In a flotation pilot plant for instance, a variographic experiment carried out on the overflow of a hydrocyclone and involving increments extracted from the stream at a 4.5 second interval disclosed an important periodic term with a period of about 16 seconds. In such a situation, two solutions can be thought of in order to reduce the weighting error resulting from such periodic fluctuations of the rate of flow :

- either implementing a stratified scheme which remains the safest solution,
- or implementing a systematic scheme after regulating the rate of flow by means of a small surge tank equipped with a mixer.

13.7. CONCLUSIONS

The main conclusion of this chapter, which will be illustrated by the example presented in chapter 15, is that the rate of flow of the stream to be sampled should always be regulated in some way prior to its sampling.

With solids, there are two main types of regulating devices working according to two models :

- regulation of the weight delivered per unit time (see Colijn, 1974),
- regulation of the volume delivered per unit time (most feeders).

The first type is more expensive but more efficient. The second type is cheaper but usually results in weighting errors ranging from barely acceptable to definitely unacceptable. No regulation at all is likely to result in unacceptable weighting errors.

With pulps, the regulation is always of the volumetric type.

The reader should never forget, however, that regulating devices are very often liable to introduce periodic fluctuations, due to a sort of pendulum effect, with the consequence that stratified sampling remains the safest of all sampling schemes, as already pointed out in the conclusion of chapter 12.

CHAPTER 14

PRACTICAL IMPLEMENTATION OF THE CONTINUOUS MODEL VARIOGRAPHIC EXPERIMENT

14.1. INTRODUCTION

This chapter is dedicated to the practical implementation of the theoretical results of chapter 6. Its purpose is to show how a variographic experiment should be organized if we want it to be efficient. Examples will then be presented in order to show how to deduce the mathematical expression of the model variogram from the results of the variographic experiment, and how to estimate the various variographic parameters in terms of which the components of the continuous selection error have been expressed (chapters 9 to 13).

As the variographic experiment is the key to the experimental estimation of the moments of the continuous selection error (which will be developed in chapter 15), this is a key chapter from a practical standpoint. We must assume here that the results obtained in chapter 6 have been well assimilated by the reader.

14.2. ORGANIZATION OF A VARIOGRAPHIC EXPERIMENT

14.2.1. Planning of the experiment :

Such an experiment can be carried out with two possible purposes :

1) estimation of the moments of the errors committed when sampling a given stream,
2) analysis of the general variability of the quality and rate of flow functions in relation for instance to the control of a transformation process.

In the first hypothesis, the lower branch of the variogram alone is to be investigated. In the second one, it is necessary to investigate also the higher branch of the variogram, up to, say, 24 hours or more. The thorough study of a bed-blending system, for instance, made it necessary to collect increments during the formation of a complete pile (56 hours) and during the reclaiming of the same pile (90 hours).

The lower branch of the variogram should cover the usual values of T_{sy} and T_{st}, which means that it is usually limited to values of θ of the order of 20 to 30 mn. The higher branch depends on the problem to be investigated.

The estimation of the parameter v'_{f1}, ordinate of the variogram for $\theta = 0$ is very important, due to the part played by this parameter in the estimation of the sampling variance as well as in the bed blending theory. It requires one or several series of increments extracted from the stream at a constant interval as small as possible (e.g. one second or a few seconds according to the problem to be solved On the other hand, if we want to explore the higher branch of the variogram up to say 24 hours, a single series of increments would lead to the extraction, reduction and assay of thousands of samples which is always economically impossible. This difficulty can be overcome by taking several independent series of increments during the same period and for instance :

- <u>First series</u> : h_1 = 2 sec = 0.03 mn. As the only purpose of this series is to provide an estimate of $v'_{f1} = w''_f(h_1)$, a series of Q_1 = 30 increments will provide enough data to calculate both the variance of the population (29 degrees of freedom) and the variogram $w''_{f1}(h_1)$ (28 degrees of freedom) which should be two independent estimates of the same quantity, for the effects of the long-range fluctuations should be imperceptible at the scale of one minute.

- <u>Second series</u> : h_2 = 1 mn. The object of this series is to cover the useful domain of the variogram, i.e. the domain containing the usual intervals or strata extents. If we want the variogram $w''_f(30\ mn)$ to be known with about 30 degrees of freedom, we shall have to extract a number Q_2 = 60 increments.

- <u>Third series</u> : h_3 = 30 mn. If we want the point $w''_f(24\ hours)$ to be known with about 30 degrees of freedom, the third series will have to be made of about 80 increments.

A variogram covering the domain ranging from 2 sec to 24 hours requires about 170 increments which is acceptable. As far as sampling errors alone are concerned, 90 increments would be enough.

14.2.2. Practical considerations :

In order to reduce the estimation variance s_f^2 the following rules should be respected :

1) All increments should be extracted by a correct sampling device (this notion will be developed in the third part of this book). Hand sampling is definitely unadvisable.

2) Increment reduction to the bulk and particle size required of the final assay sample should be carried out by a fully standardized method and whenever possible by a single operator. Assaying or analysis also.

3) The results should be tabulated clearly.

4) Make sure that the same time unit is used throughout the calculations and is clearly mentioned. We definitely recommend the use of the decimal minute : for instance, 1 mn 30 s = 1.5 mn.

5) Whenever possible, use a computer. Hand calculation of variograms is long, tedious and likely to be inaccurate.

14.3. ANALYSIS OF A SIMPLE PERIODIC VARIOGRAM

14.3.1. Organization of the variographic experiment :

This experiment was carried out nearly twenty years ago on the feed to a flotation plant. Three series of increments were taken :

- <u>First series</u> : h_1 = 2 seconds : two separate groups of 25 increments. This was awkward : a single series of 50 increments would have brought a more precise information. The corresponding variogram is represented in fig. 14.1.

- <u>Second series</u> : h_2 = 1 mn : three groups of 20 increments. Here again a single group of 60 increments would have been more efficient. The corresponding variogram is represented in fig. 14.2.

- <u>Third series</u> : h_3 = 20 mn : one group of 72 increments. The corresponding variogram is represented in fig. 14.3. (results reported in table 14.1.).

All increments were weighed and assayed for Pb and Zn. We shall discuss here the zinc variogram that presents the most typical periodic fluctuations we have ever studied. It has been selected for this reason.

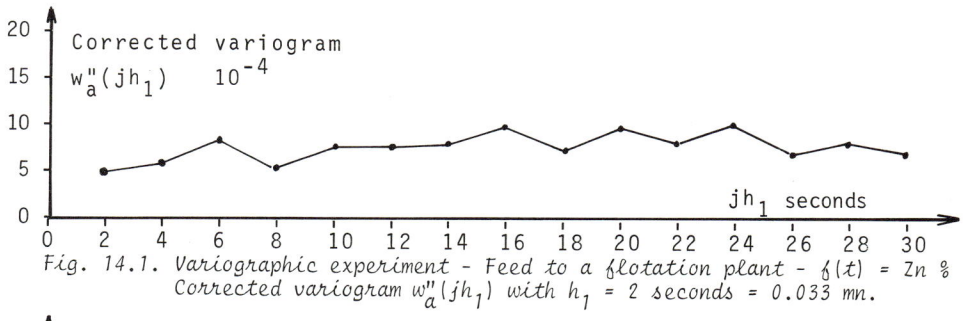

Fig. 14.1. Variographic experiment - Feed to a flotation plant - $\zeta(t)$ = Zn % Corrected variogram $w_a''(jh_1)$ with h_1 = 2 seconds = 0.033 mn.

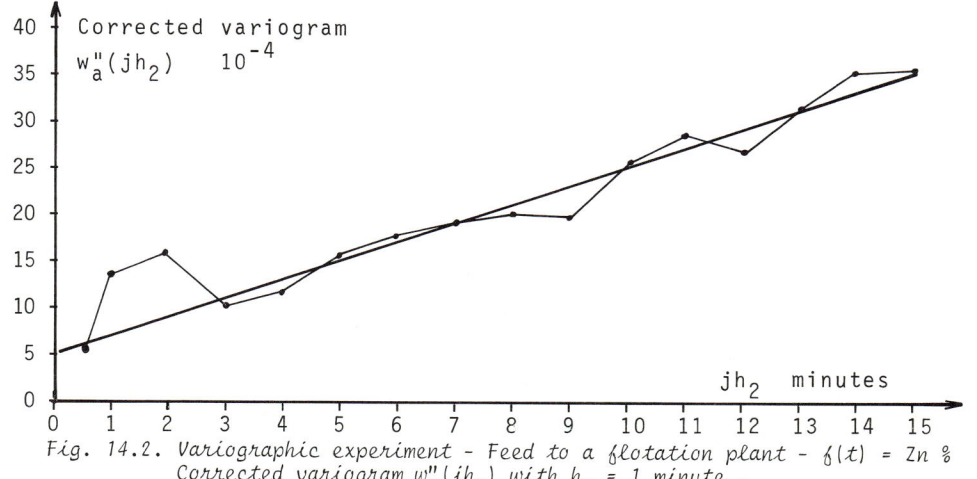

Fig. 14.2. Variographic experiment - Feed to a flotation plant - $\zeta(t)$ = Zn % Corrected variogram $w_a''(jh_2)$ with h_2 = 1 minute .

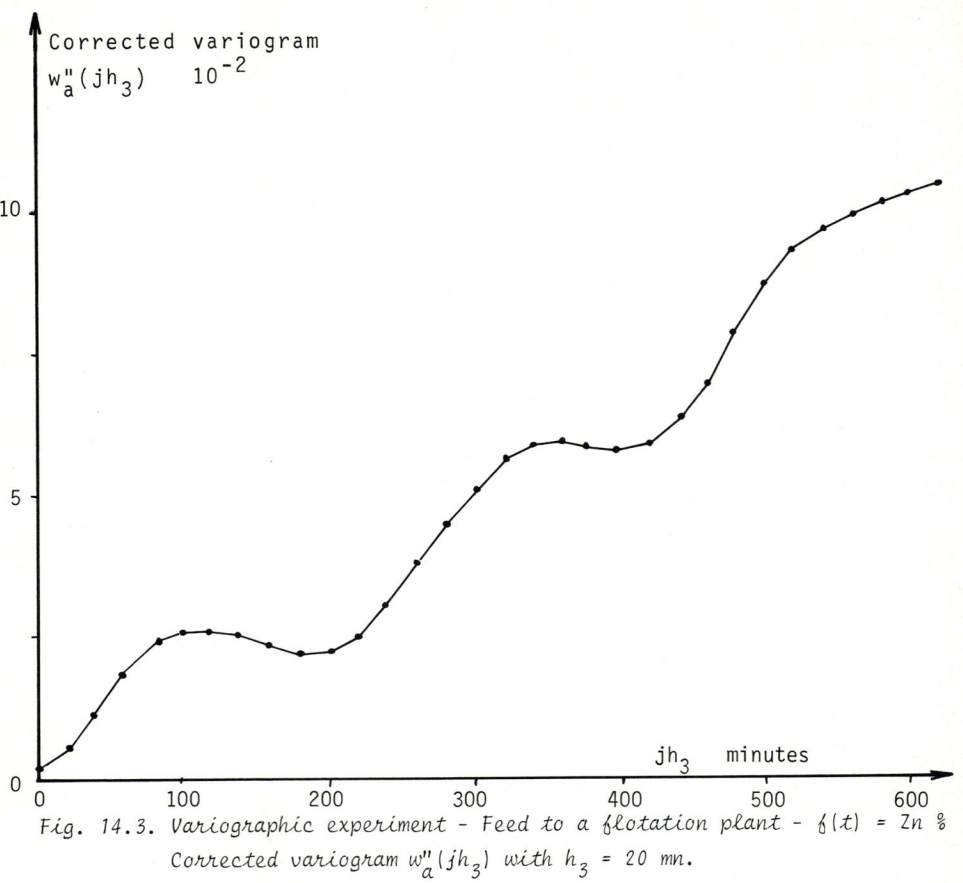

Fig. 14.3. *Variographic experiment - Feed to a flotation plant - $f(t) = Zn$ % Corrected variogram $w''_a(jh_3)$ with $h_3 = 20$ mn.*

A separate experiment had shown the absolute standard deviation of the sample reduction and Zn-assaying to be about 0.1 % Zn. With $a_L = 6.29$ % Zn, the relative standard deviation was about 1.6×10^{-2} and the relative variance $s_f^2 = s_a^2$:

$$s_a^2 = 2.5 \times 10^{-4}$$

We substracted s_a^2 from the raw experimental estimates $w'_a(jh)$ in order to obtain the corrected estimates $w''_a(jh)$. From now on we shall use only the corrected estimates $w''_a(jh)$.

14.3.2. Analysis of the first series with $h_1 = 2$ sec $= 0.033$ mn.

Fig. 14.1. shows well enough that for all practical purposes the variogram can be regarded as flat throughout the interval $0 < \theta \leq 0.5$ mn (30 seconds). The best available estimate of v'_{a1} is provided by the point $w''_a(h_1) = 5 \times 10^{-4}$ obtained for $j = 1$, $jh_1 = 0.033$ mn.

As a comparison, the relative variance of the population of Zn assays belonging to these two groups of 25 increments was 5.4×10^{-4}. The point $w''_a(15h_1)$ is 5×10^{-4} i.e. equal to $w''_a(h_1)$.

14.3.3. Analysis of the second series with $h_2 = 1$ mn :

Fig. 14.2. shows that somewhere between 0.5 and 1 mn the flat tendency of the variogram disappears. The first point ($\theta = 0.5$ mn) is borrowed from the first series and serves as a link between both series. The threshold of the variogram as defined in section 6.7.2. is $\theta_{ao} = 0.5$ mn. In other words, θ_{ao} is the highest point on the θ-axis for which the difference $w_a''(\theta) - w_a''(h_1)$ is negative. Still in other words, θ_{ao} is the point of the θ-axis beyond which the influence of the long-range term of the variogram begins to be noticeable. This is the useful domain of the variogram as far as sampling is concerned ($1 < \theta < 15$ mn). For all practical purposes we would retain a rectilinear representation such as :

$$v_a'(\theta) = v_{a1}' + v_{a2}'\theta \quad \text{with} \quad v_{a1}' = 5 \times 10^{-4} \text{ (section 14.3.2)} \quad \text{and} \quad v_{a2}' = 2 \times 10^{-4} (\text{mn}^{-1})$$

The value of v_{a2}' is the slope of the straight line representing the model variogram in fig. 14.2.

$$v_a'(\theta) = (5 + 2\theta) \times 10^{-4}$$

It would be pointless to look for a more sophisticated representation of the variogram or for a more precise estimation of the parameters v_{a1}' and v_{a2}'.

14.3.4. Analysis of the third series with $h_3 = 20$ mn :

14.3.4.1. Checking the periodic nature of the variogram : this is obvious in fig. 14.3. but we shall use this example to illustrate the test of the differential described in section 6.7.3. Table 14.1 shows :

- column 1 : the value of j (j = 1, 2, ... 36)
- column 2 : the value of jh_3 (mn)
- column 3 : the corrected variogram $w_a''(jh_3)$
- column 4 : the successive differences : $\Delta_j = w_a''(j+1)h - w_a''(jh)$. The quotient Δ_j/h_3 is an estimate of the differential $dv_a/d\theta$ for the value $\theta = (2j + 1) h_3 / 2 = 10 (2j + 1)$ mn.

These estimates of the differential have been plotted against θ (fig. 14.4.). The graph shows a sequence of fairly regularly spaced maxima and minima which is characteristic of a periodic variogram. As a term of comparison, we have plotted in fig. 14.5. the same differences obtained from a non-periodic variogram (as a matter of fact, the same variogram after elimination of the periodic term).

14.3.4.2. First estimation of T_p and v_{a3} : this is the application of the method described in section 6.7.5.2. The lower branch of the corrected variogram $w_a''(jh_3)$ has been enlarged in fig. 14.6. for $0 \leq \theta \leq 400$ mn (curve Γ). The enveloping curves Γ_1 and Γ_2 are supposed to be roughly drawn. In fact they have been accurately drawn as we shall explain later on, but at this point of the demonstration, a rough drawing would be enough. The abscissa of point P_5 (first non-zero contact between

Table 14.1. Analysis of a periodic variogram - Feed to a flotation plant - $f(t) = Z$
Series $h_3 = 20$ mn - Test of the differential.

j	jh_3 mn	$w_a''(jh_3)$ 10^{-4}	Δ_j 10^{-4}	j	jh_3 mn	$w_a''(jh_3)$ 10^{-4}	Δ_j 10^{-4}
-	0.033	5	+ 31		Continuation		
1	20	36	+ 65	19	380	641	- 35
2	40	101	+ 83	20	400	606	- 2
3	60	184	+ 60	21	420	604	+ 21
4	80	244	+ 39	22	440	625	+ 60
5	100	283	+ 6	23	460	685	+ 89
6	120	289	- 8	24	480	774	+ 98
7	140	281	- 20	25	500	868	+ 71
8	160	261	- 11	26	520	939	+ 44
9	180	250	+ 1	27	540	983	+ 14
10	200	251	+ 24	28	560	997	+ 13
11	220	275	+ 46	29	580	1010	+ 8
12	240	321	+ 57	30	600	1018	+ 5
13	260	378	+ 72	31	620	1023	+ 15
14	280	450	+ 76	32	640	1038	+ 32
15	300	526	+ 69	33	660	1070	+ 40
16	320	595	+ 38	34	680	1110	+ 60
17	340	633	+ 18	35	700	1170	+ 62
18	360	651	- 10	36	720	1232	

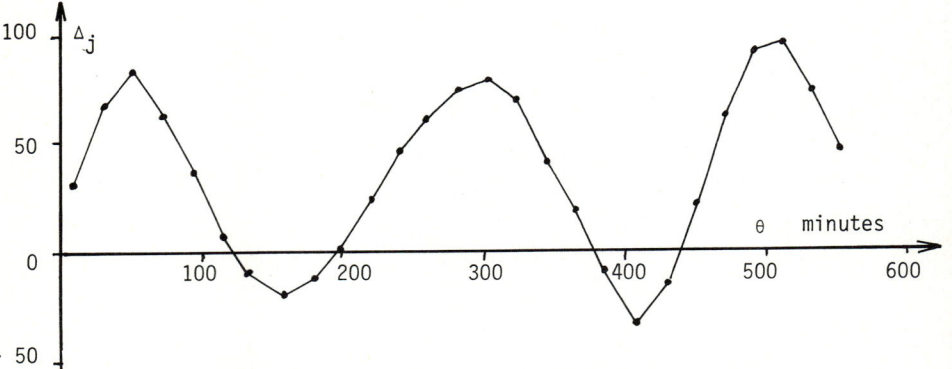

Fig. 14.4. Variographic experiment - Test of the differential -
Example of a periodic variogram : the variogram $w_a''(jh_3)$

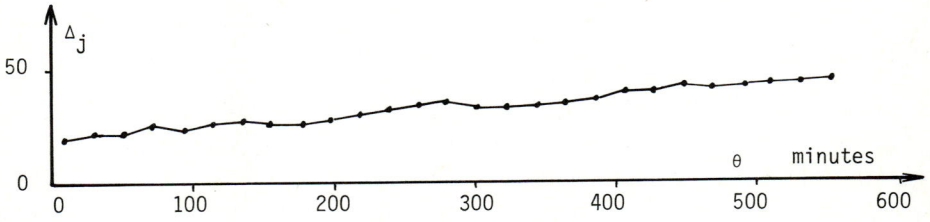

Fig. 14.5. Variographic experiment - Test of the differential -
Example of a non-periodic variogram : the component $w_{a2}''(jh_3)$

Γ and Γ_1) provides a first estimate of T_p :

200 mn < T_p < 220 mn

We would then retain T'_p = 210 mn. The vertical distance between the curves Γ_1 and Γ_2 provides a rough estimate of $2v_{a3}$:

$160 \times 10^{-4} < 2v_{a3} < 170 \times 10^{-4}$

We would retain $v'_{a3} = 83 \times 10^{-4}$.

For all practical purposes this simple method can be recommended. We shall nevertheless present a more accurate method.

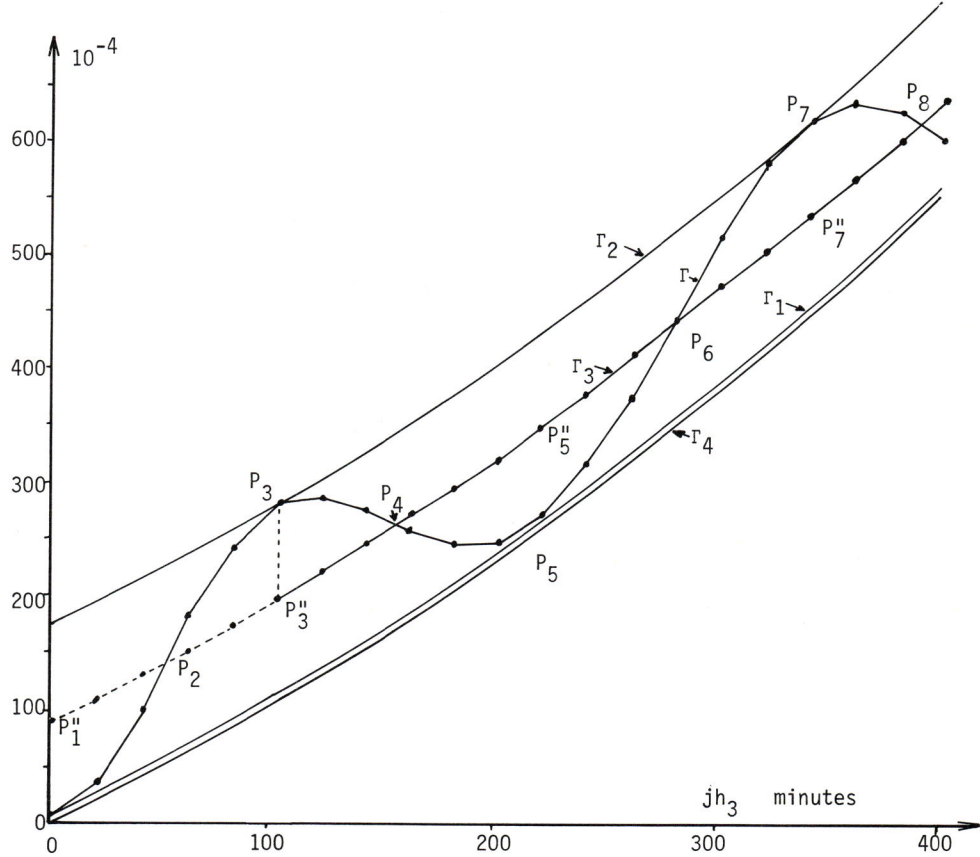

Curve Γ : corrected variogram $w''_a(jh_3)$
Curve Γ_1 : envelope of the minima : estimate of $v_{a1} + v_{a2}(\theta)$
Curve Γ_2 : envelope of the maxima : estimate of $v_{a1} + v_{a2}(\theta) + 2v_{a3}$
Curve Γ_3 : shifting mean of the corrected variogram : $W_a(jh_3)$
Curve Γ_4 : difference $W_a(jh_3) - (v'_{a1} + v'_{a3})$: estimate of $v_{a2}(\theta)$

Fig. 14.6. *Variographic experiment - Example of analysis of a periodic variogram.*

14.3.4.3. Second estimation of T_p and v_{a3} : this is the application of the method described in section 6.7.5.3.

Calculations : by graphic estimation, we know that $10\ h_3 < T_p < 11\ h_3$. We shall therefore calculate the shifting mean of $K + 1 = 11$ consecutive values of the corrected variogram (we need K to be even). We have reported in table 14.2.:

Column 1 : j   with   j = 1, 2, ... 20
Column 2 : jh_3 (minutes)
Column 3 : $w_a''(jh_3)$: corrected variogram (extracted from table 14.1.)
Column 4 : $W_a(jh_3)$: shifting mean of 11 consecutive values of $w_a''(jh_3)$. This shifting mean can be calculated only for j larger than or equal to 5. The values given between brackets for smaller values of j were obtained by graphic interpolation.
Column 5 : $w_{a3}'(jh_3) = W_a(jh_3) - w_a''(jh_3)$: estimate of $v_{a3} \cos \dfrac{2\pi jh_3}{T_p} = v_{a3} - v_{a3}(jh)$
Column 6 : $w_{a2}''(jh_3) = W_a(jh_3) - (v_{a1}' + v_{a3}')$

Table 14.2. Analysis of a periodic variogram - Calculations.

j	jh_3 minutes	$w_a''(jh_3)$ 10^{-4}	$W_a(jh_3)$ 10^{-4}	$w_{a3}'(jh_3)$ 10^{-4}	$w_{a2}''(jh_3)$ 10^{-4}
-	0	5	(91)	(+ 86)	(0)
1	20	36	(111)	(+ 75)	(20)
2	40	101	(131)	(+ 30)	(40)
3	60	184	(153)	(- 31)	(62)
4	80	244	(177)	(- 67)	(86)
5	100	283	199	- 84	108
6	120	289	223	- 66	132
7	140	281	249	- 32	158
8	160	261	274	+ 13	183
9	180	250	299	+ 49	208
10	200	251	324	+ 73	233
11	220	275	353	+ 78	262
12	240	321	384	+ 63	293
13	260	378	418	+ 40	327
14	280	450	452	+ 2	361
15	300	526	484	- 42	393
16	320	595	516	- 79	425
17	340	633	548	- 85	457
18	360	651	581	- 70	490
19	380	641	617	- 24	526
20	400	606	655	+ 51	564

Accurate estimation of v_{a3} : the curve represented in fig. 14.7. has been obtained by plotting $w_{a3}'(jh_3)$ against jh_3. It is an estimate of $v_{a3} \cos \dfrac{2\pi jh_3}{T_p}$. The points P_4', P_6' and P_8', intersections of the curve with the axis of the abscissae can be determined with great accuracy. The abscissae of the points $P_5' = (P_4' + P_6')/2$ and $P_7' = (P_6' + P_8')/2$, second minimum and second maximum of the curve can also be determined with accuracy. Their ordinates are estimates of $+ v_{a3}$ and $- v_{a3}$ respectively. The best available estimate of v_{a3} is therefore the half-difference between the ordinates of P_5' and P_7'. We find $v_{a3}' = 86 \times 10^{-4}$. This value has been used to calculate the figures of column 6 of table 14.2. It should be a more accurate estimate than the first one (83×10^{-4}). To be true, it is exceptional to encounter such a nearly perfect sinusoid.

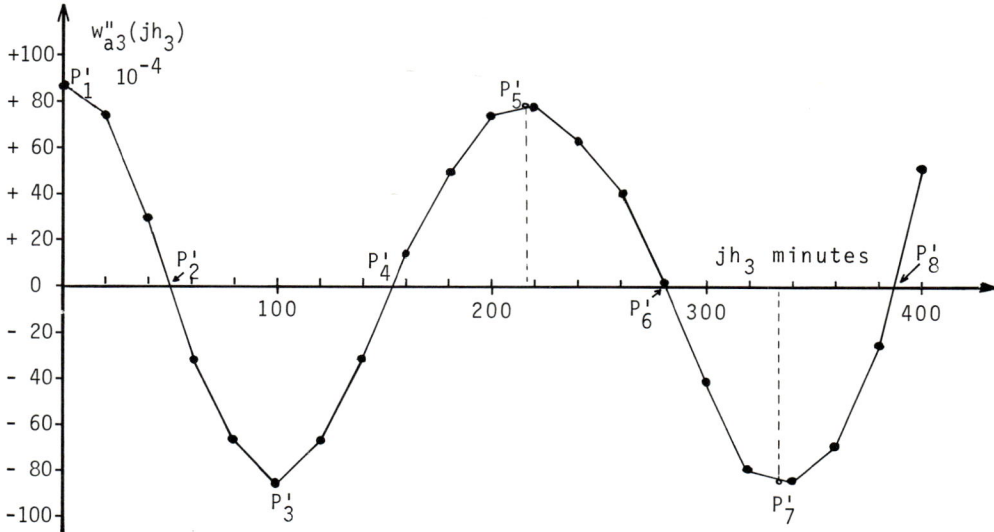

Fig. 14.7. *Variographic experiment - Value of* $w_{a3}'(jh_3) = w_a(jh_3) - w_a'''(jh_3)$
Estimate of $v_{a3} - v_{a3}(jh)$.

Accurate estimation of T_p : the knowledge of v_{a3}' makes it possible to place the point P_1'' of fig. 14.6. characterized by a zero abscissa and by the ordinate :

$$v_{a1}' + v_{a3}' = 91 \times 10^{-4}$$

Let's recall that the points belonging to the branch $P_3'' P_8''$ of the curve Γ_3 have been calculated but that the points of the branch $P_1'' P_3''$ cannot be objectively defined. Knowing however the points P_1'' and P_3'', this branch has been drawn by following the natural curvature of the upper branch and the ordinates of the four points situated between P_1'' and P_3'' have been graphically estimated and reported between brackets in column 4 of table 14.2.. These figures have been used to calculate the values of $w_{a3}'(jh_3)$. There are several ways of estimating the period of the sinusoid represented in fig. 14.7. The abscissae of P_2' (50 mn), P_4' (155 mn), P_6' (280 mn) and P_8' (386 mn) are estimates of $\frac{1}{4}T_p$, $\frac{3}{4}T_p$, $\frac{5}{4}T_p$ and $\frac{7}{4}T_p$ respectively. The corresponding estimates of T_p are:

200 mn , 207 mn , 224 mn and 220 mn respectively. If we want our model to follow reality as closely as possible, we must break up the curve into four branches and adopt slightly different periods for each of them :

$0 < \theta \leq 50$ mn : $T_{P1} = 4\ 0'P_2' = 200$ mn
$50 < \theta \leq 155$ mn : $T_{P2} = 2\ P_2'P_4' = 210$ mn
$155 < \theta \leq 280$ mn : $T_{P3} = 2\ P_4'P_6' = 250$ mn
$280 < \theta \leq 400$ mn : $T_{P4} = 2\ P_6'P_8' = 212$ mn

14.3.4.4. Expression of $v_{a3}'(\theta)$: (section 6.7.5.4.) : the general expression is

$$v_{a3}'(\theta) = (1 - \cos\frac{2\pi\theta}{T_P})\ v_{a3}'$$

For each of the four branches of the curve we have :

$0 < \theta \leq 50$ mn : $v_{a3}'(\theta) = (1 - \cos 2\pi\frac{\theta}{200})\ 86 \times 10^{-4} = (1 - \cos 0.0314\ \theta)86\times 10^{-}$

$50 < \theta \leq 155$ mn : $v_{a3}'(\theta) = (1 + \sin 2\pi\frac{\theta - 50}{210})\ 86 \times 10^{-4}$

$155 < \theta \leq 280$ mn : $v_{a3}'(\theta) = (1 - \sin 2\pi\frac{\theta - 155}{250})\ 86 \times 10^{-4}$

$280 < \theta \leq 400$ mn : $v_{a3}'(\theta) = (1 + \cos 2\pi\frac{\theta - 280}{212})\ 86 \times 10^{-4}$

In the lower branch, we can retain (with 1.8 θ expressed in degrees) :

$$v_{a3}'(\theta) = (1 - \cos 1.8\ \theta)\ 86 \times 10^{-4}$$

14.3.4.5. Expression of $v_{a2}'(\theta)$: (section 6.7.5.5.) : curve Γ_4 of fig. 14.6. represents the function $w_{a2}''(jh_3)$ the value of which is given in column 6 of table 14.2. Its shape suggests a parabolic representation of the type :

$$v_{a2}'(\theta) = v_{a2}'\theta + v_{a2}''\theta^2$$

If such a representation is valid, then the points $\dfrac{w_{a2}''(jh_3)}{jh_3}$ plotted against jh_3 should match a straight line of equation :

$$\frac{v_{a2}'(\theta)}{\theta} = v_{a2}' + v_{a2}''\theta$$

This method is illustrated by fig. 14.8. The method of the least squares provide the values of v_{a2}' and v_{a2}'' and we obtain :

$$v_{a2}'(\theta) = (0.97\ \theta + 0.0011\ \theta^2) \times 10^{-4} \qquad \text{(with } \theta \text{ in minutes)}$$

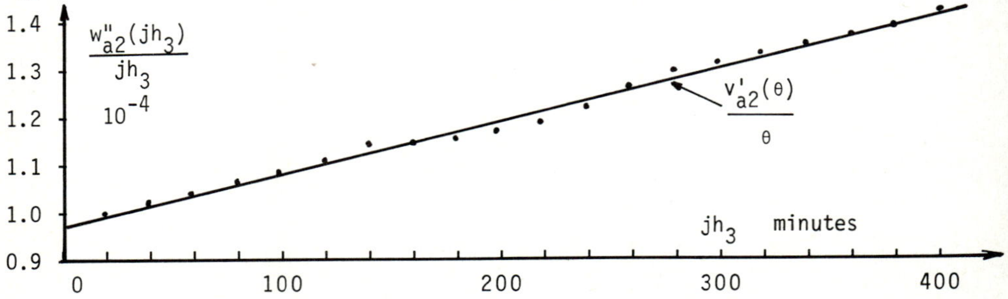

Fig. 14.8. Variographic experiment - Straight representation of $w_{a2}''(jh_3)/jh_3$

14.3.4.6. General expression of $v'_a(\theta)$: (section 6.7.5.6.) : the best available estimator of the continuous variogram $v_a(\theta)$ is (in the useful domain) :

$v'_a(\theta) = v'_{a1}(\theta) + v'_{a2}(\theta) + v'_{a3}(\theta)$ with   (θ in minutes and angles in degrees) :

$v'_{a1}(\theta) = v'_{a1} = 5 \times 10^{-4}$

$v'_{a2}(\theta) = v'_{a2}\theta + v''_{a2}\theta^2 = (0.97\,\theta + 0.0011\,\theta^2) \times 10^{-4}$

$v'_{a3}(\theta) = (1 - \cos 2\pi\theta/T_p)\,v'_{a3} = (1 - \cos 1.8\,\theta)\,86 \times 10^{-4}$

$v'_a(\theta) = (91 + 0.97\theta + 0.0011\,\theta^2 - 86 \cos 1.8\,\theta) \times 10^{-4}$

In order to recapitulate all components of $v'_a(\theta)$ and to check the closeness of agreement between model and experimental data, we have gathered in table 14.3. the following data :

Column 1 : j
Column 2 : $\theta = jh_3$
Column 3 : $v'_{a1} = 5 \times 10^{-4}$ = constant
Column 4 : $v'_{a2}(\theta) = (0.97\,\theta + 0.0011\,\theta^2) \times 10^{-4}$
Column 5 : $v'_{a3}(\theta)$
Column 6 : $v'_a(\theta) = v'_{a1} + v'_{a2}(\theta) + v'_{a3}(\theta)$: model variogram
Column 7 : $w''_a(jh_3)$: corrected experimental variogram
Column 8 : $v'_a(\theta) - w''_a(jh_3)$: difference between model and experimental data.

Table 14.3. Model variogram $v'_a(\theta)$ and its components

j	$\theta = jh_3$ mn	v'_{a1} 10^{-4}	$v'_{a2}(\theta)$ 10^{-4}	$v'_{a3}(\theta)$ 10^{-4}	$v'_a(\theta)$ 10^{-4}	$w''_a(\theta)$ 10^{-4}	$v'_a(\theta) - w''_a(\theta)$ 10^{-4}
-	0	5	0	0	5	5	0
1	20	5	20	16	41	36	+ 5
2	40	5	40	59	104	101	+ 3
3	60	5	62	111	178	184	- 6
4	80	5	84	153	242	244	- 2
5	100	5	108	172	285	283	+ 2
6	120	5	132	160	297	289	+ 8
7	140	5	157	123	285	281	+ 4
8	160	5	183	75	263	261	+ 2
9	180	5	210	36	251	250	+ 1
10	200	5	238	8	251	251	0
11	220	5	266	0	271	275	- 4
12	240	5	296	13	314	321	- 7
13	260	5	326	43	374	378	- 4
14	280	5	357	84	446	450	- 4
15	300	5	390	134	529	526	+ 3
16	320	5	423	166	594	595	- 1
17	340	5	457	170	622	633	- 11
18	360	5	491	146	642	651	- 9
19	380	5	527	101	633	641	- 8
20	400	5	564	51	620	606	+ 14

The 20 deviations recorded in column 8 of table 14.3. have a mean equal to 0.25×10^{-4} and a standard deviation equal to 5.65×10^{-4}. We therefore consider that the agreement between model and experimental data is very satisfactory.

14.3.4.7. *Variographic parameters* : (section 6.7.7.) : in the useful domain of θ the variogram $v'_a(\theta)$ is completely determined by its general expression (section 14.3.4.6.) and by the five variographic parameters :

$v'_{a1} = 5 \times 10^{-4}$ (dimensionless)
$v'_{a2} = 0.97 \times 10^{-4}$ (mn)$^{-1}$
$v''_{a2} = 0.0011 \times 10^{-4}$ (mn)$^{-2}$
$v'_{a3} = 86 \times 10^{-4}$ (dimensionless)
$T'_P = 200$ (mn)

The relative importance of the parameter v'_{a3} calls for our special attention. The amplitude a_3 of the periodic phenomenon is $a_3 = a_L \sqrt{2 v'_{a3}}$. In this example,
$a_L = 6.29$ % Zn
$a_3 = 0.82$ % Zn
$a_3/a_L = 0.13$ which is an exceptionally high value.

14.4. ANALYSIS OF A NON-PERIODIC VARIOGRAM

Fig. 14.9. represents the variogram of the feed to a copper flotation plant (60 increments taken at a two minute interval). Table 14.4. shows how such a variogram can be broken up into a sum of two terms. Another series of increments had shown (for θ = 5 sec = 0.083 mn) the parameter v'_{a1} to be equal to 8×10^{-4}.

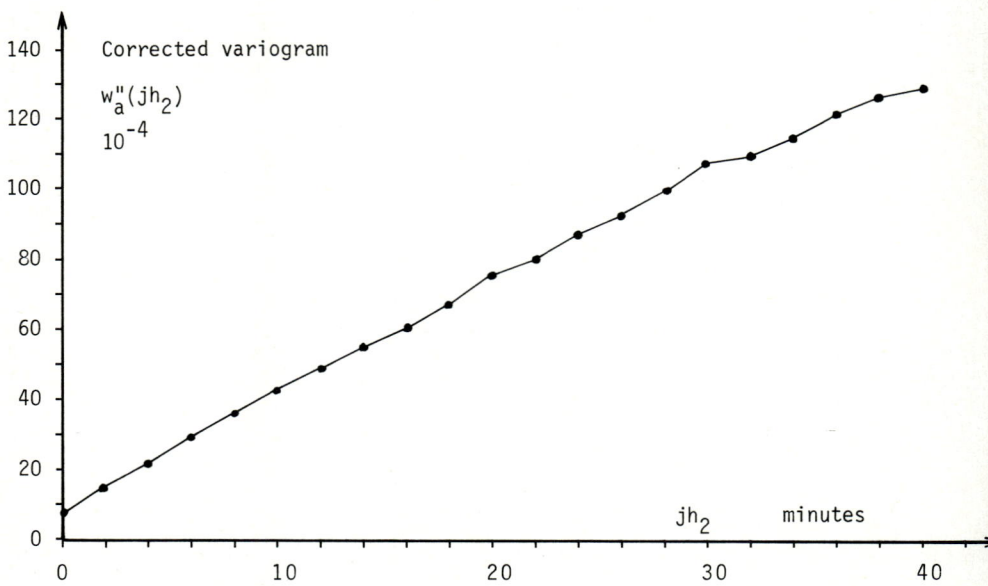

Fig. 14.9. Variographic experiment - Example of a non-periodic variogram - Feed to a copper flotation plant - $a(t) = Cu$ % .

Column 1 : j   (j = 1, 2, ... 20)
Column 2 : jh_2 = 2j minutes
Column 3 : $w_a''(jh_2)$: corrected variogram
Column 4 : v_{a1}' = 8 x 10^{-4}
Column 5 : $w_{a2}''(jh_2) = w_a''(jh_2) - v_{a1}'$ = column 3 - column 4
Column 6 : $w_{a2}''(jh_2)/jh_2$ = column 5/column 2

Table 14.4. *Analysis of a non-periodic variogram* .

j	jh_2 mn	$w_a''(jh_2)$ 10^{-4}	v_{a1}' 10^{-4}	$w_{a2}''(jh_2)$ 10^{-4}	$w_{a2}''(jh_2)/jh_2$ 10^{-4}
-	0	8	8	0	0
1	2	15	8	7	3.40
2	4	21	8	13	3.31
3	6	28	8	20	3.38
4	8	35	8	27	3.32
5	10	41	8	33	3.33
6	12	47	8	39	3.29
7	14	53	8	45	3.25
8	16	59	8	51	3.20
9	18	66	8	58	3.21
10	20	73	8	65	3.24
11	22	77	8	69	3.16
12	24	85	8	77	3.20
13	26	89	8	81	3.13
14	28	96	8	88	3.15
15	30	104	8	96	3.20
16	32	106	8	98	3.05
17	34	113	8	105	3.10
18	36	120	8	112	3.12
19	38	125	8	117	3.09
20	40	127	8	119	2.98

We might have opted for a rectilinear variogram (fig. 14.9.) but fig.14.10 shows a slightly decreasing tendency of $w_{a2}''(jh_2)/jh_2$ which leads to the expression :

$v_{a2}'(\theta) = (3.4\ \theta - 0.009\ \theta^2) \times 10^{-4}$
$v_a'(\theta) = (8 + 3.4\ \theta - 0.009\ \theta^2) \times 10^{-4}$ valid in the useful domain of θ.

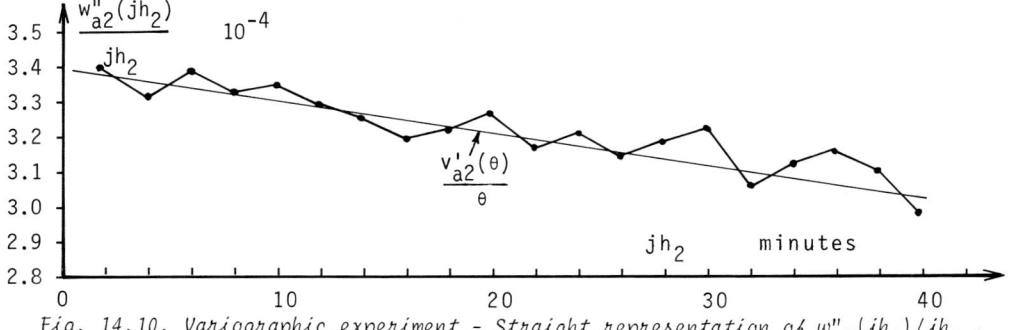

Fig. 14.10. *Variographic experiment - Straight representation of $w_{a2}''(jh_2)/jh_2$* .

CHAPTER 15

PRACTICAL IMPLEMENTATION OF THE CONTINUOUS MODEL
ERROR ESTIMATION

15.1. INTRODUCTION

The purpose of this chapter is to illustrate chapters 8 to 13 and to show on two examples how to calculate the variances of the continuous selection error CE and of its components QE_1, QE_2 and WE in terms of the variographic parameters characterizing the material to be sampled, for each of the three reference selection schemes and various values of the free parameter.

15.2. EXAMPLE No. 1

This first example is borrowed from the metal mining industry and concerns a pulp of finely ground material, the feed to a lead flotation plant. The sampling was usually carried out by means of an automatic sampler but for this particular experiment (performed in 1962) which was supposed not to disturb the production, the increments were taken by hand, which is definitely unadvisable.

15.2.1. The material to be sampled :

A variographic experiment was carried out according to the procedure detailed in chapter 14 and provided the following variograms :

$v_M(\theta) = (319 + 5.32 \, \theta) \times 10^{-4}$ (valid between 0 and 60 mn)
$v_A(\theta) = (341 + 11.07 \, \theta) \times 10^{-4}$ (valid between 0 and 60 mn)
$v_a(\theta) = (40 + 3.12 \, \theta) \times 10^{-4}$ (valid between 0 and 140 mn)

$v'_{M1} = 319 \times 10^{-4}$ $v'_{M2} = 5.32 \times 10^{-4}$ (mn)$^{-1}$ $v''_{M2} = 0$
$v'_{A1} = 341 \times 10^{-4}$ $v'_{A2} = 11.07 \times 10^{-4}$ (mn)$^{-1}$ $v''_{A2} = 0$
$v'_{a1} = 40 \times 10^{-4}$ $v'_{a2} = 3.12 \times 10^{-4}$ (mn)$^{-1}$ $v''_{a2} = 0$

The following parameters were also calculated from the population of increments :

$u^2(M_q) = 536 \times 10^{-4}$
$u^2(A_q) = 1124 \times 10^{-4}$
$u^2(a_q) = 5.2 \times 10^{-4}$

$m(a_q) = 0.0158$ or 1.58 % Pb
$\rho(A_q, M_q) = 0.8$

We shall assume in the next sections that $\rho(A_S, M_S) = \rho(A_q, M_q) = 0.8$

The dispersion of the increment weights (± 46 %) is partly due to actual fluctuations of the rate of flow and for a larger part to the fact that the increments were taken by hand and by various operators. With mechanical samplers, the dispersion due to the second cause is suppressed and the dispersion of increment weights actually reflects the fluctuations of the rate of flow. A sample is to be collected at the end of each 8-hour shift (T_L = 480 mn).

15.2.2. Relative variance of the sample weight M_S (illustration of section 8.3.3.2)

$$u^2_{sy}(M_S) = \frac{1}{T_L}(v'_{M1} T_{sy} + \frac{v'_{M2}}{6} T^2_{sy}) = (66.46 \, T_{sy} + 0.185 \, T^2_{sy}) \times 10^{-6}$$

$$u^2_{st}(M_S) = \frac{1}{T_L}(v'_{M1} T_{st} + \frac{v'_{M2}}{3} T^2_{st}) = (66.46 \, T_{st} + 0.370 \, T^2_{st}) \times 10^{-6}$$

$$u^2_{ra}(M_S) = \frac{1}{Q_{ra}} u^2(M_q) = \frac{536 \times 10^{-4}}{Q_{ra}}$$

The values of $u^2(M_S)$ and $u(M_S)$ are presented in table 15.1. for the three selection schemes and for values of T_{sy} and T_{st} ranging from 1 to 40 mn, or for values of Q_{ra} ranging from 480 to 12.

Table 15.1. Example No.1 - Variance and standard deviation of the sample weight

Q_{ra} or Q	T_{sy} T_{st}	Systematic		Stratified		Random	
	mn	$u^2_{sy}(M_S)$ 10^{-6}	$u_{sy}(M_S)$ 10^{-3}	$u^2_{st}(M_S)$ 10^{-6}	$u_{st}(M_S)$ 10^{-3}	$u^2_{ra}(M_S)$ 10^{-6}	$u_{ra}(M_S)$ 10^{-3}
480	1	66.6	8.16	66.8	8.17	111.8	10.5
240	2	133.7	11.56	134.4	11.59	223.5	14.9
160	3	201.0	14.18	202.7	14.24	335.2	18.3
120	4	268.8	16.40	271.8	16.49	447.0	21.1
96	5	336.9	18.36	341.6	18.48	558.7	23.6
80	6	405.4	20.14	412.1	20.30	660.5	25.8
60	8	543.5	23.31	555.4	23.57	894.0	29.9
48	10	683.1	26.14	701.6	26.49	1117.5	33.4
24	20	1403.2	37.46	1477.2	38.43	2235.0	47.2
16	30	2160.3	46.48	2326.8	48.24	3352.4	57.9
12	40	2954.4	54.35	3250.4	57.01	4469.9	66.8

15.2.3. Relative variance of the weight A_S of critical component in the sample (illustration of section 8.4.3.3.) :

$$u^2_{sy}(A_S) = \frac{1}{T_L}(v'_{A1} T_{sy} + \frac{v'_{a2}}{6} T^2_{sy}) = (71.04 \, T_{sy} + 0.38 \, T^2_{sy}) \times 10^{-6}$$

$$u^2_{st}(A_S) = \frac{1}{T_L}(v'_{A1} T_{st} + \frac{v'_{a2}}{3} T^2_{st}) = (71.04 \, T_{st} + 0.77 \, T^2_{st}) \times 10^{-6}$$

$$u^2_{ra}(A_S) = \frac{1}{Q_{ra}} u^2(A_q) = \frac{1124 \times 10^{-4}}{Q_{ra}}$$

The values of $u^2(A_S)$ and $u(A_S)$ are calculated in table 15.2. for the three reference selection schemes and for the same values of T_{sy}, T_{st} and Q_{ra} as in table 15.1.

Table 15.2. Example No.1 - Variance and standard deviation of the weight of lead in the sample

Q_{ra} or Q	T_{sy} T_{st} mn	Systematic		Stratified		Random	
		$u^2_{sy}(A_S)$ 10^{-6}	$u_{sy}(A_S)$ 10^{-3}	$u^2_{st}(A_S)$ 10^{-6}	$u_{st}(A_S)$ 10^{-3}	$u^2_{ra}(A_S)$ 10^{-6}	$u_{ra}(A_S)$ 10^{-3}
480	1	71.4	8.45	71.8	8.47	234.2	15.30
240	2	143.6	11.98	145.2	12.05	468.4	21.64
160	3	216.6	14.72	220.0	14.83	702.7	26.51
120	4	290.3	17.04	296.5	17.22	936.9	30.61
96	5	364.8	19.10	374.4	19.35	1171.1	34.22
80	6	440.1	20.98	453.9	21.31	1405.3	37.49
60	8	592.9	24.35	617.5	24.85	1873.8	43.29
48	10	748.8	27.37	787.3	28.06	2342.2	48.40
24	20	1574.6	39.68	1728.3	41.57	4684.4	68.44
16	30	2477.2	49.77	2823.1	53.13	7026.6	83.82
12	40	3456.7	58.79	4071.7	63.81	9368.8	96.79

15.2.4. Variance of the continuous selection error CE :

By definition, irrespective of the selection scheme, we can write :

$$\sigma^2(CE) = u^2(a_S) = u^2(A_S) + u^2(M_S) - 2\rho(A_S, M_S) u(A_S) u(M_S)$$

Table 15.3. has been calculated from the results of tables 15.1. and 15.2.

Table 15.3. Example No.1 - Variance and standard deviation of the continuous selection error - Confidence interval of the Pb content

Q_{ra} or Q	T_{sy} T_{st} mn	Systematic			Stratified			Random		
		σ^2_{sy} 10^{-6}	σ_{sy} 10^{-3}	$\pm 2\sigma(a_S)$ % Pb	σ^2_{st} 10^{-6}	σ_{st} 10^{-3}	$\pm 2\sigma(a_S)$ % Pb	σ^2_{ra} 10^{-6}	σ_{ra} 10^{-3}	$\pm 2\sigma(a_S)$ % Pb
480	1	28	5.3	0.016	28	5.3	0.016	87	9.3	0.030
240	2	56	7.5	0.024	56	7.5	0.024	174	13.2	0.042
160	3	84	9.2	0.029	85	9.2	0.029	261	16.2	0.051
120	4	112	10.6	0.033	114	10.7	0.033	348	18.7	0.059
96	5	141	11.9	0.038	144	12.0	0.038	436	20.9	0.066
80	6	170	13.0	0.041	174	13.2	0.041	524	22.9	0.072
60	8	228	15.1	0.048	236	15.4	0.049	697	26.4	0.083
48	10	288	17.0	0.054	300	17.3	0.055	871	29.5	0.093
24	20	599	24.5	0.077	649	25.5	0.081	1742	41.7	0.132
16	30	936	30.6	0.097	1049	32.4	0.102	2613	51.1	0.162
12	40	1298	36.0	0.114	1501	38.8	0.123	3485	59.0	0.187

This table gives also the value of the 95 % probability confidence interval.

15.2.5. Mean of the continuous selection error CE :

By definition, irrespective of the selection scheme :

$$m(CE) = B_c(a_S) = u^2(M_S) - \rho(A_S, M_S) \, u(A_S) \, u(M_S)$$

Table 15.4. has been calculated from the results of tables 15.1 and 15.2.

Table 15.4. Example No.1 - Mean and coefficient of bias of the continuous selection error CE

Q_{ra} or Q	T_{sy} T_{st}	Systematic		Stratified		Random	
		m(CE)	b(CE)	m(CE)	b(CE)	m(CE)	b(CE)
	mn	10^{-6}	10^{-3}	10^{-6}	10^{-3}	10^{-6}	10^{-3}
480	1	+ 11	2.2	+ 11	2.2	- 18	1.9
240	2	+ 23	3.1	+ 23	3.0	- 35	2.7
160	3	+ 34	3.7	+ 34	3.7	- 53	3.3
120	4	+ 45	4.3	+ 45	4.2	- 71	3.8
96	5	+ 56	4.8	+ 55	4.6	- 88	4.2
80	6	+ 67	5.2	+ 66	5.0	- 110	4.8
60	8	+ 89	5.9	+ 87	5.6	- 141	5.4
48	10	+ 111	6.5	+ 107	6.2	- 177	6.0
24	20	+ 214	8.7	+ 199	7.8	- 354	8.5
16	30	+ 310	10.1	+ 276	8.5	- 530	10.4
12	40	+ 398	11.0	+ 340	8.8	- 707	12.0

This table confirms that the values of m(CE) as well as those of the coefficien of bias b(CE) = |m(CE)|/σ(CE) are always negligible, which confirms the statement made in chapter 8 that the first approximation m(CE) = 0 is always acceptable.

15.2.6. Confidence interval of the Pb content of the sample S :

The 95 % probability confidence interval of the lead content is :

$$\pm 2 \, \sigma(a_S) = \pm 2 \times 1.58 \times \sigma(CE) \, \% \, Pb$$

It has been calculated and reported in table 15.3. for the three reference selection schemes. If a ± 0.05 % Pb confidence interval can be regarded as acceptabl then it is easy to see that such a reproducibility can be achieved either by systematic sampling with interval 8 mn, or by stratified sampling with strata extent 8 mn, or by random sampling with at least 180 increments per 8-hour shift. Systematic and stratified sampling are practically equivalent but random sampling requires three times as many increments, which involves an important increase in the cost of sample reduction.

15.2.7. Breaking up of the total variance $\sigma^2(CE)$:

According to the results of chapter 9, $\sigma^2(CE)$ can be broken up as follows :

$$\sigma^2(CE) = \sigma^2(QE) + \sigma^2(WE) = \sigma^2(QE_1) + \sigma^2(QE_2) + \sigma^2(WE) \qquad \text{with :}$$

$$\sigma_{sy}^2(QE) = \frac{1}{T_L}(v'_{a1} T_{sy} + \frac{v'_{a2}}{6} T_{sy}^2) = (8.33 T_{sy} + 0.108 T_{sy}^2) \times 10^{-6}$$

$$\sigma_{sy}^2(QE_1) = 8.33 \times 10^{-6} T_{sy} \quad \text{and} \quad \sigma_{sy}^2(QE_2) = 0.108 \times 10^{-6} T_{sy}^2$$

$$\sigma_{st}^2(QE) = \frac{1}{T_L}(v'_{a1} T_{st} + \frac{v'_{a2}}{3} T_{st}^2) = (8.33 T_{st} + 0.217 T_{st}^2) \times 10^{-6}$$

$$\sigma_{st}^2(QE_1) = 8.33 \times 10^{-6} T_{st} \quad \text{and} \quad \sigma_{st}^2(QE_2) = 0.217 \times 10^{-6} T_{st}^2$$

The corresponding results are given in table 15.5. As the variograms are not valid throughout the domain $(T_L) = 0$ to 480 mn, this breaking up cannot be applied to the random scheme.

Table 15.5. *Example No.1 - Partition of the continuous selection error CE Variance of QE_1, QE_2, QE and WE.*

Q_{ra} or Q	T_{sy} or T_{st}	Systematic				Stratified			
		QE_1 10^{-6}	QE_2 10^{-6}	QE 10^{-6}	WE 10^{-6}	QE_1 10^{-6}	QE_2 10^{-6}	QE 10^{-6}	WE 10^{-6}
	mn								
480	1	8	0	8	19	8	0	8	19
240	2	17	0	17	39	17	1	18	39
160	3	25	1	26	58	25	2	27	58
120	4	33	2	35	77	33	4	37	77
96	5	42	3	45	96	42	5	47	97
80	6	50	4	54	116	50	8	58	116
60	8	67	7	74	155	67	14	81	155
48	10	83	11	94	193	83	22	105	195
24	20	167	43	210	389	167	87	254	396
16	30	250	97	347	589	250	195	445	604
12	40	333	174	507	791	333	347	680	821

15.2.8. Interpretation of the results :

15.2.8.1. Comparison between the three selection schemes : the systematic and stratified selection schemes are practically equivalent, except for very low numbers of increments (say smaller than 50). As an example, for a number Q = 60 (i.e. for T_{sy} or T_{st} = 8 mn) the confidence intervals on the lead content are 0.048 and 0.049 % Pb respectively, whereas with the random scheme and the same number of increments it is equal to 0.072 % Pb.

15.2.8.2. Relative importance of $\sigma^2(QE_1)$, $\sigma^2(QE_2)$ and $\sigma^2(WE)$: table 15.5. confirms that the long-range fluctuation error QE_2 remains small as compared with the short-range fluctuation error QE_1 except when the number of increments falls below 30, irrespective of the selection scheme. The weighting variance represents between 55 and 70 % of the total variance for fluctuations of the rate of flow of ± 46 %.

15.3. EXAMPLE No. 2

This example concerns the raw mix fed to the blending pile of a cement plant.

15.3.1. The material to be sampled :

Upon entering the bed-blending system, the - 35 mm raw mix is sampled by means of a correct cross-stream sampler and an hourly sample is prepared in order to control the average composition of the pile. Several elements are to be analysed, the most relevant being CaO, SiO_2, Fe_2O_3, Al_2O_3 and MgO. A variographic experiment was realized according to the procedure detailed in chapter 14 and the variograms were drawn for all elements. The most important being CaO, we shall present the calculation of the sampling errors incurred on the CaO content for the three selection schemes and various values of the free parameters T_{sy}, T_{st} and Q_{ra}.

The number of increments making up an hourly sample cannot be too small, which limits the useful domain of the variogram. For the three variograms, we retained a parabolic representation valid for all values of θ smaller than 10 mn.

$$v'_a(\theta) = (195 + 282.5\,\theta - 15.85\,\theta^2) \times 10^{-6}$$

$$v'_M(\theta) = (1584 + 3350\,\theta - 179.7\,\theta^2) \times 10^{-6}$$

$$v'_A(\theta) = (1590 + 4383\,\theta - 221.1\,\theta^2) \times 10^{-6}$$

$v'_{a1} = 195 \times 10^{-6}$ $v'_{a2} = 282.5 \times 10^{-6}\,(mn)^{-1}$ $v''_{a2} = -15.85 \times 10^{-6}\,(mn)^{-2}$

$v'_{M1} = 1584 \times 10^{-6}$ $v'_{M2} = 3350 \times 10^{-6}\,(mn)^{-1}$ $v''_{M2} = -179.7 \times 10^{-6}\,(mn)^{-2}$

$v'_{A1} = 1590 \times 10^{-6}$ $v'_{A2} = 4383 \times 10^{-6}\,(mn)^{-1}$ $v''_{A2} = -221.1 \times 10^{-6}\,(mn)^{-2}$

The following parameters were also calculated from the population of increments

$u^2(a_q) = 2072 \times 10^{-6}$

$u^2(M_q) = 13002 \times 10^{-6}$

$u^2(A_q) = 17441 \times 10^{-6}$

$\rho(A_q, M_q) = +\,0.9396$ We shall assume that $\rho(A_S, M_S) = \rho(A_q, M_q) = +\,0.94$

15.3.2. Relative variance of the sample weight M_S : (section 8.3.3.2.) :

For the three selection schemes, we shall use the following formulas ($T_L = 60$

$$u^2_{sy}(M_S) = \frac{1}{T_L}(v'_{M1}\,T_{sy} + \frac{v'_{M2}}{6}\,T^2_{sy}) = (26.4\,T_{sy} + 9.31\,T^2_{sy}) \times 10^{-6}$$

$$u^2_{st}(M_S) = \frac{1}{T_L}(v'_{M1}\,T_{st} + \frac{v'_{M2}}{3}\,T^2_{st} + \frac{v''_{M2}}{6}\,T^3_{st}) = (26.4\,T_{st} + 18.6\,T^2_{st} - 0.50\,T^3_{st}) \times 1$$

$$u^2_{ra}(M_S) = \frac{u^2(M_q)}{Q_{ra}} = \frac{13002}{Q_{ra}} \times 10^{-6}$$

Table 15.6 gives the value of $u^2(M_S)$ and $u(M_S)$ for the three selection schemes, for values of T_{sy} and T_{st} ranging from 0.5 to 6 mn or for values of Q_{ra} and Q ranging from 120 to 10.

Table 15.6. Example No. 2 - Variance and standard deviation of the sample weight M_S

Q_{ra} or Q	T_{sy} T_{st}	Systematic		Stratified		Random	
		$u^2_{sy}(M_S)$	$u_{sy}(M_S)$	$u^2_{st}(M_S)$	$u_{st}(M_S)$	$u^2_{ra}(M_S)$	$u_{ra}(M_S)$
	mn	10^{-6}	10^{-3}	10^{-6}	10^{-3}	10^{-6}	10^{-3}
120	0.50	15.53	3.94	17.79	4.22	108.35	10.41
100	0.60	19.19	4.38	22.43	4.74	130.02	11.40
80	0.75	25.03	5.00	30.06	5.48	162.52	12.75
60	1.00	35.71	5.98	44.51	6.67	216.70	14.72
40	1.50	60.54	7.78	79.79	8.93	325.04	18.03
30	2.00	90.03	9.49	123.26	11.10	433.39	20.82
24	2.50	124.17	11.14	174.53	13.21	541.74	23.28
20	3.00	162.96	12.77	233.24	15.27	650.09	25.50
15	4.00	254.50	15.95	371.46	19.27	866.78	29.44
12	5.00	364.66	19.10	534.92	23.13	1083.48	32.92
10	6.00	493.43	22.21	720.63	26.84	1300.18	36.06

15.3.3. Relative variance of the weight A_S of CaO in the sample (section 8.4.3.3.)

$$u^2_{sy}(A_S) = \frac{1}{T_L}(v'_{A1} T_{sy} + \frac{v'_{A2}}{6} T^2_{sy}) = (26.50 T_{sy} + 12.18 T^2_{sy}) \times 10^{-6}$$

$$u^2_{st}(A_S) = \frac{1}{T_L}(v'_{A1} T_{st} + \frac{v'_{A2}}{3} T^2_{st} + \frac{v''_{A2}}{6} T^3_{st}) = (26.5 T_{st} + 24.35 T^2_{st} - 0.61 T^3_{st}) \times 10^{-6}$$

$$u^2_{ra}(A_S) = \frac{1}{Q_{ra}} u^2(A_q) = \frac{17441}{Q_{ra}} \times 10^{-6}$$

Table 15.7 : Example No.2 - Variance and stand. dev. of the weight A_S of CaO in sample

Q_{ra} or Q	T_{sy} T_{st}	Systematic		Stratified		Random	
		$u^2_{sy}(A_S)$	$u_{sy}(A_S)$	$u^2_{st}(A_S)$	$u_{st}(A_S)$	$u^2_{ra}(A_S)$	$u_{ra}(A_S)$
	mn	10^{-6}	10^{-3}	10^{-6}	10^{-3}	10^{-6}	10^{-3}
120	0.50	16.29	4.04	19.26	4.39	145.34	12.06
100	0.60	20.28	4.50	24.53	4.95	174.41	13.21
80	0.75	26.72	5.17	33.31	5.77	218.01	14.77
60	1.00	38.67	6.22	50.23	7.09	290.69	17.05
40	1.50	67.14	8.19	92.46	9.62	436.03	20.88
30	2.00	101.69	10.08	145.48	12.06	581.37	24.11
24	2.50	142.34	11.93	208.83	14.45	726.72	26.96
20	3.00	189.06	13.75	282.04	16.79	872.06	29.53
15	4.00	300.78	17.34	456.25	21.36	1162.75	34.10
12	5.00	436.84	20.90	664.42	25.78	1453.43	38.12
10	6.00	597.25	24.44	902.85	30.05	1744.12	41.76

15.3.4. Variance of the continuous selection error CE (section 8.7.4.)

By definition and irrespective of the selection scheme :

$$\sigma^2(CE) = u^2(A_S) + u^2(M_S) - 2\rho(A_S, M_S) u(A_S) u(M_S)$$

Table 15.8 is obtained from the figures of tables 15.6 and 15.7. It gives the variance and standard deviation of CE and the confidence interval of the content a_S

Table 15.8. Example No. 2 - Variance and standard deviation of the continuous selection error CE - Confidence interval of the CaO content a_S.

Q_{ra} or Q	T_{sy} or T_{st}	Systematic			Stratified			Random		
		σ^2_{sy}	σ_{sy}	$\pm 2\sigma(a_S)$	σ^2_{st}	σ_{st}	$\pm 2\sigma(a_S)$	σ^2_{ra}	σ_{ra}	$\pm 2\sigma($
	mn	10^{-6}	10^{-3}	% CaO	10^{-6}	10^{-3}	% CaO	10^{-6}	10^{-3}	% C
120	0.50	1.93	1.39	0.11	2.27	1.51	0.12	17.87	4.23	0.
100	0.60	2.40	1.55	0.12	2.88	1.70	0.13	21.44	4.63	0.
80	0.75	3.15	1.78	0.14	3.91	1.98	0.16	26.81	5.18	0.
60	1.00	4.55	2.13	0.17	5.89	2.43	0.19	35.74	5.98	0.
40	1.50	7.87	2.81	0.22	10.84	3.29	0.26	53.61	7.32	0.
30	2.00	11.91	3.45	0.27	17.10	4.13	0.33	71.48	8.45	0.
24	2.50	16.68	4.08	0.32	24.60	4.96	0.39	89.35	9.45	0.
20	3.00	22.17	4.71	0.37	33.30	5.77	0.46	107.22	10.35	0.
15	4.00	35.35	5.95	0.47	54.08	7.35	0.58	142.96	11.96	0.
12	5.00	51.47	7.17	0.57	79.03	8.89	0.70	178.71	13.37	1.
10	6.00	70.53	8.40	0.67	107.69	10.38	0.82	214.45	14.64	1.

15.3.5. Confidence interval of the CaO content :

The grade of the mix is about 41 % CaO. The 95 % probability confidence interv is :

$\pm 2 \sigma(a_S) = \pm 2 \times 41 \times \sigma(CE)$ % CaO and is expressed in table 15.8. As periodic va riations are likely to occur, the responsible of the project wisely chose to carr out a stratified scheme. As furthermore a ± 0.2 % CaO confidence interval was to regarded as a maximum on the hourly sample, the value T_{st} = 1 mn was selected at primary sampling stage (± 0.19 % CaO in table 15.8). A pile being completed in 50 60 hours, the grade of the pile is known within $\pm 0.19 / \sqrt{56} = \pm 0.025$ % CaO.

Thanks to immediate reduction and analysis of the hourly samples, the average composition of the pile is known with precision at every moment, making it possib to correct any deviation from the average composition required. Accurate and repr ducible sampling are as necessary as blending and homogenisation if a uniform raw mix is to be fed to the kiln.

15.3.6. Mean of the continuous selection error CE (section 8.7.3.) :

In first approximation, we know that this mean is zero. In second approximation, irrespective of the selection scheme :

$$m(CE) = B_C(a_S) = u^2(M_S) - \rho(A_S, M_S) u(A_S) u(M_S)$$

Table 15.9 is obtained from the figures of tables 15.6 and 15.7. It gives the mean m(CE) and the coefficient of bias b(CE) = $|m(CE)|$ / $\sigma(CE)$. The coefficient of bias is everywhere smaller than 1 % and in the useful range than 0.1 % which confirms that the bias is perfectly negligible.

Table 15.9. Example No.2 - Mean and coefficient of bias of the continuous selection error CE.

Q_{ra} or Q	T_{sy} T_{st} mn	Systematic m(CE) 10^{-6}	Systematic b(CE) 10^{-3}	Stratified m(CE) 10^{-6}	Stratified b(CE) 10^{-3}	Random m(CE) 10^{-6}	Random b(CE) 10^{-3}
120	0.50	+ 0.58	0.42	+ 0.40	0.26	- 9.56	2.26
100	0.60	+ 0.65	0.42	+ 0.39	0.23	- 11.47	2.48
80	0.75	+ 0.73	0.41	+ 0.33	0.17	- 14.34	2.77
60	1.00	+ 0.79	0.37	+ 0.08	0.03	- 19.12	3.20
40	1.50	+ 0.64	0.23	- 2.91	0.28	- 28.69	3.92
30	2.00	+ 0.12	0.04	- 2.56	0.62	- 38.25	4.52
24	2.50	- 0.75	0.18	- 4.85	0.98	- 47.81	5.06
20	3.00	- 1.97	0.42	- 7.75	1.34	- 57.37	5.54
15	4.00	- 5.47	0.76	-15.35	2.09	- 76.50	6.40
12	5.00	-10.36	1.44	-25.24	2.84	- 95.62	7.15
10	6.00	-16.65	1.98	-37.26	3.59	-114.75	7.84

15.3.7. Breaking up of the continuous selection variance $\sigma^2(CE)$:

As no periodic fluctuations have been detected at the scale 0-10 mn, we may write :

$$\sigma^2(CE) = \sigma^2(QE) + \sigma^2(WE) = \sigma^2(QE_1) + \sigma^2(QE_2) + \sigma^2(WE)$$

The value of the components of $\sigma^2(CE)$ are given in tables 15.10, 15.11 and 15.12. They have been computed from the following formulas :

$$\sigma^2_{sy}(QE) = \frac{1}{T_L} (v'_{a1} T_{sy} + \frac{v'_{a2}}{6} T^2_{sy}) = (3.25\, T_{sy} + 0.785\, T^2_{sy}) \times 10^{-6}$$

$$\sigma^2_{sy}(QE_1) = 3.25 \times 10^{-6}\, T_{sy} \quad \text{and} \quad \sigma^2_{sy}(QE_2) = 0.785 \times 10^{-6}\, T^2_{sy}$$

$$\sigma^2_{sy}(WE) = \sigma^2_{sy}(CE) - \sigma^2_{sy}(QE)$$

$$\sigma^2_{st}(QE) = \frac{1}{T_L}(v'_{a1}T_{st} + \frac{v'_{a2}}{3}T^2_{st} + \frac{v''_{a2}}{6}T^3_{st}) = (3.25\ T_{st} + 1.569\ T^2_{st} - 0.044\ T^3_{st}) \times 10^{-6}$$

$$\sigma^2_{st}(QE_1) = 3.25 \times 10^{-6}\ T_{st} \quad \text{and} \quad \sigma^2_{st}(QE_2) = 1.569 \times 10^{-6}\ T^2_{st} - 0.044 \times 10^{-6}\ T^3_{st}$$

$$\sigma^2_{st}(WE) = \sigma^2_{st}(CE) - \sigma^2_{st}(QE)$$

$$\sigma^2_{ra}(QE) = \frac{\sigma^2(a_q)}{Q_{ra}} = \frac{2072 \times 10^{-6}}{Q_{ra}}$$

$$\sigma^2_{ra}(QE_1) = \sigma^2_{sy}(QE_1) = \sigma^2_{st}(QE_1) = \frac{v'_{a1}}{Q_{ra}} = \frac{195 \times 10^{-6}}{Q_{ra}}$$

$$\sigma^2_{ra}(QE_2) = \sigma^2_{ra}(QE) - \sigma^2_{ra}(QE_1) \quad \text{and} \quad \sigma^2_{ra}(WE) = \sigma^2_{ra}(CE) - \sigma^2_{ra}(QE)$$

Table 15.10. *Breaking up of the continuous selection variance σ^2 (CE) - Unit 10^{-} Systematic sampling. Hourly samples : T_L = 60 mn.*

T_{sy}	Q	$\sigma^2(CE)$	$\sigma^2(CE) = \sigma^2(WE) + \sigma^2(QE)$		$\sigma^2(QE) = \sigma^2(QE_1) + \sigma^2(QE_2)$	
			$\sigma^2(WE)$	$\sigma^2(QE)$	$\sigma^2(QE_1)$	$\sigma^2(QE_2)$
0.50	120	1.93	0.11	1.82	1.62	0.20
0.60	100	2.40	0.17	2.23	1.95	0.28
0.75	80	3.15	0.27	2.88	2.44	0.44
1.00	60	4.55	0.51	4.04	3.25	0.78
1.50	40	7.87	1.23	6.64	4.87	1.77
2.00	30	11.91	2.27	9.64	6.50	3.14
2.50	24	16.68	3.65	13.03	8.13	4.90
3.00	20	22.17	5.36	16.81	9.75	7.06
4.00	15	35.35	9.80	25.55	13.00	12.55
5.00	12	51.47	15.60	35.87	16.25	19.62
6.00	10	70.53	22.78	47.75	19.50	28.25

Table 15.11. *Stratified sampling. Hourly samples : T_L = 60 mn.*

0.50	120	2.26	0.25	2.01	1.62	0.39
0.60	100	2.88	0.38	2.50	1.95	0.55
0.75	80	3.91	0.61	3.30	2.44	0.86
1.00	60	5.89	1.11	4.78	3.25	1.53
1.50	40	10.84	2.58	8.26	4.88	3.38
2.00	30	17.10	4.67	12.43	6.50	5.93
2.50	24	24.60	7.35	17.25	8.13	9.12
3.00	20	33.30	10.62	22.68	9.75	12.93
4.00	15	54.08	18.79	35.29	13.00	22.29
5.00	12	79.03	29.05	49.98	16.25	33.73
6.00	10	107.69	41.21	66.48	19.50	46.98

Table 15.12. Random sampling. Hourly samples : T_L = 60 mn.

Q	σ^2(CE)	σ^2(CE) = σ^2(WE) + σ^2(QE)		σ^2(QE) = σ^2(QE$_1$) + σ^2(QE$_2$)	
		σ^2(WE)	σ^2(QE)	σ^2(QE$_1$)	σ^2(QE$_2$)
120	17.87	0.60	17.27	1.63	15.64
100	21.44	0.72	20.72	1.95	18.77
80	26.81	0.91	25.90	2.44	23.46
60	35.74	1.21	34.53	3.25	31.28
40	53.61	1.81	51.80	4.87	46.93
30	71.48	2.41	69.07	6.50	62.57
24	89.36	3.02	86.34	8.13	78.21
20	107.22	3.62	103.60	9.75	93.85
15	142.97	4.83	138.14	13.00	125.14
12	178.71	6.04	172.67	16.25	156.42
10	214.44	7.24	207.20	19.50	187.70

15.3.8. Interpretation of the results :

15.3.8.1. Comparison of the three selection schemes : as already observed in our first example, the systematic and stratified selection schemes are practically equivalent throughout the useful range of the number Q of increments but the random scheme is much less reproducible. For Q = 60 for instance (table 15.8) the confidence intervals are ± 0.17, ± 0.19 and ± 0.47 respectively. This observation is quite general and definitely rules out the random scheme as a practical solution to any flowing stream sampling problem. This example shows that the slight loss of reproducibility observed when implementing a stratified scheme instead of a systematic one is negligible and is more than compensated by the safety which is gained when periodic fluctuations are likely to occur. Assume for instance that a very moderate sinusoidal variation with amplitude a_3 = 1 % CaO and period T_p = 1 mn adds to the fluctuations already observed, it will be responsible for an additional variance σ^2(QE$_3$) the maximum of which is, for T_{sy} or T_{st} = T_p = 1 mn and for Q = 60 :

σ^2_{sy}(QE$_3$) = 297 x 10^{-6} σ^2_{st}(QE$_3$) = 4.95 x 10^{-6}

to be compared with :

σ^2_{sy}(QE$_1$) = 3.25 x 10^{-6} σ^2_{st}(QE$_1$) = 3.25 x 10^{-6}

σ^2_{sy}(QE$_2$) = 0.78 x 10^{-6} σ^2_{st}(QE$_2$) = 1.53 x 10^{-6}

σ^2_{sy}(WE) = 0.51 x 10^{-6} σ^2_{st}(WE) = 1.11 x 10^{-6}

σ^2_{sy}(CE) = 4.54 x 10^{-6} (without periodic fluctuations) σ^2_{st}(CE) = 5.89 x 10^{-6}

σ^2_{sy}(CE') = 301.54 x 10^{-6} (with periodic fluctuations) σ^2_{st}(CE') = 10.84 x 10^{-6}

With systematic sampling, the periodic fluctuations increase the confidence interval from ± 0.17 to 1.42 % CaO, while with stratified sampling the increase is from ± 0.19 to ± 0.27 % CaO, which makes a serious difference.

15.3.8.2. Relative importance of QE_1, QE_2 and WE : Tables 15.10. and 15.11. show that with systematic and stratified sampling, the long-range fluctuation error QE_2 remains smaller than the short-range error QE_1 as long as the number Q of increments is larger than 30, while table 15.12 shows that with random sampling the error QE_2 is on the contrary unacceptably large, irrespective of the number Q of increments. It is therefore at the level of the long-range fluctuation error QE_2 that the random scheme looses its representativity.

As regards the weighting error WE, we remember that it was very important in our first example (section 15.2. - table 15.5 page 147). The fluctuations of the increment weight, responsible for WE, were partly due to fluctuations of the rate of flow (true weighting error) but for a much larger part to the lack of uniformity of the cutter speed associated with hand sampling (artificial weighting error). In the present example, on the contrary, the increments were extracted by means of a correct sampling device that suppresses the artificial part of WE. We observe that the true weighting error remains smaller and even much smaller than the total quality fluctuation error QE, irrespective of the selection scheme. In the useful range of the number Q of increments, the weighting error WE is negligible.

Our experience confirms that this conclusion remains valid as long as the rate of flow is regulated prior to sampling, preferably by means of a weigh feeder.

THIRD PART

FROM THE CONTINUOUS MODEL TO THE DISCRETE REALITY

MATERIALIZATION OF THE PUNCTUAL INCREMENTS

> *Sampling must be correct. If money is to be spent on sampling, the first end to be achieved is to eliminate all sources of sampling incorrectness. This is the only way of eradicating sampling biases.*

When applied to the discretized functions defined in section 5.3.4.3. the continuous model presented in the second part of this book covers the whole of the increment sampling process, i.e. the complete sequence :

- point selection,
- increment delimitation,
- increment extraction,

on condition, however, that the increment materialization steps, i.e. increment delimitation and extraction be "correctly" carried out or in other words on condition that these two independent operations give all particles of the lot a uniform probability of being selected.

This third part is dedicated to a detailed analysis of the materialization of the punctual increments and aims at establishing the conditions that must be fulfilled in order for a sampling device to carry out a correct increment delimitation and a correct increment extraction. When these conditions are not respected, we show that two and only two new errors add to the continuous selection error CE studied in the preceding chapters, namely :

- increment delimitation error DE,
- increment extraction error EE,

components of the materialization error ME :

$$ME = DE + EE \qquad \text{and} \qquad SE = CE + ME$$

In this introduction to the study of the materialization errors we would like to emphasize a few very important points :

1) Inspection of thousands of sampling devices distributed all over the world shows that about 75 % of these introduce either a delimitation or an extraction error and more often than not both of these.

2) Materialization errors affect the mean as well as the variance of the total sampling error. Incorrect sampling devices are responsible for a lack of accuracy as well as for a lack of reproducibility. They are definitely unreliable : biases as large as 20 % have been recorded more than once, a 5 % bias is not uncommon. A 1 to 2 % bias is frequent and can be detected only by the eye of an expert.

3) Although the main causes of materialization errors have been publicly denounced on many occasions and for a long time, a number of sampling equipment manufacturers go on designing and building, a number of users go on utilizing incorrect sampling devices and we can see no improvement whatever in this worrying situation

4) The only efficient strategy with materialization errors is one of preventive elimination by exclusively designing, manufacturing, selling or using sampling methods or devices ensuring a correct delimitation and a correct extraction.

5) Correct sampling devices are in no way more expensive than incorrect ones. There should therefore be no reason not to prefer them, except ignorance.

6) Increment delimitation and extraction errors can be much more important than the continuous selection error CE and any of its components.

These few remarks should point out the practical and economical importance of the next three chapters :

Chapter 16 : Components of the materialization error ME
Chapter 17 : Increment delimitation error DE
Chapter 18 : Increment extraction error EE.

CHAPTER 16

COMPONENTS OF THE MATERIALIZATION ERROR ME

16.1. INTRODUCTION

We showed in section 4.6. that the increment sampling process could be, from a logical standpoint, broken up into a sequence of four elementary and independent steps :

1) <u>Point selection</u> : A certain number of points are selected throughout the domain (D_L) occupied by the lot L to be sampled, according to a certain selection scheme. These points are the "<u>punctual increments</u>".

2) <u>Increment delimitation</u> : In its relative motion through (D_L) the sampling tool or device delimits the geometrical boundaries of the domain (D_I) of the "<u>extended increments</u>" of uniform extent D_I. This purely geometrical operation is not supposed to take the particulate structure of the material into consideration.

3) <u>Increment extraction</u> : In its material penetration through the material making up the lot L, the sampling device extracts groups of fragments, the "<u>material increments</u>" coinciding more or less closely with the set of particles the centre of gravity of which falls within the boundaries of the extended increments.

4) <u>Increment reunion</u> : The set obtained by reunion of the punctual, extended or material increments is the "<u>punctual, extended or material sample</u>" retained to represent the lot.

According to the definition given in section 5.3.4. the continuous selection model applied to the punctual functions $\mu_P(t)$, $\alpha_P(t)$ and $a_P(t)$ covers the point selection step alone. Applied to the extended functions $\mu_E(t)$, $\alpha_E(t)$ and $a_E(t)$ it covers the sequence: point selection + increment delimitation. Applied to the fragmental functions $\mu_F(t)$, $\alpha_F(t)$ and $a_F(t)$ it covers the whole sequence: point selection + increment delimitation + increment extraction.

There is however a condition for the last two statements to be true : that the increment delimitation and extraction actually carried out be performed in accordance with the definitions of the extended and fragmental functions.

This chapter deals with the study of the materialization of the punctual increments. Its purpose is to analyse the sequence of logical steps leading from the punctual increments on which is founded the selection model to the fragmental increments making up the sample actually extracted from the lot, to make a census of the possible deviations from the definitions of the extended and fragmental functions and to define the new errors liable to take place as a result of such deviations.

In order to clearly illustrate our logical approach we shall concentrate our demonstrations on the problem set by the sampling of one-dimensional flowing stream sampled at the discharge of a belt conveyor by means of a traveling cutter. The generalization of our conclusions to two- and three-dimensional object is obvious and will be dealt with in chapter 29.

16.2. INCREMENT SAMPLING OF FLOWING STREAMS

16.2.1. The one-dimensional model :

The one-dimensional model retained for the representation of a flowing stream of particulate material supposes that matter is condensed on the extension axis of the lot by "projection" in a direction parallel to a reference plane (Λ), according to the process described in fig. 16.1. where (Λ) is perpandicular to the extension axis but the angle made by (Λ) and the axis is irrelevant provided the plane is not parallel to the axis.

The lot is actually made of the material contained within the boundaries of the domain (D_L) schematized by a cylinder in fig 16.1.(1) and delimited by two planes Λ_0 and Λ_L of the (Λ) family. By projection, the three-dimensional domain (D_L) is condensed in a one-dimensional domain (T_L) of the extension or time axis.

16.2.2. The punctual increment I_p :

Fig. 16.1. shows the lot L in three different ways :

1) In a three-dimensional perspective showing the boundaries of the domain (D_L)

2) In a two-dimensional cross-section showing the particulate structure of the material to be sampled,

3) In a one-dimensional model reduced to the segment (T_L) occupied by the lot L projected on the time axis.

The three representations show the physical meaning of the punctual increment I_p which is made of the material condensed at point t_q of the time axis. I_p is the material contained in the elementary slice of matter flowing between the instants t_q and t_q + dt. This slice is represented by the plane Λ_q.

16.2.3. The model extended increment I_E :

Fig. 16.2. shows the physical definition of the "model extended increment" I_E. The extension process of the punctual increment I_p consists in :

1) selecting a certain time extent T_I ,

2) defining the domain (T_{Iq}) occupied on the time axis by the extended increment

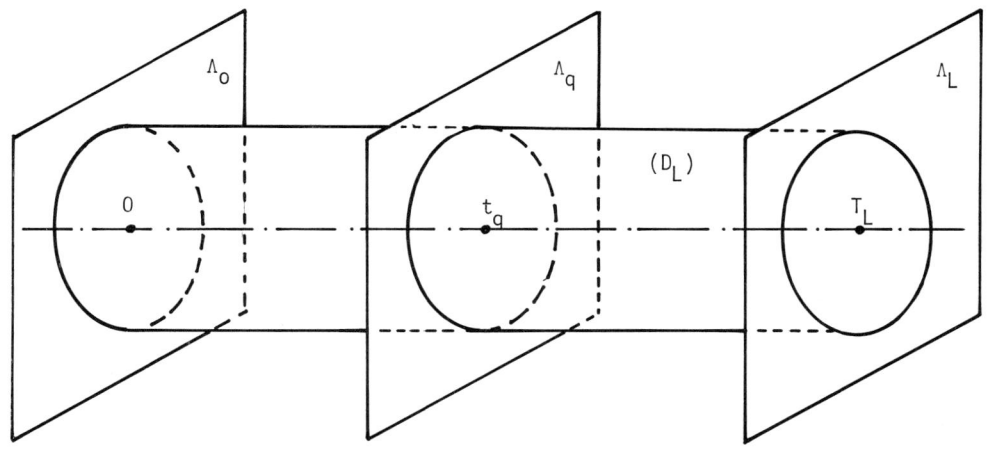

1) Three-dimensional perspective

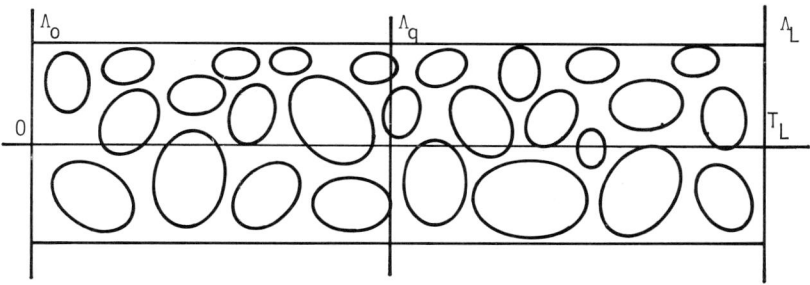

2) Two-dimensional longitudinal cross-section

3) One-dimensional model

Fig. 16.1. *Definition of the punctual increment I_q*

According to this definition :
$$(T_{Iq}) \equiv \{ t \mid t_q - \frac{T_I}{2} < t \leq t_q + \frac{T_I}{2} \}$$
the extended increment I_E is made of the matter contained between the planes Λ_1 and Λ_2 of the (Λ) family containing the points :
$$t_1 = t_q - \frac{T_I}{2} \quad \text{and} \quad t_2 = t_q + \frac{T_I}{2}$$

The two-dimensional cross-section of fig. 16.2.(2). shows that the extension process is a purely geometrical operation which does not take the particulate structure of the material into consideration. The matter belonging to the "model extended increment" I_E is hachured in fig. 16.2.(2). It strictly corresponds to the definition of the extended functions defined in section 5.3.4.2. We shall observe that all elements of the transversal cross-section of the stream are cut during the same time T_I. If an increment is taken from the stream at a constant interval

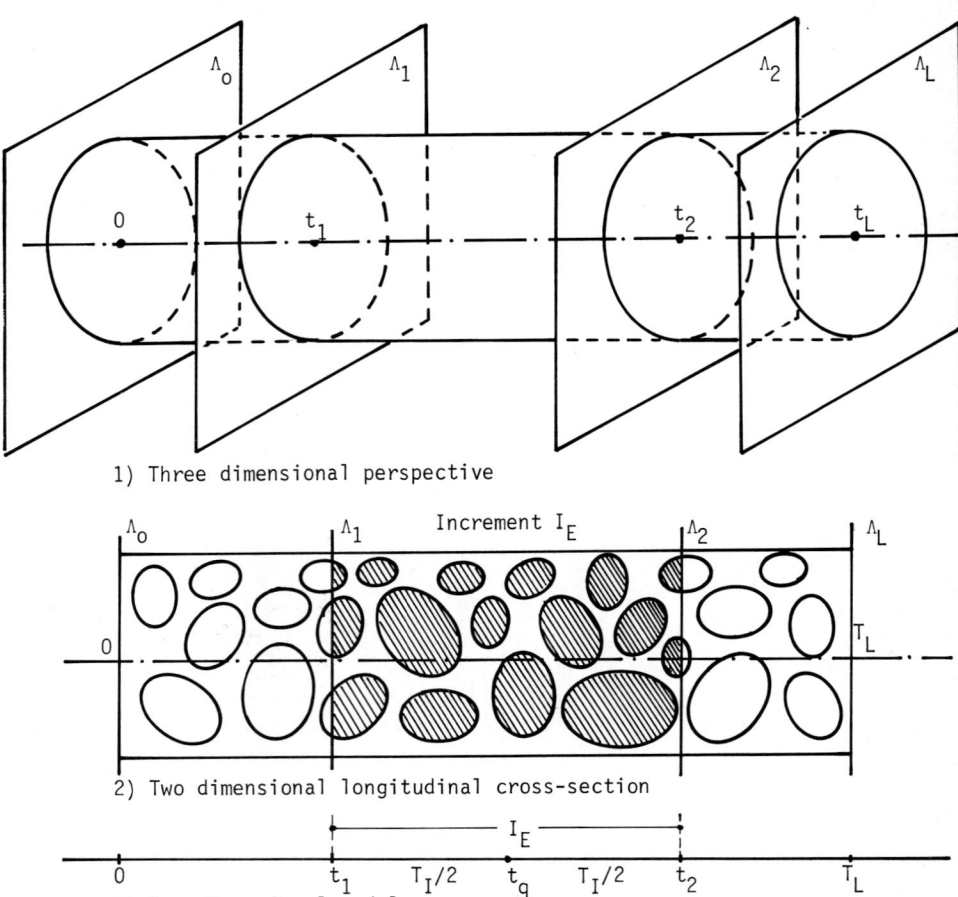

Fig.16.2. *Definition of the model extended increment I_E (enlarged)*.

T_{sy}, the probability for a given element of matter (we are not yet speaking of particles) to fall between the boundaries of the model extended increment is obviously uniform and equal to T_I/T_{sy}, irrespective of the position of this element in the stream cross-section : the model extended increment achieves a "correct increment delimitation". Now, the increment actually cut through (D_L) by the cutter may differ from the model extended increment :

16.2.4. The actual extended increment I'_E - Increment delimitation error DE :

The model extended increment is delimited by two parallel planes Λ_1 and Λ_2 but if the characteristics of the sampling device do not respect a certain number of conditions that will be reviewed in chapter 17, the "actual extended increment" I'_E will be delimited by two surfaces, for instance Λ'_1 and Λ'_2, which are no longer plane nor parallel. Fig. 16.3. shows how the actual extended increment deviates from its model. This deviation is better shown up in the longitudinal cross-section

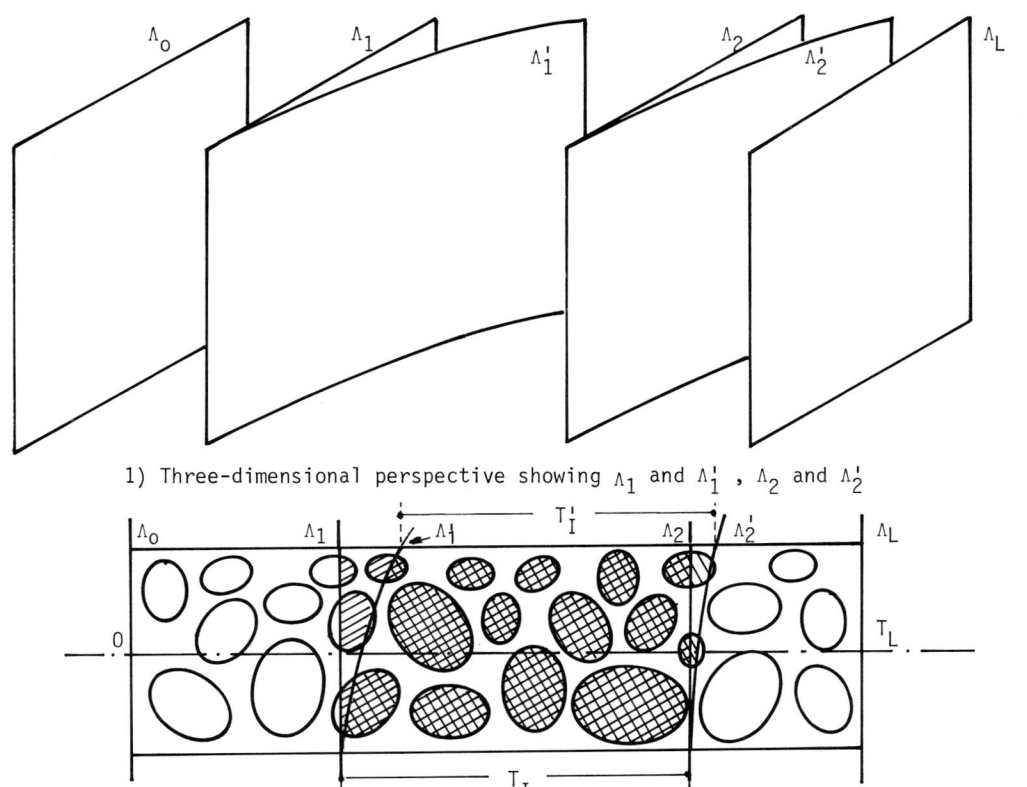

1) Three-dimensional perspective showing Λ_1 and Λ'_1 , Λ_2 and Λ'_2

2) Two-dimensional cross-section showing the model extended increment ($\Lambda_1 \Lambda_2$) and the actual extended increment ($\Lambda'_1 \Lambda'_2$).

Fig. 16.3. *Definition of the actual extended increment I'_E - Delimitation error DE*

of fig. 16.3.(2). where the actual extended increment I'_E is hachured from left to right (top to bottom) whilst the model extended increment is hachured from right to left. When the surfaces Λ'_1 and Λ'_2 can be superposed by a translation parallel to the extension axis, the two curvilinear triangles $P_1 P'_1 P''_1$ and $P_2 P'_2 P''_2$ shown on the cross-section are identical and the actual increment is equivalent to the model increment : the selection probability remains uniform for all elements of the stream cross-section. But when these surfaces are not superposable by a translation parallel to the extension axis, the selection probability is no longer uniform and the sampling ratio T_I/T_{sy} which was formerly a constant is now a function of the position in the stream cross-section. We can see for instance in fig. 16.3.(2). that the upper part of the stream is diverted during a time T'_I smaller than T_I , diversion time of the lower part : in such a case, the actual extended increment achieves an "incorrect increment delimitation". If some segregation takes place throughout the transversal stream cross-section (as suggested in our figures) or in other words if the position of a particle in the stream is a function of its physical properties (diameter, density, shape) as we can frequently observe, then the actual extended increment is systematically different from its model. This deviation from the model, which is not accounted for by the continuous selection model, generates a new error : the "increment delimitation error" DE. Its mean is likely to be non-zero and is called "increment delimitation bias". Chapter 17 is dedicated to the study of the increment delimitation error and to how to suppress the always dangerous delimitation bias. When the delimitation process is correct, the increment delimitation error is identically zero.

16.2.5. The model fragmental increment I_F :

Let's assume for a moment that the manufacturer has read and assimilated the rules of delimitation correctness : the extended increment actually cut through the stream is identical with its model of fig. 16.2. There remains now to take the particulate structure of the material into consideration and we may wonder what kind of relationship should exist between an extended increment and its fragmental equivalent.

Let's observe a falling stream of particulate material sampled at the discharge of a belt conveyor by means of a cutter traveling across the stream. In this process most fragments by-pass the cutter, others fall directly into the cutter and a few particles bounce on one of the cutter edges. These are the fragments that in fig. 16.4.(1). are cut by one of the planes Λ_1 or Λ_2 and the problem reduces itself to a simple question : how does a fragment rebound when hitting a cutter edge ? This problem will be discussed in detail in chapter 18 but we must anticipate by stating that under ideal conditions (to be defined later) a fragment hitting a cutter edge will bounce on the side of the edge that contains its centre of gravity. This "rebounding rule" or "rule of the centre of gravity" illustrated by fig. 16.4.(3). has been retained in section 5.3.4.3. for the definition of the fragmental functions. These functions describe the properties of the group of fragments the centre of gravity of which falls within the boundaries of the (model) extended increment.

We shall accordingly define the "model fragmental increment" I_F as the set of particles, the centre of gravity of which falls between the planes Λ_1 and Λ_2. These particles are hachured in fig. 16.4.(2). The probability for a given particle to belong to the model fragmental increment solely depends on the abscissa of its centre of gravity and is taken into account by the point selection scheme. More specifically, it does not depend on its physical properties (diameter, density, shape) and may be regarded as uniform if the selection scheme is correct. The model fragmental increment defines the rules of "correct increment extraction". Extraction correctness and respect of the rebounding rule are synonymous.

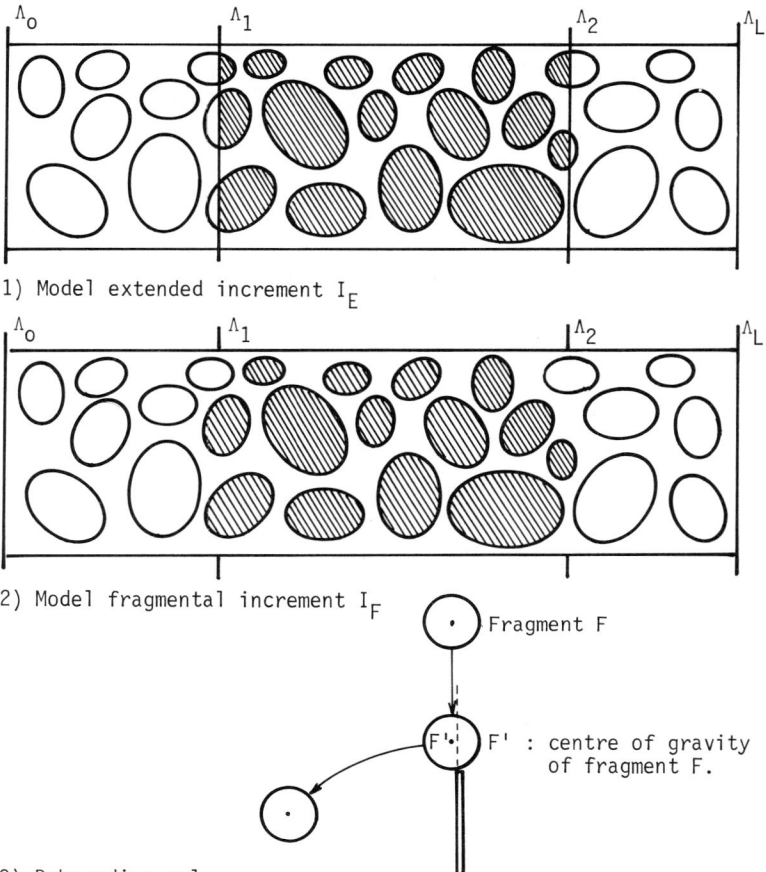

Fig. 16.4. *Definition of the model fragmental increment I_F - Rebounding rule*

16.2.6. The actual fragmental increment I'_F - Increment extraction error EE :

The "<u>actual fragmental increment</u>" I'_F is the set of particles actually extracted from the stream by the sampling device. It differs from the model fragmental increment if the rebounding rule is not respected. Fig. 16.5.(1) shows the composition of the model fragmental increment and fig. 16.5.(2) the actual fragmental increment. Fragments marked by an x which belonged to the model increment do not eventually belong to the actual increment, due to the non-respect of the rule of the centre of gravity. This deviation from the model generates a new error which has not been taken into account by the continuous selection model : the "<u>increment extraction error</u>" EE. Its mean is usually non-zero and is called "<u>increment extraction bias</u>". Chapter 18 is dedicated to a study of the extraction error and on how to suppress the always dangerous extraction bias. When the extraction process is correct, the increment extraction error EE is identically zero.

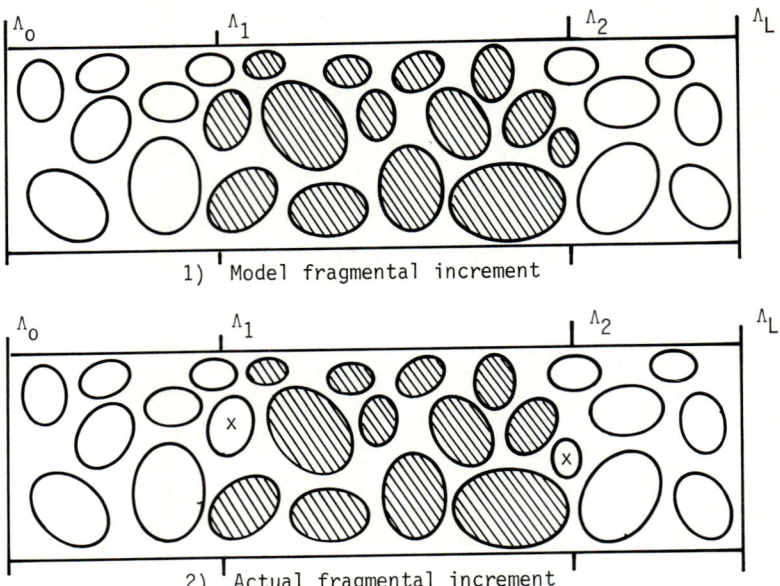

Fig. 16.5. *Definition of the actual fragmental increment I'_F - Extraction error E*

16.3. RECAPITULATION

In our logical analysis of the steps encountered in our progress from the punctual increment taken into account by the continuous selection model to the material increment actually extracted by the sampling cutter, we have defined :

- the punctual increment I_P
- the model extended increment I_E
- the actual extended increment I'_E
- the model fragmental increment I_F
- the actual fragmental increment I'_F.

We have also defined two and only two new errors which had not been taken into account by the continuous selection model dealt with in the second part of this book :

- <u>the increment delimitation error DE</u> that may be incurred when the actual extended increment I'_E differs from the model extended increment I_E or in other words when the delimitation is incorrect.

- <u>the increment extraction error EE</u> that may be incurred when the actual fragmental increment I'_F differs from the model fragmental increment I_F or in other words when the extraction is incorrect.

If we assume both the increment delimitation and extraction to be correct, then both errors ED and EE are identically nil. Then but only then the total sampling error SE incurred when representing the lot L by a sample S obtained by reunion of increments such as I_F is equal to the continuous selection error CE defined in chapter 8 :

Correct increment delimitation and extraction : SE ≡ CE

But when the increment delimitation and extraction are incorrect, a materialization error ME takes place which is the sum of the delimitation and extraction errors DE and EE :

Incorrect increment delimitation and extraction : SE = CE + ME = CE + DE + EE

The properties of the delimitation error are studied in chapter 17 and those of the extraction error in chapter 18.

16.4. SELECTION, EXTRACTION AND SAMPLING PROBABILITIES

16.4.1. Density of selection probability :

We shall assume that the point selection has been correctly carried out, i.e. that all points of the domain (T_L) occupied by the lot L on the time axis are selected with a uniform density π_o of selection probability. According to known results (chapter 7) we have :

- systematic selection : $\pi_o = 1/T_{sy}$
- stratified selection : $\pi_o = 1/T_{st}$
- random selection : $\pi_o = Q_{ra}/T_L$

16.4.2. Selection probability :

We shall now call "selection probability P_{Ei} of the fragment F_i" the probability for the centre of gravity of F_i to fall within the boundaries of the actual extended increment I'_E or in other words the probability for the fragment F_i to belong to the model fragmental increment I_F.

Assuming the increment delimitation to be correct, I'_E coincides with its model I_E of uniform extent T_I. Then, irrespective of the fragment F_i, the probability P_{Ei} is uniformly equal to $P_o = \pi_o T_I$:

- systematic selection : $P_{Ei} \equiv P_o = T_I/T_{sy}$
- stratified selection : $P_{Ei} \equiv P_o = T_I/T_{st}$
- random selection : $P_{Ei} \equiv P_o = Q_{ra} T_I/T_L$.

16.4.3. Extraction probability :

We shall now call "extraction probability P_{Fi} of the fragment F_i belonging already to the model fragmental increment I_F" the probability for F_i to be actually extracted from the lot L and recovered in the actual fragmental increment I'_F.

Assuming the increment extraction to be correct, I'_F coincides with its model I_F.

Then, by definition and irrespective of the fragment F_i, the probability P_{Fi} turns out to be uniformly equal to unity :

Correct extraction : $P_{Fi} \equiv 1$

16.4.4. Sampling probability :

We shall define as "sampling probability P_i of the fragment F_i" the probability for a given fragment F_i of L to be extracted and to belong to the sample S obtained by reunion of Q actual fragmental increments I'_F.

Assuming that no fragment extraneous to I_F will eventually fall in I'_F (which is always observed in practice), the presence of F_i in I'_F results of a sequence of two independent random events :

- presence of F_i in I_F (or presence of the centre of gravity of F_i in I'_E) : probability P_{Ei} .

- presence of F_i (already assumed to belong to I_F) in I'_F .

Then, by definition :

$$P_i = P_{Ei} \, P_{Fi}$$

If and only if increment delimitation and extraction are both correct :

$P_{Ei} \equiv P_o$ and $P_{Fi} \equiv 1$ hence $P_i = P_{Ei} \, P_{Fi} \equiv P_o$ with :

- systematic selection : $P_i \equiv P_o = T_I/T_{sy}$
- stratified selection : $P_i \equiv P_o = T_I/T_{st}$
- random selection : $P_i \equiv P_o = Q_{ra} T_I/T_L$.

When increment delimitation or/and extraction are incorrect, then P_i is a function of the physical properties of the fragment F_i and more specifically of its diameter, of its density and of its shape. Incorrect sampling is likely to be biased as we shall see in the fourth part of this book.

CHAPTER 17

INCREMENT DELIMITATION ERROR DE

17.1. DEFINITION

According to the definition given in section 16.2.4. an "increment delimitation error" DE takes place as soon as the increment delimitation ceases to be correct. Increment delimitation is said to be correct (definition of correctness) if and only if all elements of the transversal cross-section are intercepted by the sampling cutter during the same length of time. When this condition is fulfilled, the continuous selection model is applicable to the extended functions and no new error is involved : the increment delimitation error is identically nil.

This chapter deals mainly with a study of the conditions of delimitation correctness when sampling a flowing stream falling at the discharge of a belt or of a feeder (sections 17.2. to 17.5.) but we shall also discuss the conditions of correctness when sampling non-falling one-dimensional objects (sampling on a stopped belt often used as a reference sampling method assumed to be unbiased - section 17.6.) or two-dimensional objects (section 17.7.).

17.2. FALLING STREAM SAMPLING - CORRECTNESS CONDITIONS INVOLVING THE CUTTER GEOMETRY

It is convenient to distinguish three categories of cutters :

1) Straight path cutters : their geometry is correct if and only if the cutter edges are parallel, irrespective of their angle with the stream.

2) Circular path cutters : their geometry is correct if and only if the cutter edges are radial, i.e. intersecting on the revolution axis of the cutter, irrespective of their angle with the axis.

3) Other cutters : during their motion across the stream a few mechanical samplers do actually generate a curve which is neither a straight line nor a circle but the prototype of this category is hand sampling, the trajectory of which is neither straight nor circular nor even defined. Since the correct geometry of a cutter depends on its trajectory, it can be easily shown that there is no correct cutter geometry when the cutter path is neither straight nor circular.

17.2.1. Straight path cross-stream cutters - correct geometry :

Fig. 17.1. shows a rectangular cutter opening. Assuming all other correctness conditions to be respected, the increment actually delimited through the stream is an inclined parallelogram : the cutter geometry is correct as all elements of

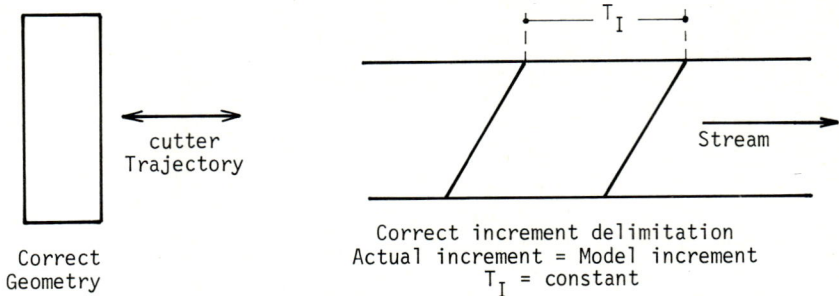

Fig. 17.1. *Straight path cross-stream cutters with parallel edges - Correct geometry.*

the stream cross-section are intercepted during the same length of time. The sampling probability is uniform for all particles, irrespective of their position in the stream cross-section.

The usual tendency of the manufacturers is to respect this simple rule if only for the sake of simplicity and low cost. We shall see in section 17.2.4. that it is unlikely to be for the sake of theoretical considerations.

17.2.2. Straight path cross-stream cutters - deviations from correctness :

Fig. 17.2. shows what happens to originally correct cutters when they are too lightly built and submitted to repeated hammering by high rates of flow of coarse materials. The original rectangle widens in the middle and takes the general shape of a cask. The corresponding increment delimitation is incorrect as shown in the figure. The remedy is obvious : sampling equipment must be mechanically adapted to the tonnage to be sampled and systematically inspected by the maintenance team for checking eventual deformations.

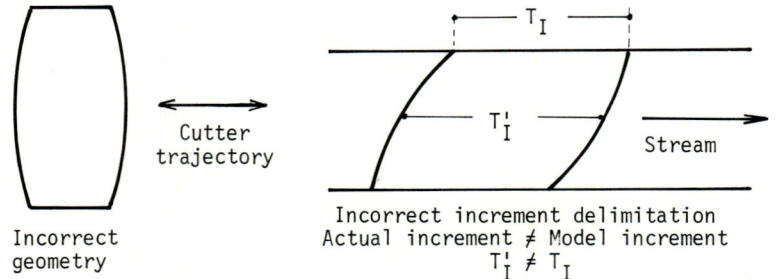

Fig. 17.2. *Straight path cross-stream cutters - weak construction.*

Fig. 17.3. shows a typical example of pulp sampling by means of a usually narrow (nearly always too narrow but this is another problem which shall be dealt with in chapter 18) cutter.

It happens that wood fibers present in the pulp build up on the cutter edges and form a kind of felt pad cemented by clay particles. Since these cutters are very often enclosed in blind boxes, it is not until the cutter opening is completely obstructed and no sample recovered that the operator understands that something has gone wrong. Here again the solution is obvious though too often ignored : systematic inspection and cleaning, for instance once a day, by the maintenance team. Worn out cutters present the same kind of defect and the solution is the same as above.

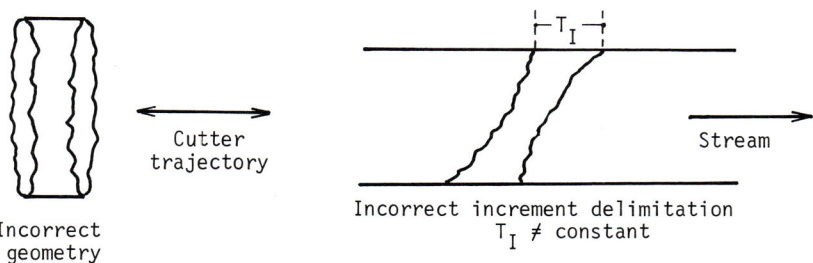

Fig. 17.3. Straight path cross-stream pulp cutters - cutter opening partly obstructed.

17.2.3. Straight path cutters of other types - deviation from correctness :

Fig. 17.4. represents a **flap sampler** in vertical cross-section. Such samplers can be regarded, from a theoretical standpoint, as single edge cutters. They cut trapezoidal increments always containing a larger part of one side of the stream (the lower side in our figure). These devices are always incorrect and there is no remedy : the conception is basically wrong. Such samplers should be avoided.

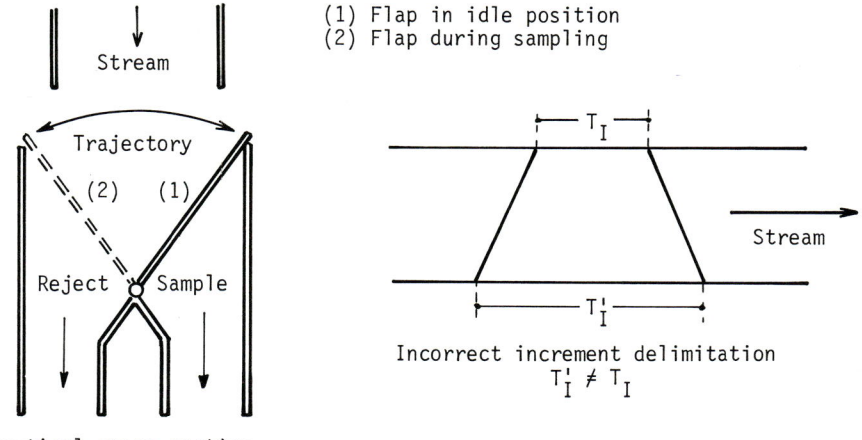

Fig. 17.4. Flap samplers - always incorrect

Akin to flap samplers from a theoretical standpoint, the <u>flexible hose pulp samplers</u> present the same defect, as shown in fig. 17.5. Such devices are cheap and can be found in small flotation plants or in pilot plants. The usual model represented in fig.17.5. carries out an incorrect delimitation and should be avoided.

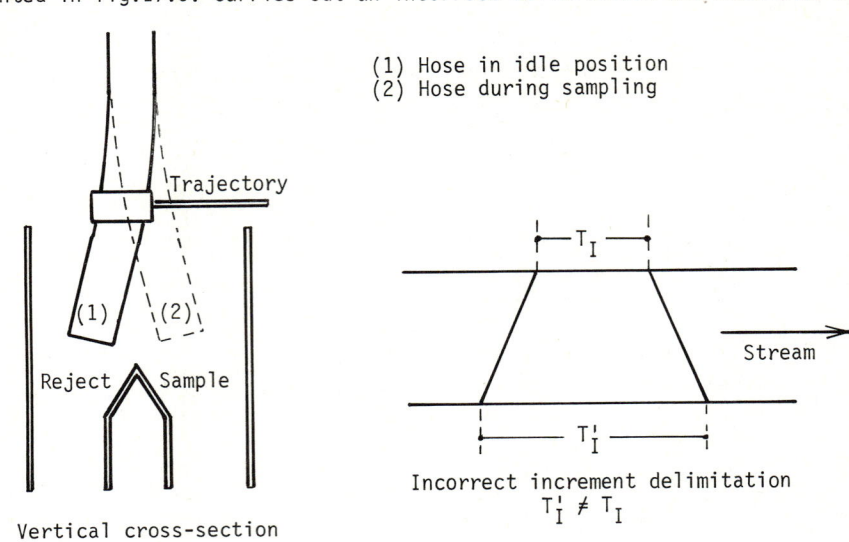

Fig. 17.5. *Flexible hose pulp samplers - usual model - incorrect lay-out.*

But contrary to flap samplers, flexible hose samplers can be easily corrected, at least as far as the geometry is concerned. An improved model is shown in fig. 17

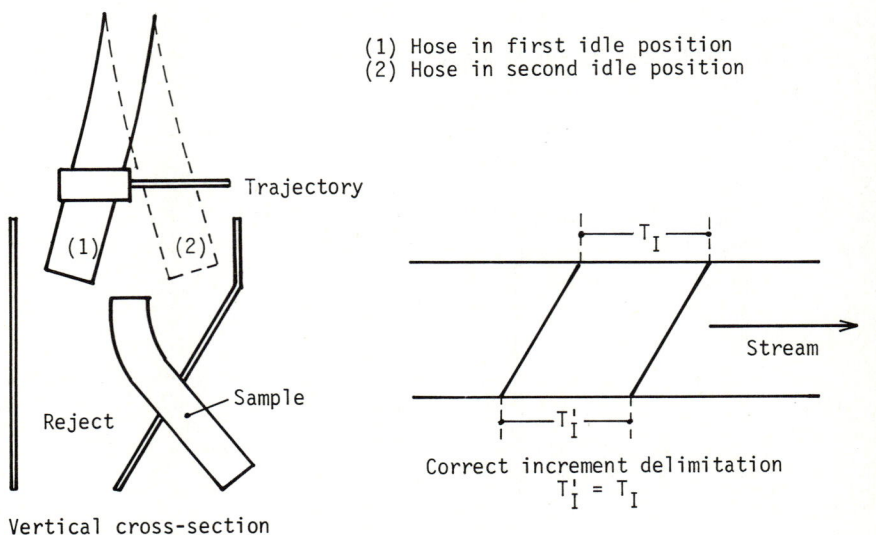

Fig. 17.6. *Flexible hose pulp samplers - improved model - correct lay-out .*

It can be regarded as a cross stream cutter with a stationary cutter and a moving feeder. On condition that the cutter opening be rectangular and that other delimitation correctness rules be respected, this simple sampler performs a correct increment delimitation. The temptation to match a circular hose by a circular cutter should not be yielded to. The trouble with flexible hose samplers is that, as far as we know, they never respect the uniform speed condition that will be presented in a further section.

17.2.4. Circular path cutters - correct geometry :

There are a great number of samplers falling within this category. These were very popular at the turn of the century and still today remain popular in many parts of the mining world. They all have in common to revolve around a rotation axis which can be vertical, horizontal or inclined.

With circular path cutters, a correct increment delimitation is achieved if and only if the cutter edges converge towards the rotation axis. Fig. 17.7 shows the shape of a correct cutter opening and the corresponding increment cut through the stream by the cutter.

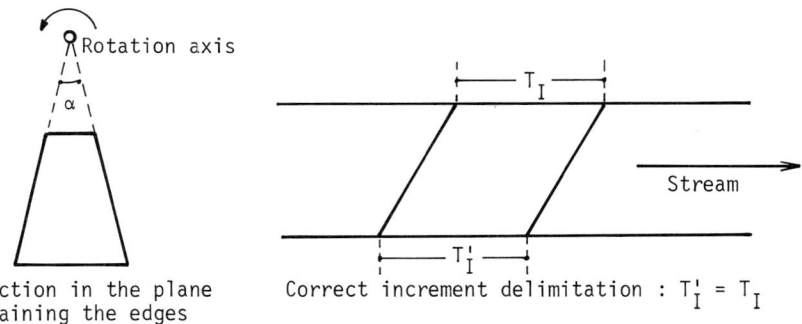

Projection in the plane   Correct increment delimitation : $T_I' = T_I$
containing the edges

Fig. 17.7. *Circular path cutters - correct geometry*.

But, whereas it was natural for the manufacturers to build a rectangular cutter opening in straight path samplers, it seems that the trapezoidal shape delimited by radial cutter edges does not look as "natural" as it should and a number of deviations from the correctness rule can be observed.

17.2.5. Circular path cutters - deviations from correctness :

The most frequent of all deviations is the rectangular opening, which was the standard of correctness with straight path cutters but can be regarded as the classical example of delimitation incorrectness with circular path cutters. It must be observed here that the sampling ratio of a given element E of the stream cross-section, a certain distance r from the axis, is proportional to the angle α_r of the cutter opening at this distance of the axis. This is shown in fig. 17.8. If W denotes the uniform width of the rectangular opening, then the sampling ratio and with it the sampling probability is proportional to arc sin W/2r, i.e. the sampling probability is non-uniform throughout the stream cross-section and is a function of the distance r from the rotation axis. In the example represented in fig. 17.8. the sampling ratio is nearly three times larger for the side of the stream nearest the axis than for the farthest side. We met with such a device in harbour sampling facilities loading huge tonnages of iron ores from a large African mine. The sampler was made by a world famous mining equipment manufacturer. As far as we know, it is still operating though an important bias had been detected and proved beyond any possible doubt. Unfortunately, rectangular shape seems to be the rule rather than the exception with circular path cutters.

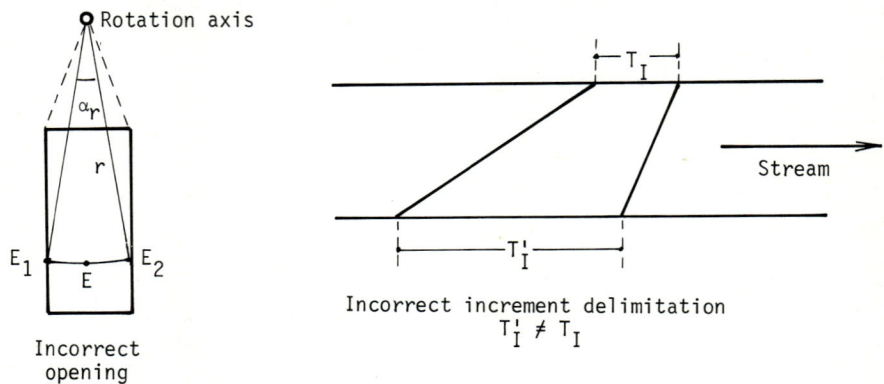

Fig. 17.8. Circular path cutters - Incorrect rectangular opening.

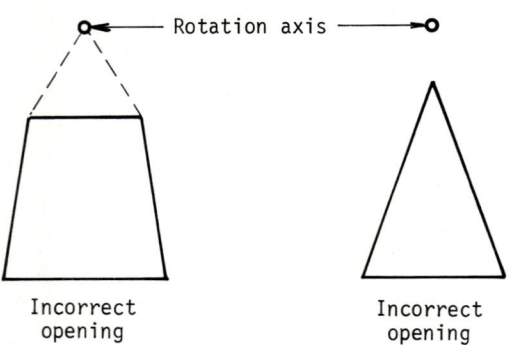

Fig. 17.9. Circular path cutters - Other incorrect openings.

Fig. 17.9. shows two other incorrect cutter openings that have been actually found in small to medium mining Companies, which had found appropriate and cheap to design and build their own sampling devices through ignorance of the dangers of so doing. Their only excuse, which is certainly a good one, is that few are the sampling equipment manufacturers who supply 100 % correct sampling devices, even among the world famous.

17.3. FALLING STREAM SAMPLING - CORRECTNESS CONDITIONS INVOLVING THE CUTTER SPEED

These conditions are simple (proof has been given in Gy, 1975, chapter 24) :

1) The cutter velocity should remain constant during its travel across the stream,

2) The cutter velocity should remain uniform from one increment to the next.

17.3.1. Achievement of the uniformity of the cutter velocity during its travel across the stream :

We must distinguish here two categories of samplers :

- samplers in continuous motion,
- samplers working intermittently, set into motion by a timer and automatically stopped at the end of their travel through the stream.

Falling into the <u>first category</u> are a number of circular path samplers and a few straight path samplers. These have in common to perform a usually large sampling ratio (say from 2 to 20 %). They usually reach their cruising speed very quickly and maintain it consistently. With such devices, it may always be safely assumed that the cutter velocity remains constant during its travel across the stream.

Falling into the <u>second category</u>, we find all intermittent samplers. Schematically, they work in the following way :

1) a timer sets the cutter and its carriage into motion from an idle position which we shall assume to be out of the stream and some distance away.

2) the cutter crosses the stream and takes an increment,

3) at the end of its travel, the cutter drive is automatically switched off and the carriage braked to a standstill. Alternately, at the end of its travel, the cutter motion is reversed, the cutter crosses the stream a second time and travels back to its starting point where its drive is switched off and its carriage stopped.

The same sequence repeats itself according to a given timing program, systematic or random stratified (see chapter 7).

We must now distinguish the various driving systems in use : electric, pneumatic, hydraulic, magnetic or manual.

17.3.1.1. Electric drive :
electric motors, usually coupled to a reduction gear are by far the most usual, at least in large and medium plants. They are cheap, convenient and highly reliable. Electric drive is the only system achieving a constant speed during the cutter travel across the stream. Now, several conditions must be fulfilled if a correct delimitation is to be performed :

1) the idle positions of the cutter should be far enough from the stream, for the cutter carriage to have enough time to reach its cruising speed prior to entering the stream. The adequate distance depends on many factors such as type and power of the motor, weight of the cutter carriage, etc ... In reciprocating samplers cutting the stream once from left to right, once from right to left, the same distance is used for speeding up and for slowing down and braking, no deceleration being allowed before all parts of the cutter are out of the stream. For a small pulp sampler, the reasonable minimum distance is about 0.2 to 0.3 m but for heavy bucket samplers (more than one ton dead-weight) such as those currently used for sampling high rates of flow (over 8000 t/h) of coarse materials (200 mm) requiring heavy-duty machinery, large bucket openings and huge capacity, this minimum distance can be as large as three meters. Fig. 17.10. shows the increment delimitation achieved by an otherwise correct cutter still accelerating while crossing the stream. Precise measurements have shown in some instances the speed ratio to be 1 to 1.5 between the beginning and the end of stream crossing.

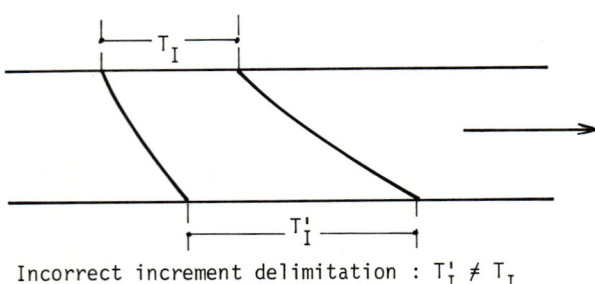

Incorrect increment delimitation : $T'_I \neq T_I$

Fig. 17.10. Incorrect delimitation achieved by a cutter still accelerating while crossing the stream - Idle position too near the stream.

2) Assuming the preceding condition to be fulfilled, the cutter should not be slowed down while crossing the stream. Such a defect may appear when sampling high tonnages of coarse materials imposing a very heavy load on the motor. The solution consists in using a motor with sufficient power reserve. Fig. 17.11. shows the increment delimitation achieved by an otherwise correct cutter slowing down progressively when entering the stream and accelerating again when the load decreases. Rough estimations showed that the speed reduction can reach 25 %.

3) Speed reduction gears should be so constructed as to prevent jerky motion.

<u>17.3.1.2. Pneumatic, hydraulic, magnetic or manual drives</u> : those are to be found in small or medium mines and plants and are likely to equip home-made clever sampling devices. None of these drives respects the constant speed condition. They should accordingly be avoided. They could certainly be improved by means of some regulating device or other but the cost of improvement would probably surpass the

cost of superseding them by new electric drives.

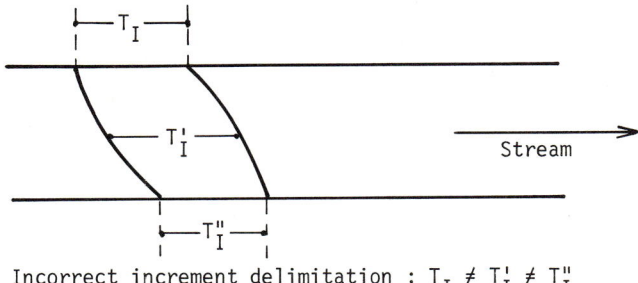

Incorrect increment delimitation : $T_I \neq T'_I \neq T''_I$

Fig. 17.11. Incorrect delimitation achieved by a cutter slowed down by the load while crossing the stream - Insufficient horsepower of the electric drive.

17.3.2. Achievement of a uniform cutter velocity from one increment to the next :

Here again, we must distinguish electric drives from the others.

17.3.2.1. Electric drive : of all possible drives, electric drive is the aptest for the achievement of a uniform speed from one increment to the next. Slight speed modulations may result from fluctuations of the characteristics of the electric current, sizeable in industrial areas and for instance in a big plant when high H.P. motors are being started. Alteration of motor speed may also result from lack of maintenance or wear.

Anyway, the delimitation error incurred as the result of such fluctuations is not liable to be other than negligible. For all practical purposes this error can be assimilated to the weighting error (see section 13.5.) if only for the reason that in a variographic experiment it is recorded as such.

17.3.2.2. Pneumatic, hydraulic, magnetic or manual drives :

The observations made in section 17.3.1.2. remain valid here.

17.4. FALLING STREAM SAMPLING - CORRECTNESS CONDITIONS INVOLVING THE GENERAL LAY-OUT

Several conditions fall under this heading :

17.4.1. Correct lay-out of the sampling device :

Fig. 17.12. shows the correct lay-out of a reciprocating straight path cross-stream sampler : the rectangles $A_1B_1C_1D_1$ and $A_2B_2C_2D_2$ represent the cutter opening in the two idle positions of the sampler. The dotted lines define the boundaries of the surface actually generated by the cutter opening in the horizontal plane. The stream cross-section is well delimited and falls right towards the centre of the rectangle $A_2B_1C_1D_2$, far enough from the sides B_1C_1 and A_2D_2. The vertical cross-section shows that the stream is liberated just above the surface generated by the cutter during its travel across the stream.

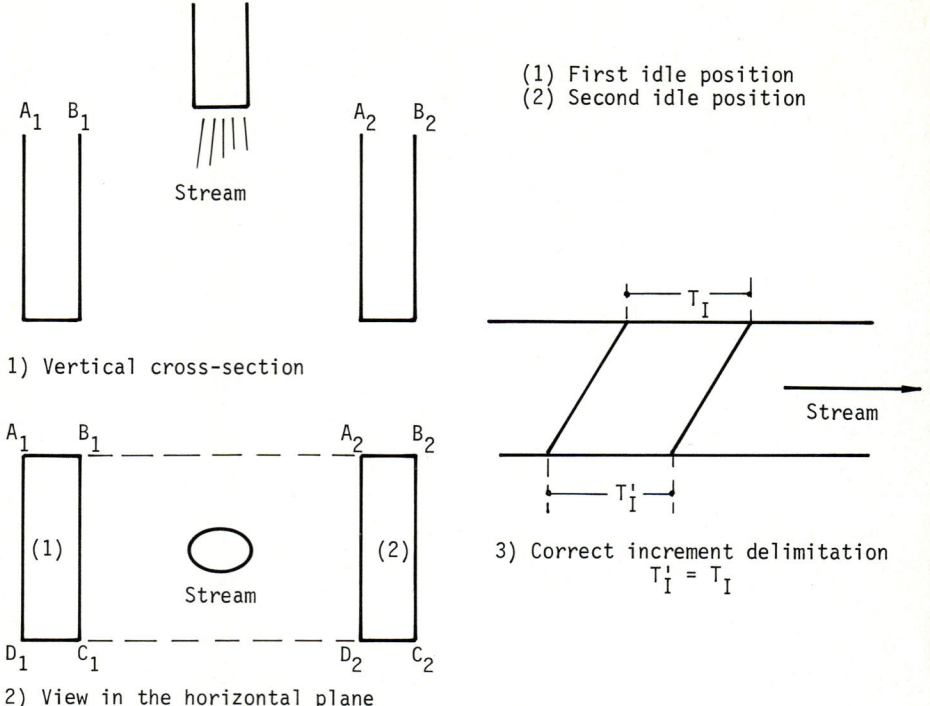

Fig. 17.12. *Correct general lay-out of the sampling device.*

17.4.2. Incorrect lay-out : part of the stream escapes sampling :

Fig. 17.13. shows an example of such a faulty lay-out. The stream falls astride the boundary B_1A_2 of the area generated by the cutter opening. The hachured part of the stream (which may represent only a few loose fragments) fails to be sampled We have seen such a defect at the discharge of a belt conveyor, the speed of which had been increased without consideration to the parabola described by the stream. A deflector may be the remedy. It is cheaper anyway than the displacement of eithe the conveyor head pulley or the sampling device.

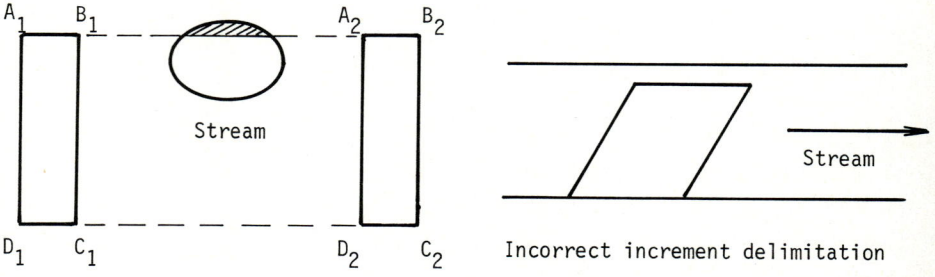

Fig. 17.13. *Incorrect lay-out. Part of the stream escapes sampling.*

17.4.3. Incorrect lay-out : part of the stream enters the cutter in idle position :

Fig. 17.14. shows how this can happen. One of the idle positions has been set too near the stream. A fringe of it (sometimes no more than a few loose coarse fragments) falls into the cutter between two increments. We can see on the right part of the figure that the right side of the stream, however small it may be, is sampled with a 50 % sampling ratio.

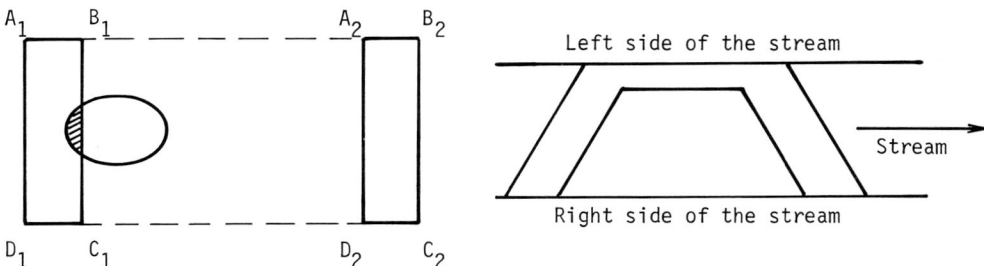

Fig.17.14. Incorrect lay-out : part of the stream falls into the cutter in idle position.

17.4.4. Incorrect lay-out : reciprocating sampler with a non-idle reversing position :

With the same respective dispositions of the stream and of the sampler as in fig. 17.14. , fig. 17.15. shows what happens when the cutter reverses its motion whilst it is not completely out of the stream. In such a case, the sampling ratio is a little smaller for the left side of the stream than for the rest of the stream cross-section.

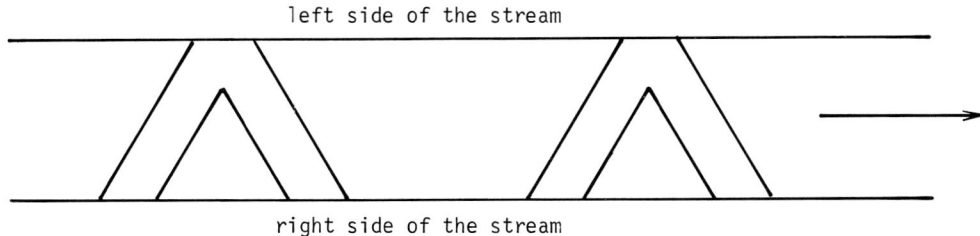

Fig. 17.15. Incorrect lay-out : reciprocating sampler with a non-idle reversing position.

17.4.5. Sampling dry materials containing fines :

When the stream contains dry fine particles, its sampling is always tricky. The numerous errors likely to take place in such a case are illustrated in fig. 17.16.

representing a cloud of dust invading the plant around the sampling device.

We can observe that some of this dust escapes sampling and falls outside the rectangle $A_2B_1C_1D_2$ (defect analysed in section 17.4.2.), that some of this dust enters the cutter in either idle positions and falls inside the rectangles $A_1B_1C_1D_1$ and $A_2B_2C_2D_2$ (defect analysed in section 17.4.3.)

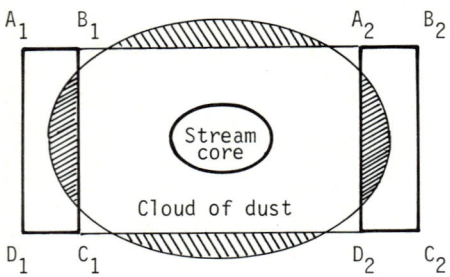

Fig. 17.16. Sampling dry fine materials - The dust problem.

The well known solution which consists in enclosing the head pulley of the conveyor, the sampler and its ancillary equipment in a neat tight box is certainly the pet solution of most designers of sampling plants. It may please the eye and save the environment, it does not solve the problem set by the requirements for delimitation correctness. When in addition they connect the box (as they always do) to one of these powerful dust collecting systems, the draft is likely to achieve a near-perfect sample dedusting (the errors associated with this dedusting effect fall within the province of increment extraction errors which will be studied in the next chapter).

Now, what can be done ? There seems to be no perfect solution. Here are a few suggestions :

1) Never use a belt conveyor to feed dry fine materials : the free fall is likely to be too high. Use for instance screw feeders discharging 1 to 2 centimeters above the plane generated by the cutter opening,

2) Collect the sampling reject in a small hopper from which the material can be extracted by means of another screw feeder or rotor-type vane feeder,

3) In both idle positions, cover the cutter by means of protecting caps preventing dust from entering the cutter between increments,

4) Use dust collectors with great discretion so as to create a slight downward draft in the reject hopper.

Such a solution is schematized in fig 17. 17.

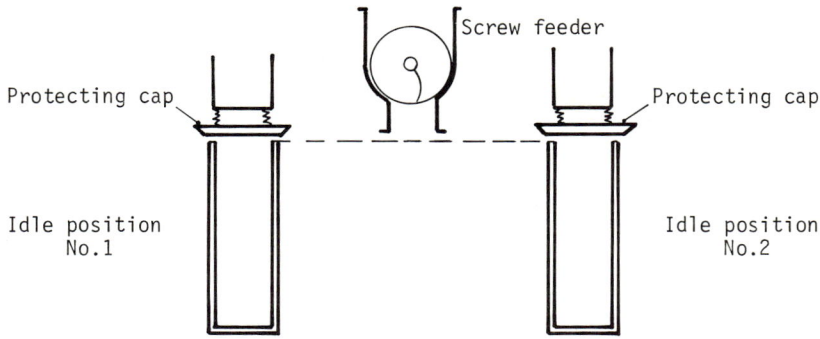

Vertical cross-section of feeder, cutter and protecting caps

Fig. 17.17. *Sampling dry fine materials - a possible solution to the dust problem.*

17.5. FALLING STREAM SAMPLING - RECAPITULATION OF THE CONDITIONS OF DELIMITATION CORRECTNESS

For the sake of convenience, the conditions of increment delimitation correctness are summarized in table 17.1.

Table 17.1. *Falling stream sampling - Increment delimitation correctness :*

Increment delimitation is correct and the increment delimitation error DE is cancelled if and only if the following conditions are simultaneously fulfilled.

Cutter geometry	Cutter velocity	Sampler lay-out
1) <u>Straight trajectory</u> : edges should be parallel. 2) <u>Circular trajectory</u> : edges should be radial. 3) <u>Undefined trajectory</u>: no correct geometry. 4) <u>Defined but non-straight non-circular trajectory</u> : no correct geometry. 5) <u>All types</u> : Cutter openings should be periodically checked for deformation and wear and cleaned.	1) <u>Electric drive</u> : a) idle positions of the cutter should be a certain distance from the stream. b) motor should be generously calculated 2) <u>Hydraulic, pneumatic, magnetic or manual drive</u>: The uniform velocity condition is practically never respected. Should be avoided.	1) <u>Stream discharge</u> : At the level of the cutter opening, the stream should be made to pass through a well delimited area. 2) <u>Idle positions</u> : Idle or reversing positions of the cutter should be far away from the stream. 3) <u>Positioning</u> : The stream should fall well inside the area generated by the cutter. 4) <u>Dust</u> : Caps should protect cutter openings.

17.6. STOPPED BELT SAMPLING

Stopped belt sampling is often resorted to and is recommended by certain national and international standards as the reference sampling method when checking a falling stream sampler for a possible bias. This check, which is based on the postulate that the reference sampling method is itself unbiased, will be studied in chapter 32 but our purpose in this section is to state the conditions that should be fulfilled in order for stopped belt sampling to be correct and therefore practically unbiased.

Increment delimitation, in stopped belt sampling, is correct if and only if the actual extended increment is delimited by parallel planes. As this sampling method is usually carried out by means of a metallic frame made of two parallel steel plates, the delimitation is to be regarded as correct. This is shown in fig. 17.18.

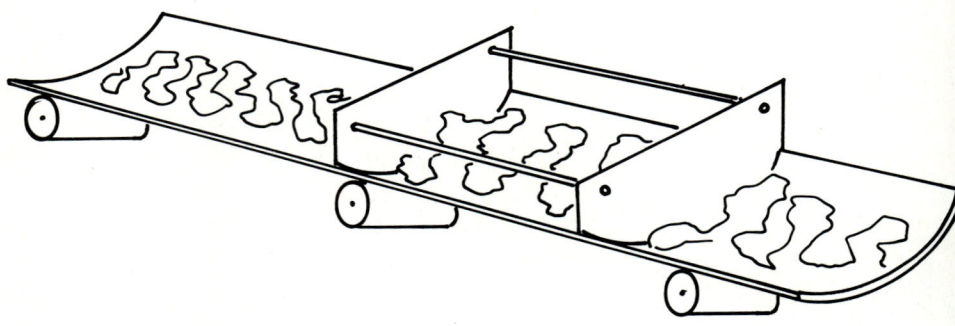

Fig. 17.18. Stopped belt sampling - Correct delimitation.

17.7. TWO-DIMENSIONAL SAMPLING

Two-dimensional sampling is resorted to when sampling for instance truckloads of fine materials by means of pipe samplers, probes or augers. The object to be sampled must be uniformly thick throughout the horizontal extension of the lot for the two-dimensional model to be efficient.

The two-dimensional isotropic extended increment is a circle. The corresponding three-dimensional model increment is a cylinder built up on the circular basis and usually perpendicular to it. Practically all two-dimensional sampling devices have a circular uniform cross-section, except for a conical head devised in order to make easier the penetration into the solid mass. We know an example of two-dimensional sampler with a square cross-section, devised for the sampling of truckloads of sugar beets. As far as increment delimitation is concerned, this sampler is corre

Fig. 17.19. shows that with cylindrical devices, the conical head is responsible for the deeper part of the truckload to be incompletely represented and to be sampled with a sampling ratio systematically low (left part of the figure). Furthermore, it is often difficult and sometimes impossible for the sampler to penetrate the material down to the bottom of the truck (right part of the figure). In both cases, the missing part of the increment, deviation of the actual sample from its model, is represented in dotted lines.

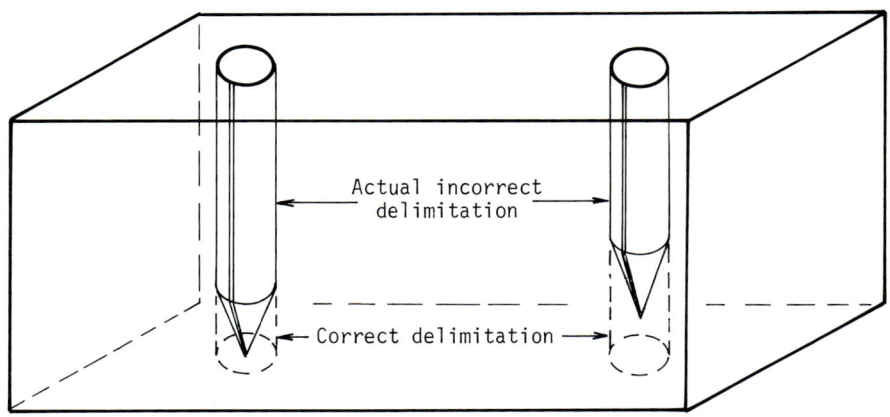

Fig. 17.19. Two-dimensional sampling - Impossibility to achieve a correct increment delimitation.

From this section, we shall retain that two-dimensional sampling by means of pipe samplers or the like can practically never achieve a correct delimitation. This is one of several reasons for which this kind of sampling can hardly be relied upon. As an example of the dangers of such samplers we shall mention the sampling of galena flotation concentrates uniformly loaded on rectangular lorries by means of a pipe sampler of the type schematized in fig. 17.19. The sampling took place at the smelter facilities in the same time as wet weighing. The samples were used both for assaying and for moisture determination. As the lorry had travelled on very rough tracks in a desert area for about one hour, due to gravity and vibrations, a large part of the water had migrated towards the bottom of the lorry whereas in the same time the top layer had dried under the sun. The vertical segregation of moisture was near-perfect, the largest part of water being concentrated in the 10 cm bottom layer. The pipe sampler failing to sample correctly this part of the concentrate load, the samples actually extracted had an average moisture content much lower than the actual moisture content to be estimated. This is a typical example of delimitation error, associated with an important segregation, magnifying the effects of this error.

17.8. CHANCES OF OBTAINING UNBIASED SAMPLES THROUGH INCORRECT DELIMITATION

Consider a lot L made of N_L fragments F_i with : a_i critical content, M_i mass and P_i selection probability of F_i. We shall show in section 20.5.4.1. that the bias resulting from the non-uniformity of P_i is proportional to $(a_o - a_L)$ with :

$$a_o = \frac{\sum_i a_i M_i P_i}{\sum_i M_i P_i} \quad \text{and} \quad a_L = \frac{\sum_i a_i M_i}{\sum_i M_i}$$

The general solution to the equation : $a_o = a_L$ that cancels the bias consists in suppressing the correlation $Cor(P_i/a_i, M_i)$ between the selection probability P_i and the "personality" (a_i, M_i) of the fragment F_i. Let's call X_i the position of the fragment F_i in the domain (D_L) occupied by L. The correlation $Cor(P_i/a_i, M_i)$ can be regarded as the logical product of three elementary correlations :

1 - $Cor(P_i/X_i)$: between selection probability P_i and position X_i of fragment F_i,

2 - $Cor(X_i/M_i)$: between position X_i and physical properties of F_i, namely its diameter d_i, density λ_i and shape factor f_i defining the mass $M_i = f_i d_i^3 \lambda_i$.

3 - $Cor(M_i/a_i)$: between physical properties and mineralogical composition of F_i.

Symbolically, we can write :

$Cor(P_i/a_i, M_i) = Cor(P_i/X_i) \times Cor(X_i/M_i) \times Cor(M_i/a_i)$

To cancel the product, it is necessary and sufficient to cancel at least one of the three factors. We can therefore think of three mathematical solutions :

1 - $Cor(P_i/X_i) = 0$: this equation is satisfied if and only if the delimitation is correct. We can achieve delimitation correctness and we know how to achieve it.

2 - $Cor(X_i/M_i) = 0$: this equation is satisfied if and only if the distribution is homogeneous but experience shows that homogeneity is never achieved in practice.

3 - $Cor(M_i/a_i) = 0$: this correlation is an intangible datum of the problem and cannot be cancelled at will. We are never entitled to assume it is zero.

We therefore reach the conclusion that the only safe way to obtain unbiased samples is to achieve a correct increment delimitation.

17.9. COST OF CORRECT INCREMENT DELIMITATION

Sampling devices respecting the rules of delimitation correctness are in no way more expensive than incorrect samplers. They entail neither larger sample weights nor larger sample reduction cost. When they fail to respect rules as simple as those reviewed in this chapter, manufacturers and engineering firms have no excuse except ignorance. But is ignorance an excuse for those whose duty is to know and whom are relied upon by their customers ?

CHAPTER 18

INCREMENT EXTRACTION ERROR EE

18.1. DEFINITION

According to the definition given in section 16.2.6. an "increment extraction error" EE takes place as soon as the increment extraction ceases to be correct. Increment extraction is said to be correct if and only if the "rebounding rule" or "rule of the centre of gravity" is respected. When this condition is fulfilled, assuming the conditions of delimitation correctness to be respected too (see chapter 17), the continuous selection model is applicable to the fragmental functions and no new error is involved : the increment extraction error is identically nil.

This chapter deals with a study of the conditions of increment extraction correctness when sampling a falling stream.

18.2. ANALYSIS OF THE REBOUNDING RULE

18.2.1. Definitions and notations :

We shall use the following notations :

F : a spherical fragment, with diameter d, falling vertically without spinning,
F' : centre of gravity of F. It is assumed to coincide with the geometrical centre of the sphere.
$\vec{V}_F$: velocity vector of F, relatively to earth. Its magnitude V_F is a function of time.
C : a rectangular cutter moving on a straight horizontal path.
C_L : the leading edge of C.
C_T : the trailing edge of C. C_L and C_T are assumed to be straight, horizontal, parallel, on the same level and perpendicular to the path of C.
W : the cutter width, distance between the cutter edges.
H : the horizontal plane generated by the movement of C_L and C_T.
$\vec{V}_C$: velocity vector of C, relatively to earth. Its magnitude V_C is assumed to be uniform.
$\vec{V}_{FC}$: velocity vector of F, relatively to the cutter C. By definition :

$$\vec{V}_{FC} = \vec{V}_F - \vec{V}_C \quad \text{and} \quad V_{FC}^2 = V_F^2 + V_C^2$$

$\vec{V}_{FC}$ is a function of time, both in direction and magnitude (see fig. 18.1.).

A geometrical study of the collision of F (speed vector $\vec{V}_F$) with a moving cutter edge (speed vector $\vec{V}_C$) is equivalent to that of the collision of F (speed vector $\vec{V}_{FC}$) with a still cutter edge (speed vector $\vec{0}$). The speed vectors $\vec{V}_F$ and $\vec{V}_{FC}$ are applied to the centre of gravity F' of F.

X : intersection of the vector $-\vec{V}_C$ with the surface of F,
X' : intersection of XF' with the surface of F,
Y : intersection of the vector $\vec{V}_F$ with the surface of F,
Z : intersection of the vector $\vec{V}_{FC}$ with the surface of F. The position of Z on the arc XY is a function of time.
t_F : time variable characterizing the chronology of the fragment F,
t_L : time variable characterizing the chronology of the leading edge C_L,
t_T : time variable characterizing the chronology of the trailing edge C_T,
t_C : collision time.

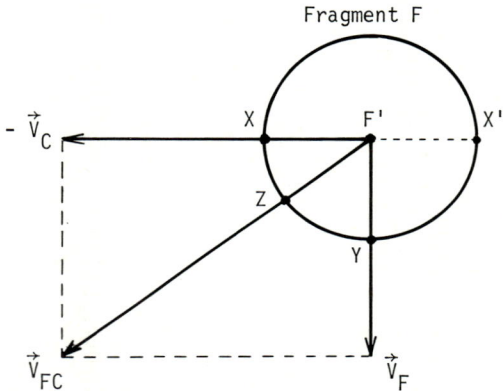

Fig. 18.1. Fragment F - Vectors $\vec{V}_F$ and $\vec{V}_{FC}$ - Vertical cross-section.

18.2.2. Chronology of the fragment F :

We shall take the following instants into consideration (fig. 18.2.) :

t_{F1} : Y crosses the plane H in Y_H
t_{F2} : Z crosses the plane H in Z_H (will be referred to as the "critical point"),
t_{F3} : X , F' and X' cross the plane H in X_H , F'_H and X'_H .

The chronology of F is characterized by the following inequalities :

$t_{F3} > t_{F2} > t_{F1}$ and $t_{F3} > t_C > t_{F1}$

18.2.3. Chronology of the leading edge C_L :

We shall take the following instants into consideration (fig. 18.3.) :

t_{L1} : C_L reaches the point X_H

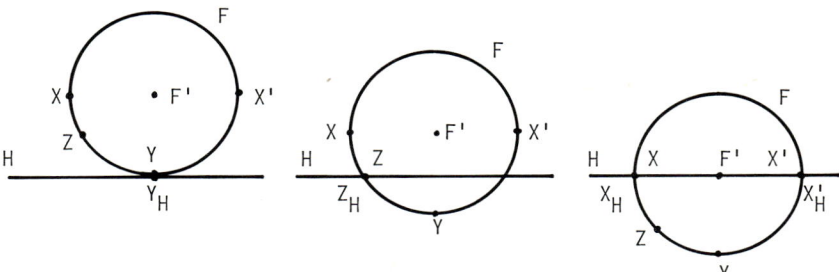

1) time t_{F1} : $Y \to Y_H$ 2) time t_{F2} : $Z \to Z_H$ 3) time t_{F3} : $X \to X_H$

Fig. 18.2. Chronology of the fragment F - Vertical cross-section.

t_{L2} : C_L reaches the point Z_H
t_{L3} : C_L reaches the point X'_H

The chronology of C_L is characterized by the following inequalities :

$t_{L3} > t_{L2} > t_{L1}$ and $t_{L3} > t_C > t_{L1}$

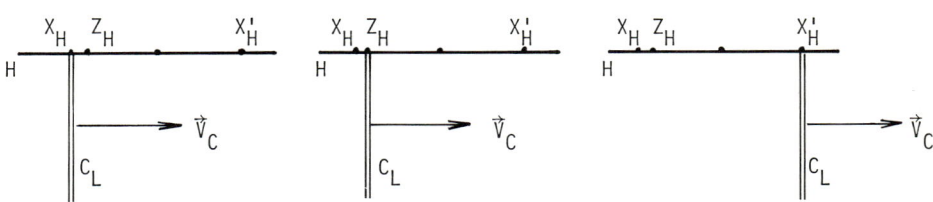

1) time t_{L1} : $C_L \to X_H$ 2) time t_{L2} : $C_L \to Z_H$ 3) time t_{L3} : $C_L \to X'_H$

Fig. 18.3. Chronology of the leading edge C_L - Vertical cross-section.

18.2.4. Chronology of the trailing edge C_T :

In the same way we shall define the instants :

t_{T1} : C_T reaches the point X_H
t_{T2} : C_T reaches the point Z_H
t_{T3} : C_T reaches the point X'_H.

The chronology of C_T is characterized by the following inequalities :

$t_{T3} > t_{T2} > t_{T1}$ and $t_{T3} > t_C > t_{T1}$

18.2.5. Collision between the fragment F and parts of the cutter C :

We shall distinguish :

- the cutter edges C_L and C_T in a proper sense (immaterial straight lines of plane H represented by points C_L and C_T of fig. 18.4.),
- the inner walls of C_L and C_T (walls facing the inner part of the cutter),
- the outer walls of C_L and C_T (walls facing the outer side of the cutter).

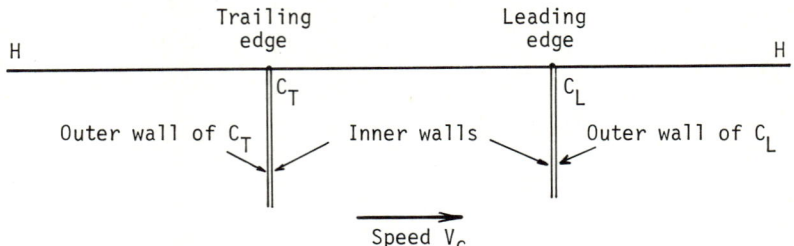

Fig. 18.4. *Edges and walls of the cutter C - Vertical cross-section.*

The falling fragment F ends its course into one of the following domains :

- the leading reject R_L : outer side of the leading edge C_L,
- the increment I : inner side of the cutter,
- the trailing reject R_T : outer side of the trailing edge C_T.

The purpose of the next sections is to analyse the extraction problem or in other words the respect of the rebounding rule by "superposing" the chronology of the fragment F to those of the cutter edges C_L and C_T.

We shall first consider what happens at the level of the leading edge C_L at time t_{F3} (F' crosses the horizontal plane H generated by the cutter edges).

18.2.6. Respective positions of F and C_L at time t_{F3} with $t_{F3} \leq t_{T3}$:

This is illustrated by fig. 18.5.

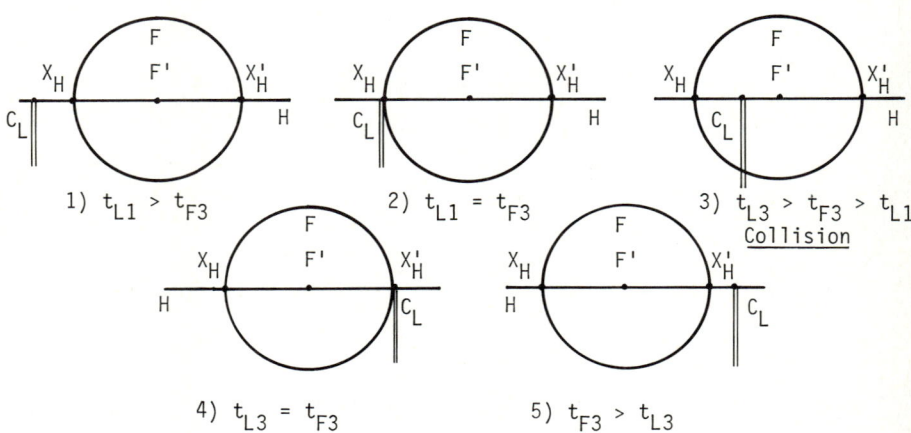

Fig. 18.5. *Respective positions of F and C_L at time t_{F3}.*

Five possibilities may arise :

1) $t_{L1} > t_{F3}$: X reaches X_H before C_L. Then, at time t_{L1}, F collides with the outer wall of C_L and eventually falls in the leading reject R_L.

2) $t_{L1} = t_{F3}$: X and C_L reach X_H in the same time. F collides with the upper part of the outer wall of C_L and eventually falls in R_L.

3) $t_{L3} > t_{F3} > t_{L1}$: X and X' cannot reach X_H and X'_H because C_L is on the way of F. Then, at some time t_C ($t_{F3} > t_C > t_{L1}$), F collides with the edge C_L and bounces in a direction that will be precised in section 18.2.7.

4) $t_{L3} = t_{F3}$: X' and C_L reach X'_H in the same time. F glides against the inner wall of C_L and eventually falls into the cutter.

5) $t_{F3} > t_{L3}$: X' reaches X'_H after C_L. F falls directly into the cutter as we have assumed in this section that $t_{T3} > t_{F3}$.

The final destination of F is uncertain only in case No.3 when it collides with the edge C_L. We shall now examine in detail what happens in this case.

18.2.7. Collision between F and C_L :

Fig 18.6. shows the three possibilities that may arise at time t_{F2} when Z crosses the plane H at point Z_H.

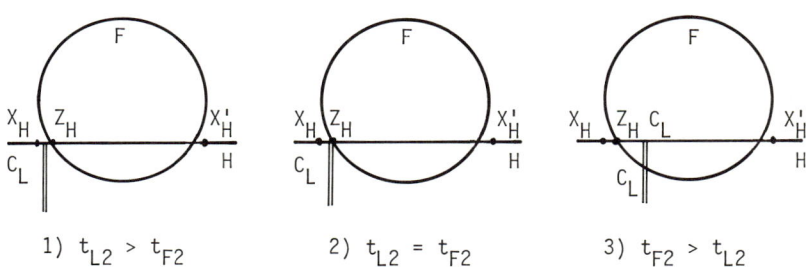

Fig. 18.6. Respective positions of F and C_L at time t_{F2} when $t_{L3} > t_{F3} > t_{F2} > t_{L1}$.

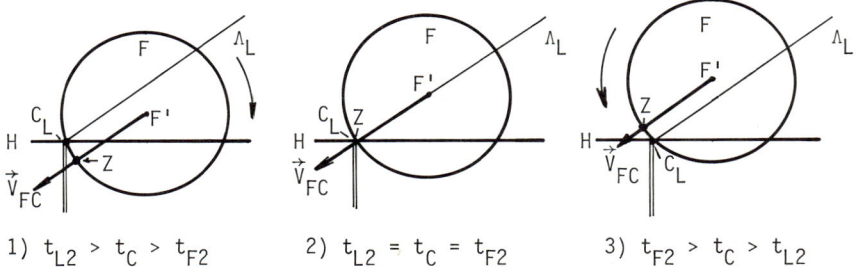

Fig. 18.7. Collision between F and C_L - Respective positions of F' and Λ_L at collision time t_C - Respective positions of Z and H - Rebounding rule.

When the fragment F collides with the edge C_L, it is submitted to two forces: the first one is applied to its centre of gravity F' and the second one to the collision point with C_L, both parallel to $\vec{V}_{FC}$. Let's consider the "critical plane" Λ_L containing the edge C_L and parallel to $\vec{V}_{FC}$ (fig. 18.7.). The rebound of F depends on the respective positions of F' and Λ_L at collision time t_C. Three possibilities may arise :

1) $t_{L2} > t_{F2}$: Z reaches Z_H before C_L. F collides with C_L at some time t_C such that $t_{L2} > t_C > t_{F2}$, as Z is already below the plane H. Fig. 18.7.(1). shows that F' and $\vec{V}_{FC}$ are on the right side of the plane Λ_L that contains the reaction force of C_L. Then, F bounces toward the leading reject R_L and spins clockwise.

2) $t_{L2} = t_{F2}$: Z and C_L reach Z_H at the same time and collide ($t_{L2} = t_C = t_{F2}$). This is shown in fig. 18.7.(2). and corresponds to the <u>first critical position</u>. If $t_{L2} = t_{F2} + \varepsilon$, then F bounces towards the outer side of C_L. If $t_{L2} = t_{F2} - \varepsilon$, then F bounces towards the inner side of C_L (see below).

3) $t_{F2} > t_{L2}$: C_L reaches Z_H before Z. F collides with C_L at some time t_C such that $t_{F2} > t_C > t_{L2}$ as Z is still above the plane H. Fig. 18.7.(shows that F' and $\vec{V}_{FC}$ are above the plane Λ_L that contains the reaction force of C_L. Then, F bounces towards the inner side of the cutter, belonging therefore to the "<u>model fragmental increment</u>". Whether or no it will eventually belong to the "<u>actual fragmental increment</u>" depends on what will happen at the level of the trailing edge C_T (section 18.4.4.).

18.2.8. Respective positions of F and C_T at time t_{F3} - collision of the first type between F and C_T :

We are discussing in this section what happens at time t_{F3} when F' crosses the plane H or comes near to it in the vicinity of the trailing edge C_T, assuming that F falls vertically. What happens at the level of C_T after F has bounced on the leading edge C_L is another problem which will be dealt with in section 18.4.4. We shall consider the plane Λ_T containing the edge C_T and parallel to Λ_L.

Both edges being assumed to be identical, C_L and C_T play symmetrical parts with the difference that the inner sides of C_L and Λ_L become the outer sides of C_T and Λ_T and vice versa.

A collision of the first type takes place between F and C_T if $t_{T3} > t_{F3} > t_{T1}$.

1) $t_{T2} > t_{F2}$: F bounces towards the inner side of the cutter (model increment),
2) $t_{T2} = t_{F2}$: F' is in plane Λ_T, which defines the second critical position,
3) $t_{F2} > t_{T2}$: F bounces towards the outer side of the cutter (trailing reject).

18.2.9. Rebounding rule and model increment :

As V_F is a function of time and varies with the elevation of F' above H, the direction of $\vec{V}_{FC}$ is also a function of time. Then, the "model fragmental increment" defined in chapter 16 is delimited by two surfaces Λ'_L and Λ'_T that can be superposed by a translation $-\vec{W}$ parallel to $\vec{V}_C$. Assuming the effects of the surrounding atmosphere to be negligible (which is acceptable with coarse fragments) the intersections of Λ'_L and Λ'_T with the vertical plane containing fig.18.8. moving with the cutter C and at the same speed $\vec{V}_C$ are parabolas. These curves are the envelopes of the relative speed vectors $\vec{V}_{FC}$ applied to the centre of gravity of a fragment F falling in critical position. This figure shows the identity of the "rebounding rule" and of the "rule of the centre of gravity".

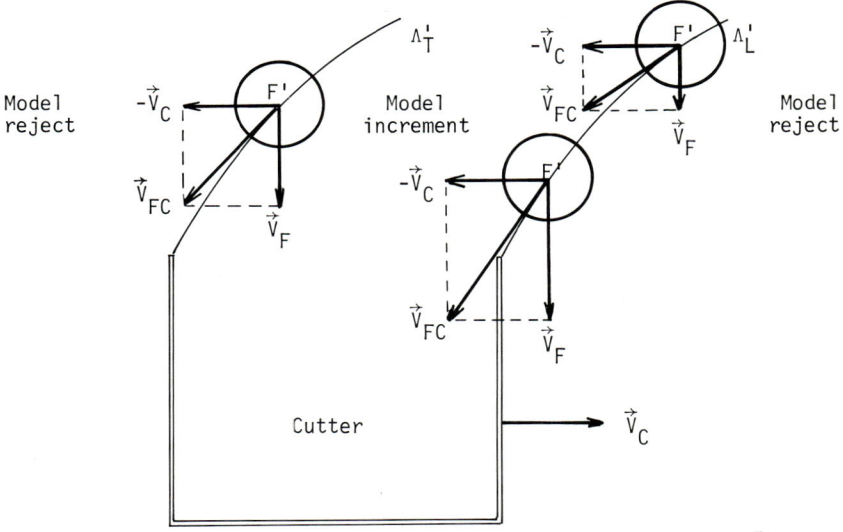

Fig 18.8. Definition of the model increment

18.3. CONDITIONS OF EXTRACTION CORRECTNESS INVOLVING THE MATERIAL TO BE SAMPLED

The rebounding rule is respected when a certain number of conditions are respected. These involve either the fragments making up the stream or the cutter characteristics. This section deals with the first group of conditions. Section 18.4. will deal with the second group.

We have more or less implicitly assumed in the preceding section that :

1) the fragments fall one by one,
2) they do not spin,
3) they fall in the plane containing our figures, irrespective of the angle made by this plane with the cutter edges.

We shall now examine how the rebounding rule may be altered when one of these conditions is not respected and more specifically when :

- the fragments fall as a more or less compact stream,
- the fragments spin,
- the speed vector $\vec{V}_F$ does not belong to a vertical plane containing $\vec{V}_C$.

18.3.1. Fragments do not fall one by one :

When a fragment belonging to a stream of particles following more or less parallel trajectories collides with one of the cutter edges and bounces, its new trajectory is more than likely to cut that of other particles and a chain reaction of inter-particle collisions takes place, which alters the probability for a fragment to respect the rebounding rule. The stream reaction depends to some extent on whether the rebounding fragment is situated at the core or at the periphery of the stream. In the core, we may safely admit that there is a good statistical equivalence between the fragments escaping extraction whilst they should fall into the cutter and those eventually found in the increment whereas they should have fallen in the sampling reject.

On the contrary, at the periphery of the stream, the reaction can but be centrifugal, as a result of which, at the level of a single edge we observe a systematic alteration of the extraction probabilities of the peripheral fragments. But as the cutter is made of two edges playing symmetrical parts, the perturbation observed on the leading edge is statistically compensated by a symmetrical perturbation on the trailing edge. Such "statistical compensations" are accounted for by the term v_{f1} of the variogram of f(t) and the error resulting from these is included in the component QE_1 of the continuous selection error CE, component which takes into account all phenomena taking place at the scale of individual particles.

We shall therefore admit that the fact that fragments fall as a stream and not one by one does not introduce a new error and does not alter the extraction probability.

18.3.2. Fragments are liable to spin :

Spinning fragments carry energy that is not accounted for by the rule of the centre of gravity. But here again the phenomenon is symmetrical at the level of the two cutters and, the demonstration being the same as above, we shall conclude that the fact that fragments are liable to spin when they collide with one of the cutter edges does not introduce a new error and does not alter the extraction probability.

18.3.3. Fragments do not fall in a vertical plane containing the vector $\vec{V}_C$:

Fig. 18.1. assumes that the vectors $\vec{V}_F$ and $\vec{V}_C$ are both contained in the vertical plane of the figure, which is not necessarily true. When F collides with the leading edge C_L at a certain angle with this plane, it will still bounce on the side of Λ_L that contains its centre of gravity. The pitch angle made by the trajectory of F may therefore be regarded as irrelevant, at least in so far as the length of the cutter is sufficient to allow all particles bouncing towards the inner side of the cutter to be eventually recovered in the increment, which will be discussed in the next section.

18.3.4. Recapitulation :

For all practical purposes we may admit that the rebounding rule is not altered when the fragments fall as a stream, when they spin or when they do not fall in a vertical plane.

18.4. CONDITIONS OF EXTRACTION CORRECTNESS INVOLVING THE CUTTER CHARACTERISTICS

The following cutter characteristics have been found to be possibly relevant :
- straightness of the cutter edges,
- thickness of the cutter edges,
- inclination of the cutter edges,
- cutter width,
- cutter speed,
- cutter depth, capacity and general construction.

These points are studied one by one in sections 18.4.1.to 18.4.6.

18.4.1. Straightness of the cutter edges :

We have implicitly assumed so far that the cutter edges were straight and perpandicular to the motion. If their horizontal projection is not straight, but if the trailing and leading edges remain superposable by a translation (or a rotation when dealing with rotating devices), then the straightness of the cutter edges is irrelevant : there is an equivalent probability for a particle rebounding towards the inner side of the leading edge (assumed to be convex) to eventually fall outside of the cutter and for a particle rebounding towards the outer side of the trailing edge (assumed to be concave) to eventually fall inside the cutter.

In fact, straight edges are the rule and curvilinear edges the exception. In the only device we know with curvilinear edges (in horizontal projection) the edges were superposable by rotation and the extraction was correct. It is interesting to note that the superposability of the cutter edges in the course of the natural motion of the cutter is nothing else than the geometrical condition of delimitation correctness (see sections 17.2.1. and 17.2.4.).

18.4.2. Thickness of the cutter edges :

We have assumed so far that the cutter edges were immaterial lines without thickness. With new samplers, sharp edges are the rule with a few notable exceptions (such as that of a cross-stream cutter we saw at unloading harbour facilities sampling high tonnages of coarse iron ores where the edges were flat and about 50mm - two inches - thick). But sharp edges do not remain sharp for a long time, especially when they are submitted to repeated hammering and the question naturally arises as to whether there is any inconvenience attached to thick, flat or rounded off edges. The rebounding rule is obviously altered but in so far as the edges play symmetrical parts, theoretical considerations show and practical observation confirms that no extraction error is incurred as the result of the thickness of cutter edges. We can however surmise that the short-range quality fluctuation error QE_1 is slightly increased but this is taken into account by the variographic experiment and by the continuous selection model.

When, however, sharp edges round off through wear, it is always advisable to check that the original shape of the cutter opening has not been altered (see section 17.2.2.).

18.4.3. Inclination of the cutter edges :

Fig. 18.9. to 18.12. represent a few types of cutters in a schematic way.

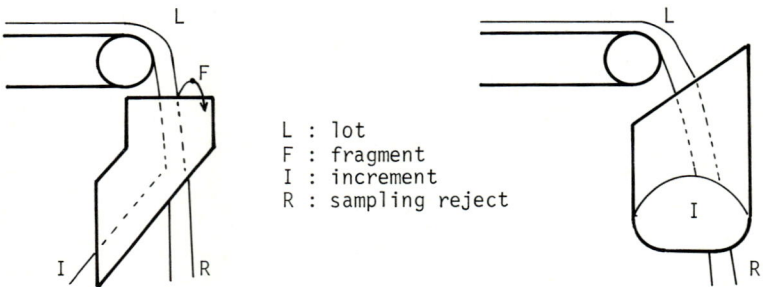

L : lot
F : fragment
I : increment
R : sampling reject

Fig. 18.9. Open discharge straight-path cutter with horizontal opening.   *Fig. 18.10. Bucket-type straight-path cutter with inclined opening.*

Fig. 18.9. shows for instance an open-discharge, straight-path cutter with horizontal edges. Such a construction performs a correct extraction on condition that the cutter be long enough (in a direction perpendicular to the motion) for a fragment such as F bouncing on one of the edges to eventually fall into the cutter if it belongs to the model increment. Too short a cutter or a cutter which is ill laid out can introduce an extraction error with a high probability for this error to be selective and affect coarse fragments more than fine ones.

Fig. 18.10. shows a bucket-type straight-path cutter with inclined edges practically perpendicular to the parabola of the stream. Such a construction is correc in so far as the length of the cutter edges is large enough (see overleaf), especi ly downwards.

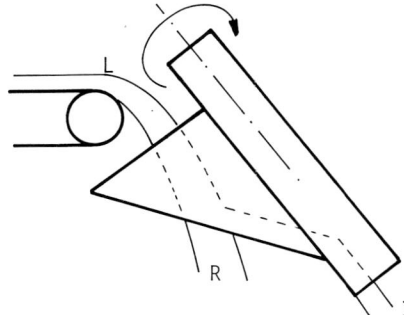 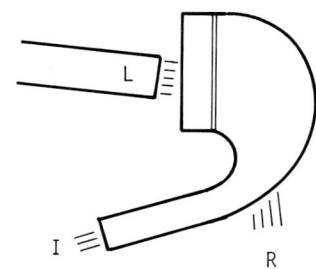

Fig. 18.11. Open-discharge circular path cutter with inclined shaft and inclined opening.

Fig. 18.12. Open-discharge straight-path pulp cutter with vertical edges.

Fig. 18.11. shows an open discharge circular path cutter with inclined shaft and inclined opening, assumed to be more or less perpandicular to the stream. The discussion is the same as with fig. 18.10.

Fig. 18.12. shows a typical open-discharge straight-path pulp cutter such as those to be found in most flotation plants for the sampling of feed, tailings and sometimes concentrates. Any particle, irrespective of its diameter, bouncing on the cutter edges is definitively lost for the increment : the rebounding rule is not respected. At the scale of a given particle, this defect amounts to a narrowing of the cutter opening. With a multi-size material, the narrowing effect varies with the particle diameter and the sampling ratio becomes a decreasing function of the particle size : the size analysis is systematically distorted towards the finer side. An extraction error and more specifically an extraction bias takes place and is the more important as the cutter width is the smaller and the size distribution the wider. This defect, often associated with too narrow cutter openings (see next section) is responsible for one of the most frequent sources of pulp sampling bias.

As a general conclusion of this section, the cutter edges should be :

1) set in such a way that the stream crosses the middle part of the area generated by the cutter edges during their motion,

2) practically perpandicular to the stream direction, with the exception of vertical cutters, sampling horizontal pulp streams which are bias generating,

3) long enough, in a direction perpandicular to the motion, to cover about three times the stream diameter or thickness,

As far as the stream is concerned, it should be sampled at a point of its trajectory where it is practically vertical.

18.4.4. Cutter width and velocity - qualitative theoretical approach :

18.4.4.1. Introduction : the effects of these two parameters cannot be dissociated as we did with the other factors. With the restrictions that will be developed in section 18.4.7., the cutter width W and velocity V_C can be regarded as irrelevant for all particles :

- falling directly into the sampling reject. These belong to the model reject and are likely to stay in the actual reject,

- falling directly into the cutter. These belong to the model increment and are likely to stay in the actual increment,

- bouncing on the leading or trailing edges towards the sampling reject. These belong to the model reject and are likely to stay in the actual reject.

The factors W and V_C become relevant only for those particles bouncing on the leading or trailing edges towards the inner side of the cutter. By definition, they belong to the model increment and there remains to find out under what conditions they will stay in the actual increment. It is intuitively obvious that if the cutter is too narrow or moves too fast, there is a non-negligible probability for a fragment bouncing for instance on the leading edge to fly over the trailing edge after a possible rebound on it and to end its course in the sampling reject. It is this possibility that we are now going to investigate, restricting our demonstration to those fragments bouncing towards the inner side of the cutter and assuming that air resistance and turbulence are both negligible. This hypothesis is justified by the fact that the problem of minimum cutter width and maximum cutter velocity is especially critical for coarse particles for which our hypothesis is acceptable.

18.4.4.2. Rebound on the leading edge :

Consider a spherical fragment F falling vertically on the leading edge C_L of a cutter C moving with a speed vector $\vec{V}_C$. Fig. 18.13. shows the trajectory of the centre of gravity F' of F in a vertical plane moving with the cutter C. This trajectory is made of two parabolic branches intersecting in F'_C, position of F' at collision time t_C. This curve is the envelope of the relative speed vector $\vec{V}'_{FC}$.

As far as the second branch of this curve is concerned, we shall call Z' the intersection of $\vec{V}'_{FC}$ with the surface of F and Z'_H the intersection of the trajectory of Z' with the horizontal plane H generated by the cutter edges. The point Z'_H is called "second critical point". We shall also call :

t_{F4} : in the chronology of F, time at which Z' reaches Z'_H ,

t_{T4} : in the chronology of the trailing edge C_T, time at which C_T reaches Z'_H.

Three possibilities may arise. They are illustrated in fig. 18.14.

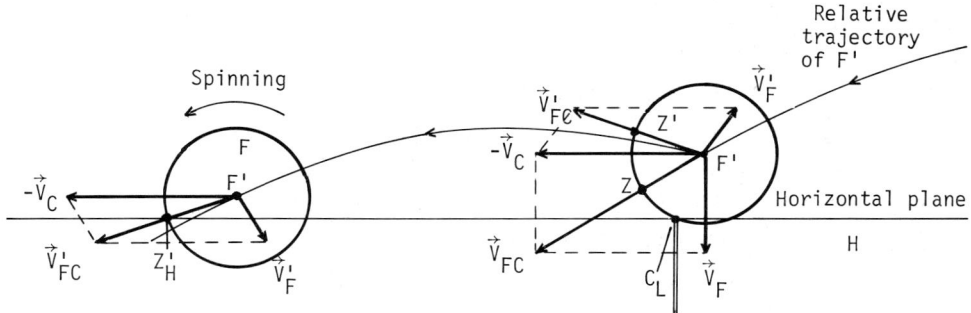

Fig. 18.13. *Trajectory of F' in a plane moving with the cutter C - Graphical definition of the critical point Z'_H*.

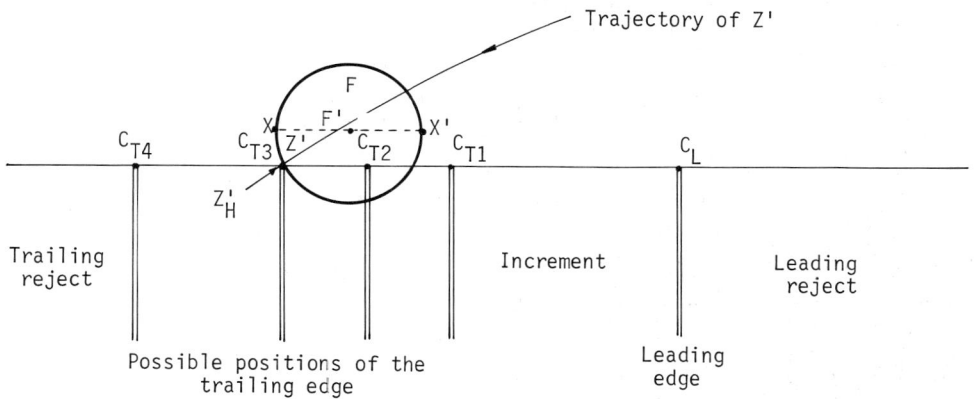

Fig. 18.14. *Trajectory of Z' - Respective positions of F and C_T at time t_{F4} when Z' is in Z'_H. Collision of the second type between F and C_T.*

1) $t_{F4} > t_{T4}$: C_T reaches the critical point Z'_H before Z'. Then, the fragment F either falls directly into the sampling reject (C_T in C_{T1}) or collides with C_T at some point of its surface between Z' and X' (C_T in C_{T2}) as a result of which it bounces towards the outer side of the cutter. In both cases F is lost for the increment to which it should belong. An extraction error takes place.

2) $t_{F4} = t_{T4}$: C_T and Z' reach the critical point Z'_H at the same time (C_T in C_{T3}). This is the <u>second critical situation</u>. If $t_{F4} = t_{T4} + \varepsilon$, F bounces towards the outer side of the cutter. If $t_{F4} = t_{T4} - \varepsilon$, F

bounces towards the inner side of the cutter (see below).

3) $t_{T4} > t_{F4}$: C_T reaches the critical point Z'_H after Z'. Then, F either falls directly into the cutter or collides with C_T at some point of its surface between X and Z' in such a way that it rebounds towards the inner side of C_T. In both cases, F eventually falls in the increment to which it belongs. No extraction error takes place.

For a given fragment F falling from a given level and for a cutter edge C_L moving with a given velocity V_C, there are as many trajectories of F' and Z' and as many critical points Z'_H as there are possible collision points between X and Z on the surface of F. We shall call Z'_{HF} the farthermost of all points Z'_H.

As the distance between Z'_{HF} and C_L is likely to increase as a function of the diameter d of F, for a given set of fragments the farthermost of all points Z'_{HF} corresponds to the coarsest particle (with diameter d_M) of the set. We shall call Z'_{HM} the corresponding critical point.

The distance between Z'_{HM} and C_L defines the "critical width" W_o of a cutter C moving at a speed V_C and sampling a material with maximum particle diameter d_M.

Then, we know that for given values of d_M and V_C :

- if $W \geq W_o$: all particles belonging to the model increment will eventually fall in the actual increment,

- if $W < W_o$: some of the particles belonging to the model increment will rebound on the leading edge and fly over the trailing edge towards the sampling reject and be lost for the actual increment. Thus, an extraction error takes place. Due to the fact that the rebounding range $Z'_{HF}C_L$ is an increasing function of d, this error will selectively affect the coarsest particles of the lot : in terms of extraction probability, we can state that :

- for all particles finer than a certain "critical diameter" d_o (function of the cutter width W), the extraction probability is equal to unity,

- for all particles coarser than d_o, the extraction probability is smaller than unity and is a decreasing function of the diameter d.

According to the definition of "correctness", the increment extraction is incorrect and we shall see in chapter 20 that the sampling is likely to be biased.

Now, when V_C increases, the trajectory of Z' obviously becomes flatter and the critical point Z'_H (for a given fragment) or its limit Z'_{HM} (for a given set of fragments with maximum diameter d_M) move further off C_L as a result of which the critical width W_o is an increasing function of V_C.

A last point should be taken into consideration : both the fragment F and the cutter C carry kinetic energy. Upon the collision between F and the leading edge C_L, part of this energy is dissipated in the form of heat, part of it is converted into the rebounding motion of F and part of it is transformed into spinning of F around its centre of gravity F'. If, as assumed in this section, F' lies on the left side of the critical surface Λ_L' (fig. 18.8.), then F rebounds towards the inner side of the cutter and the spinning is counterclockwise (arrow of fig. 18.13.). Suppose now that, after bouncing on the leading edge, F collides with the trailing edge at some time t_C' (with $t_{T4} > t_C' > t_{F4}$). F should fall into the cutter but if t_C' is near enough t_{F4}, the spinning energy of F will, upon this second collision, be partly converted into a movement of F' towards the upper left, enough for the fragment F to roll over the trailing edge and eventually fall in the trailing reject. Obviously, the larger the cutter velocity V_C, the coarser the maximum diameter d_M and the more important the spinning effect.

The analysis of a few high speed cinematographic recordings shows this spinning effect to be far from negligible with coarse fragments (say 50 mm and over).

The rolling over induced by the spinning of F (itself a consequence of the rebound of F on the leading edge) results in an increase of the critical cutter width W_o of the cutter.

18.4.4.3. Rebound on the trailing edge :

We assume in this section that F collides <u>directly</u> with the trailing edge C_T and rebounds towards the inner side of the cutter.

We can make exactly the same demonstration as in the preceding section, define a new critical point Z_H'' and a new critical width W_o' but due to the fact that the leading edge moves in the same direction as the rebounding fragment, the new critical width W_o' is always smaller and most often much smaller than W_o, as a result of which it is the rebound of F on the leading edge alone that governs the correctness of the cutter width and velocity.

Our purpose in the next sections is to disclose the relationship which must exist between the critical width W_o, the cutter velocity V_C and the diameter d_M of the coarsest particle in the material to be sampled.

18.4.4.4. Possibility of a quantitative theoretical approach :

A quantitative theoretical approach would involve a large number of factors such as the elasticity of F and C_L, the air resistance and turbulence and a few others that cannot be easily taken into consideration. Such an approach would be unlikely to provide formulas simple enough for everyday use. For this reason we have opted for an experimental approach. The conclusions of this experimental work are summarized in section 18.4.5. but as this work is relatively recent and as its results are very important, it will be described in detail in section 18.5.

18.4.5. Cutter width and velocity - Rules of extraction correctness :

We shall summarize the results of our experimental work and our recommendations as follows (see also section 18.5 and Gy - 1978) :

First rule of extraction correctness :

- for $d_M > 3$ mm : $W \geq W_o = 3\ d_M$
- for $d_M \leq 3$ mm : $W \geq W_o = 10$ mm

Second rule of extraction correctness :

Irrespective of d_M, with $W = n\ W_o$ (with $n \geq 1$) :

$V \leq V_{on} = (1 + n)\ 0.3$ m/s

Third rule of extraction correctness (economical optimum) :

The cheapest correct solution is defined by :

$W = W_o$ (irrespective of d_M)   and $V = V_o = 0.6$ m/s.

18.4.5.1. Critical cutter width W_o - Discussion :

For "coarse" materials with maximum particle diameter d_M (the definition of a coarse material being usually vague and left to the reader's imagination), there has always been a general agreement that the critical width should be proportional to d_M, whereas with "fine" materials and pulps the general idea has always been that, irrespective of the maximum particle diameter d_M, there should be an absolute minimum.

As far as coarse materials are concerned, the proportionality factor W_o/d_M varies from one author, one manufacturer or one standard to the next between 2 and 4 and the average value of 3 has been agreed upon by a majority. In one of our previous works (Gy, 1975) we proposed :

- for $d_M > 3$ mm : $W \geq W_o = 3\ d_M$

This rule was based on experiments carried out by Minemet-Industrie, Trappes, France, well known manufacturer of sampling equipment, and on various tests and observations made as a Consultant at user's facilities. Tests performed on cutters with W of the order of 2.5 and even 2.0 d_M have always shown the existence of a bias resulting from a distortion of the size analysis, the coarser size classes being partially scalped in the sample. On the contrary, test carried out on cutters with W larger than or equal to 3 d_M have never shown up any bias.

In 1977 we carried out a precise experiment whose results will be presented in section 18.5. (Gy and Marin, 1978). This experiment, involving 100 mm and 50 mm fragments (4" and 2") confirmed the rule : $W_o = 3\ d_M$.

As far as fine materials and pulps of finely ground minerals are concerned, the rule $W_o = 3\ d_M$ is obviously no longer valid : nobody would ever think of sampling a 100 mesh pulp (d_M = 0.16 mm) by means of a 0.5 mm cutter. But authors, manufacturers and standards vary as regards the absolute limit under which an extraction error is likely to take place, the proposed values ranging from 2 to 10 mm. In a previous work (Gy, 1975) we proposed :

- for $d_M \leq 3$ mm : $W \geq W_o = 10$ mm .

Dr. George Armstrong-Smith (1974) supports this recommendation by stating that in experiments carried out on the Zambian Copper Belt, the minimum width of cutter for taking pulp samples was found to be 9 mm. Just as an example of the risks incurred when utilizing too narrow cutters we shall mention the results of an experiment carried out on the feed to a lead and zinc flotation plant in Morocco (Gy, 1955). The feed was ground to about 0.16 mm (100 mesh). In two independent experiments, we recovered a certain number of increments by means of 4 mm and 8 mm cutters as well as the sampling rejects and obtained the following results by comparison of the lead contents of the samples and sampling rejects :

- <u>4 mm cutter</u> : increments :   2.64 % Pb
              rejects   :   2.75 % Pb
              difference : - 0.108 % Pb or 4 % in relative value.

A Student-Fisher test showed this difference to be significant at a 95 % probability level.

- <u>8 mm cutter</u> : increments :   3.22 % Pb
              rejects   :   3.22 % Pb
              difference : - 0.0003 % Pb or 0.01 % in relative value.

This practically zero difference was obviously non-significant.

Further experiments carried out on zinc and copper ores by means of two cutters (3 and 8 mm in the first experiment, 5 and 8 mm in the second) joined side by side and collecting simultaneous increments showed a systematic difference, significant at a 95 % probability level, which was interpreted as a bias introduced by the narrow cutter. The minimum critical width W_o = 10 mm that we recommend results from the above mentioned experiments and includes a safety factor.

Before closing this section, we must (sadly) observe that manufacturers, some of these very well known still design, build, advertize and eventually sell on a large scale sampling cutters that do not respect the first rule of extraction correctness and implementing cutter apertures such as :

- with coarse materials : $W = 2.5\ d_M$ and even $W = 2.0\ d_M$
- with fine materials   : W = 3 mm   and even W = 2 mm.

18.4.5.2. Critical cutter width W_o - Sampling very high volumes of pulp :

The minimum cutter width W_o = 10 mm proposed in the preceding section ceases to be valid when sampling high and very high rates of flow, say higher than 500 m³/h (2200 g.p.m.). In such cases, it is always advisable to take a serious safety factor, as well for W as for the other dimensions of the cutter. When designing a primary cutter for a 5000 m³/h (22,000 gpm), Minemet-Industrie rightly retained a minimum width W_o = 30 mm instead of 10 mm and multiplied by two at least the other dimensions of the cutter.

18.4.5.3. Critical cutter velocity V_o - Discussion :

Until recent years, authors on sampling used to be satisfied with recommendations as precise and useful as "speed should not be great enough to knock away pieces that should go into the sample". The recent ninth draft of a certain ISO recommendation states that "the cutting speed in relation to the speed of belt conveyor and flow rate of ore should be established so as not to introduce bias in the sample" but fails to specify this speed.

In so far as our literature survey is exhaustive, only one sampling equipment manufacturer (Minemet-Industrie, Trappes, France) has ever carried out experiments in order to define what should be the value of V_o and published the results obtained. Whereas other manufacturers go on implementing cutter speeds as high as 2 m/s (nearly 7 feet per second), Minemet established as early as 1964 that with coarse materials a bias was likely to take place with values of V_C higher than 0.5 m/s and adopted a safe V_o = 0.4 m/s. Many experiments carried out by clients showed that such a speed did not introduce any bias.

According to our experience as a Consultant and as the result of a little theoretical reasoning, we came to the conclusion that if W was actually larger than W_o, the speed V_C might be increased and we proposed (1975) the rule :

$$V_C = \frac{W}{W_o} V_o \quad \text{with} \quad V_o = 0.4 \text{ m/s}$$

But we felt that these rules needed to be backed up by undisputable experimental results. This experiment, which had been planned for years and postponed several times, was eventually carried out in June 1977 and its results published very recently (Gy and Marin, 1978). Since the conclusions of these experiments condemn the habits of most manufacturers, showing in terms of extraction probability the dangers of too narrow or too fast cutters, since furthermore our publication may have escaped the reader's attention, we found it appropriate to give a complete relation of this experiment in section 18.5. We are however conscious of the fact that this is but a preliminary work and that a small part only of our program has been realized. A number of complementary tests would be necessary to cove the whole subject. Until the results of such tests are available, we shall retain the conclusions of our latest experiment which are known to be absolutely safe :

$$V_C \leq (1 + \frac{W}{W_o})\frac{V_o}{2} \text{ with } V_o = 0.6 \text{ m/s} \quad \text{or} \quad V_C \leq V_{on} = (1 + n)\, 0.3 \text{ m/s with } n = \frac{W}{W_o}$$

Table 18.1. shows how this new rule compares with the rule proposed in our previous work (1975)

Table 18.1. Critical cutter speed V_{on} for $W = n\, W_o$ and $n \leq 1$.

$n = \dfrac{W}{W_o}$	Former rule (1975) $V_{on} = 0.4\, n$	New rule (1978) $V_{on} = 0.3\,(1 + n)$
1	0.4 m/s	0.6 m/s
2	0.8 m/s	0.9 m/s
3	1.2 m/s	1.2 m/s
4	1.6 m/s	1.5 m/s

We must, sadly again, observe that manufacturers, some of these very well known, still design, build, advertize and eventually sell on a large scale samplers that do not respect the second rule of extraction correctness and implementing for instance (sampler working on a rate of flow of 5000 t/h of coarse iron ore at loading facilities) :

- $W = 2\, d_M$ (non-respect of the first rule)
- $V_C = 2.1$ m/s (non-respect of the second rule)

This is but an example out of dozens of others. As the samples were used for commercial purposes, the users were quick to suspect an important bias which was proved beyond any possible doubt. But the transformation of the primary samplers showed how costly it was not to think of extraction correctness at the designing stage of the sampling plant.

Some national and international standards support such dangerous habits by stating for instance (emphasis added) : "in most cases it is desirable that the cutter speed is not higher than 1.5 m/s", obviously leaving open the possibility for this speed to exceed 1.5 m/s. All this but underlines the deliberate will of a number of manufacturers and standards organizations to ignore the most elementary rules and experimental evidence.

18.4.5.4. Cutter width and velocity - Economical optimum :

If μ_o is the average rate of flow and M_I the average increment weight, then :

$$M_I = \mu_o \frac{W}{V_C}$$

The larger the width W, the smaller the velocity V_C on the one hand, the heavier the increments and the costlier the sample reduction on the other. An economical optimum is therefore attained when the actual cutter width is equal to W_o and when the actual cutter speed is equal to V_o, hence the third rule stating that the economical optimum is defined by :

$$W = W_o \quad \text{and} \quad V_C = V_{o1} = V_o = 0.6 \text{ m/s.}$$

When observing this simple rule, a correct extraction is achieved and an unbiased sample at the lowest possible cost is obtained. We shall see in section 21.5.2. that this economical optimum corresponds also to the minimizing of the grouping and segregation error GE.

18.4.6. Increment integrity :

When the conditions of extraction correctness reviewed in the preceding sections are simultaneously fulfilled, every particle actually goes where it belongs according to the model - increment or sampling reject - and no extraction error EE is involved. This assumes however that the increment integrity is respected or in other words that all particles entering the cutter remain in the increment and that all particles joining the sampling reject remain in the reject. This condition splits up into two sub-conditions :

- avoidance of loss of particles belonging to the increment,
- avoidance of contamination of the increment by foreign material,

which, generally speaking, involve the cutter depth, capacity and general lay-out.

18.4.6.1. Loss of particles belonging to the increment :

A particle belonging to the increment is likely to remain in the increment unless it encounters an obstacle on which it will bounce towards the outer part of the cutter or overflow. We shall distinguish open-discharge (or chute) cutters and bucket-type cutters.

1) <u>Open-discharge or chute cutters</u> : the following rules should be observed in order that nothing could prevent the material extracted from the stream by the cutter from flowing freely towards its open discharge. At no moment should the material be allowed to accumulate on any part of the inner cutter walls or bottom. This requests :

- a minimum depth equal to about three times the maximum diameter d_M at the shallowest point of the cutter, with an absolute minimum of 0.1 m.

- a stainless steel construction with a U-shaped inclined trough with no weldings, rivets, nuts, bolts and no sharp foldings,

- a slope of at least 45° with dry materials and 60° with wet non-sticky materials,

- when sampling sticky materials such as certain ores, filter cakes etc... design oversize cutters with slopes larger than 60° and heated walls and bottom (in order to lessen water surface tension, not to dry the material, in fact, the evaporation is so small that heated cutters may be used even for moisture sampling),

- when sampling materials containing dry fines, use screw feeders to feed the sampler and rotary vane feeders down-stream in order to prevent the dedusting effect of ascending air-draught.

2) Bucket-type cutters :

- the bucket capacity should be at least twice and better three times as large as the volume of the largest possible increment, so as to prevent the last fragments entering the bucket from bouncing outside or overflowing,

- same construction as recommended for open-discharge cutters,

- the discharge system should be efficient, whether involving bottom opening or bucket tilting, and should allow no particle to remain in the bucket after discharge,

- when sampling sticky materials, particular care must be taken : the side walls should be inclined rather than vertical and heated whenever necessary.

18.4.6.2. Increment contamination :

No foreign material should be allowed to contaminate the increments. The main sources of contamination and the corresponding remedies are :

- dust : emission of dust in the sampling plant should be carefully controlled but never forget the remarks made in section 17.4.5. on the disadvantages of too efficient dust collecting systems. In its idle positions, the cutter should be covered by a protecting cap,

- abrasion and corrosion : ordinary steel should be avoided in the construction of sampling cutters in order to prevent rust from contaminating the increments. Stainless steel is suitable with most materials. With very abrasive or corrosive materials, tailor-made solutions must be specifically adapted,

- materials from a previous operation : if the cutter is so made as to prevent material from remaining in the cutter (especially bucket-type) after the taking of an increment, this source of contamination does not exist. Nor is it likely to be error generating in every day practice when sampling the same material in a routine way. This point may however arise in laboratories , pilot plants or in custom sampling plants liable to handle materials of different origins with different compositions (see also chapter 27).

18.5. CUTTER WIDTH AND VELOCITY - EXPERIMENTAL DETERMINATION OF CRITICAL VALUES

We summarized the conclusions of this experiment in section 18.4.5. but we find it advisable to provide the reader with all available experimental evidence, for him to decide whether the proposed rules are scientifically backed up, in a domain that remains controversial.

18.5.1. Introduction and notations :

This experiment was carried out in 1977 (see section 18.5.11) and its results have only recently been published (Gy and Marin, 1978). It consisted of 53 tests carried out on coarse calibrated fragments. We shall call :

L : the lot of calibrated fragments involved in a given test,
F_i : one of these fragments
N_L : the number of fragments making up the lot L : i = 1, 2, ... N_L.
W : cutter width (rectangular opening, correct for a straight path sampler). W in
V_C : cutter speed (assumed to be uniform during the cutter travel across the stream which was checked and confirmed). V_C in meters per second (m/s).
B : length of belt occupied by the N_L fragments. These are manually and uniformly deposited on the lower end of the stopped belt. B is expressed in meters (m).
D : belt loading rate or number of fragments per unit length. It is assumed to be uniform along the loaded part of the belt. By definition :

$D = N_L/B$ with D expressed in number of fragments per meter (fr/m).

V_B : belt speed (assumed to be uniform during the discharge of the load, which was checked and confirmed). V_B in (m/s).
μ : rate of flow of material to be sampled

$\mu = V_B D = V_B N_L/B$ with μ expressed in number of fragments per second (fr/s)

As both V_B and D are assumed to be constant, μ should be uniform too.
N_M : number of fragments in the model increment. By definition :

$N_M = \mu W/V_C = N_L W V_B/V_C B$

As it results from calculations, N_M is usually not an integer.
P_i : extraction probability of fragment F_i.
N_A : number of fragments in the actual increment. It is always an integer. N_A is a random variable with mean (or expected value) $m(N_A)$:

$m(N_A) = \sum_i P_i$ with i = 1, 2, .. N_L

τ : extraction ratio defined as follows :

$\tau = N_A/N_M$ it is a random variable with mean $m(\tau)$:

$m(\tau) = m(N_A)/N_M = \sum_i P_i/N_M$

18.5.2. Choice of an experimental method :

Our method is based on the following observation :

If the extraction is correct one can easily show that :

$$\sum_i P_i = N_M \qquad \text{wich involves : } m(\tau) = 1$$

If the extraction is incorrect, the extraction probability of fragments colliding with the cutter edges (and more specifically the leading edge) is significantly reduced with the consequence that :

$$\sum_i P_i < N_M \qquad \text{which involves : } m(\tau) < 1$$

Since N_M can easily be calculated and N_A enumerated, we can easily calculate the value of τ in each test and its experimental mean $\bar{\tau}$ when several tests are realized in the same experimental conditions. If this mean is significantly different from unity, we have an experimental proof that the extraction is incorrect. On the contrary, if this mean is not significantly different from unity, there is a serious presumption that the extraction is correct.

We shall use this property to estimate the critical values of the parameters. We shall call :

d_M : the maximum particle diameter. It is defined as the opening of the upper calibrating screen of the material to be sampled.

d_o : the critical particle diameter for a cutter with width W and speed V_C :
  - for $d_M \leq d_o$: $m(\tau) = 1$ (correct extraction)
  - for $d_M > d_o$: $m(\tau) < 1$ (incorrect extraction)

W_o : the critical width of a cutter sampling a material with maximum diameter d_M :
  - for $W \geq W_o$: $m(\tau) = 1$ (correct extraction)
  - for $W < W_o$: $m(\tau) < 1$ (incorrect extraction)

V_{on} : the critical speed of a cutter with a width $W = n W_o$ (with $n \geq 1$)
  - for $V_C \leq V_{on}$: $m(\tau) = 1$ (correct extraction)
  - for $V_C > V_{on}$: $m(\tau) < 1$ (incorrect extraction)

Our experiment consists in carrying out a certain number of tests for known values of d_M, W and V_C, in counting the number N_A of fragments in the increment actually recovered, in calculating the number N_M of fragments in the model increment and the extraction ratio τ and in plotting the results on various graphs representing τ as a function of either W/d_M or V_C. Both sides of $\tau = 1$ we define a ± 0.04 confidence interval corresponding to the natural fluctuations of τ due to various measurement errors. As soon as the observed value of τ falls below the lower limit (0.96) of the confidence interval, we can conclude that we have just crossed the critical value of the parameter plotted on the graph.

Among various possible methods, we chose to work on two different calibrated materials (100 - 80 mm or $d_M = 0.1$ m and 50 - 40 mm or $d_M = 0.05$ m) and to implement a sampler with adjustable cutter width (between 0.05 and 0.8 m) and adjustable cutter speed (between 0.1 and 2.0 m/s).

For the <u>first series of tests</u> which was meant to assess the critical value of the ratio W/d_M, we selected a value of V_C known to be smaller than or equal to the critical speed V_o, and tested various values of W/d_M (4.0 , 3.5 , 3.0 , 2.5 and 2.0

For the <u>second series of tests</u> which was meant to assess the critical value of the speed V_C when the actual cutter width $W = W_o = 3\ d_M$, we tested increasing values of V_C (from 0.2 to 2 m/s).

For the <u>third series of tests</u> which was meant to assess the critical value of the speed V_C when the actual cutter width is $W = n\ W_o$ with $n = 2, 3, 4$, we tested also increasing values of V_C (from 0.6 to 2.0 m/s).

18.5.3. General lay-out of the testing plant :

The testing plant which was erected for this experiment (see section 18.5.7.) is schematically represented in fig. 18.15.

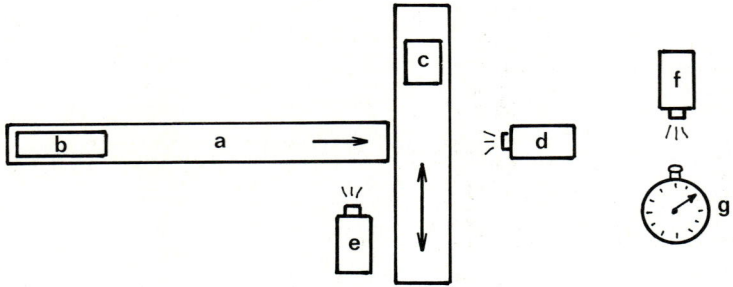

Fig. 18.15. *Critical cutter width and speed - Lay-out of the testing plant.*

This plant consisted of :

a) - horizontal conveyor belt : length 19 m, width 0.8 m, average speed 2.25 m/s,
b) - lot to be sampled deposited by hand on the stopped belt
c) - Minemet EAC sampler (prototype constructed for this experiment) equiped with an adjustable width cutter and a very precise (± 1 %) electronic speed regulator. The automotive carriage moves on a 7 m long track and is driven by means of a rack and pinion system. It is automatically switched off and stopped.
d) - Canon 814 movie camera recording the cutter during the crossing of the stream for measurement of V_C and check of its uniformity,
e) - Canon 814 movie camera recording the belt during the discharge of the lot, for measurement of V_B and check of its uniformity,

f) same cameras recording the hand of a 1/100 s chronometer (calibration of the actual speed of the cameras, expressed in frames per second - f / s - and check of its uniformity).

18.5.4. Critical value of the ratio W/d_M :

The results of our first series of tests are represented in fig. 18.16

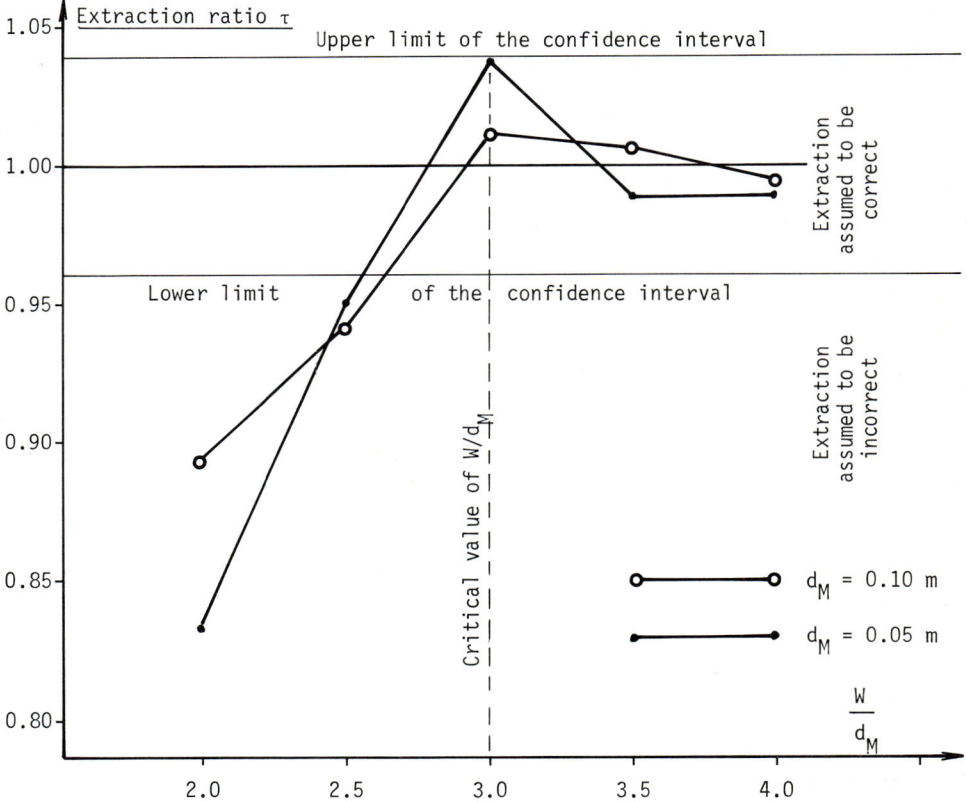

Fig. 18.16. Critical value of the ratio W/d_M.

For both tested materials (d_M = 0.1 m and d_M = 0.05 m) we selected values of the ratio W/d_M equal to 4.0 , 3.5 , 3.0 , 2.5 and 2.0. Fig. 18.16 shows that both curves fall rapidly below the lower limit 0.96 of the confidence interval of the sampling ratio τ as soon as W falls below 3.0 d_M. Our first conclusion is therefore that, at least with fragments as coarse as 100 or 50 mm, the critical value W_o of W is :

$$W \geq W_o = 3 \, d_M$$

In other words, for a cutter with a given width W, the critical particle diameter is :

$$d \leq d_o = W/3$$

18.5.5. Critical cutter speed V_{o1} when $W = W_o = 3\ d_M$:

The results of our second series of tests are represented in fig. 18.17.

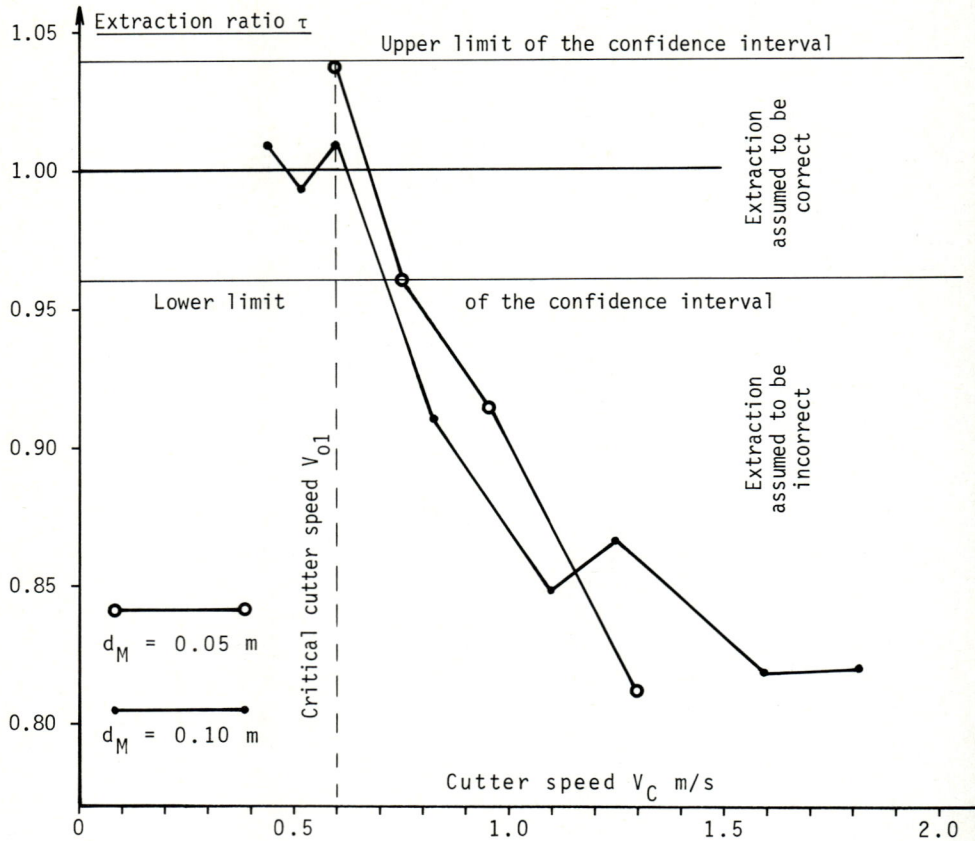

Fig. 18.17. *Critical cutter speed V_{o1} when $W = W_o = 3\ d_M$.*

Both curves show that the extraction ratio falls dangerously below the lower limit (0.96) of the confidence interval as soon as the speed exceeds 0.6 m/s.

With the coarser material (100 mm) it falls down to 0.85 (highly significant) for a speed of 1.0 m/s. With the 50 mm material, the ratio falls down to 0.81 for a speed of 1.25 m/s. We must therefore conclude that the critical speed is 0.6 m/s at least for materials such as those tested in this experiment and for actual cutter width W equal to its critical value W_o:

When $W = W_o = 3\ d_M$: $V_C \lesssim V_{o1} = 0.6$ m/s

18.5.6. Critical cutter speed V_{on} when $W = n\ W_o = 3\ n\ d_M$:

In our third series of tests, we experimented the values of n = 2, n = 3 and n = 4 and speeds ranging from 0.6 m/s to 2.0 m/s. The results of this third series of tests are represented in fig. 18.18.

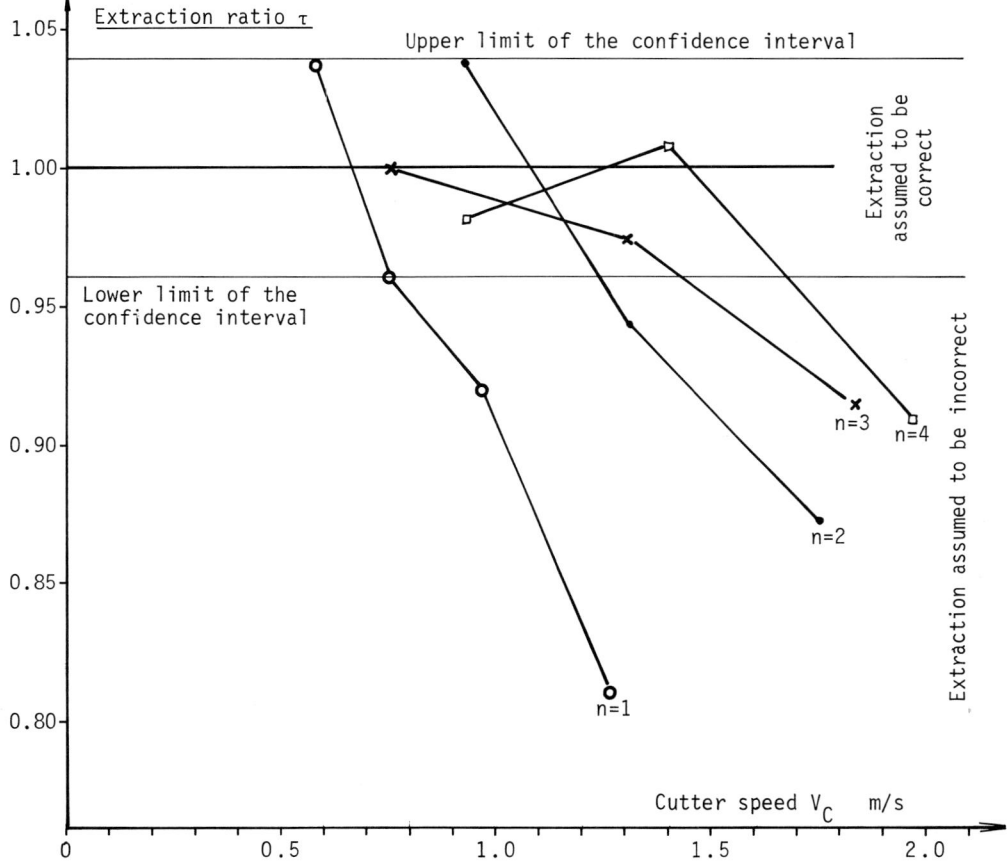

Fig. 18.18. *Critical cutter speed* V_{on} *when* $W = n W_o = 3 n d_M$.

Obviously, a much larger number of tests would have been necessary to fully investigate this problem and to ascertain unquestionable relationships. Qualitatively, however, the tendency is obvious : there is a critical speed and it increases with the value of n. The four curves cross the lower limit of the confidence interval for the following speeds :

- n = 1 : V_C = 0.72 m/s
- n = 2 : V_C = 1.20 m/s
- n = 3 : V_C = 1.45 m/s
- n = 4 : V_C = 1.70 m/s

But we know that the critical value must be lower than these values. For this reason and in order to propose a simple rule, we have retained the following cri-

tical speeds :
- $n = 1 : V_{o1} = 0.6$ m/s
- $n = 2 : V_{o2} = 0.9$ m/s $= 1.5\ V_{o1}$
- $n = 3 : V_{o3} = 1.2$ m/s $= 2.0\ V_{o1}$
- $n = 4 : V_{o4} = 1.5$ m/s $= 2.5\ V_{o1}$

and the following simple relationship :

When $W = n\ W_o = 3\ n\ d_M$: $V_C \leqq V_{on} = \dfrac{1 + n}{2} V_{o1} = (1 + n)\ 0.3$ m/s with $n \geqq 1$

18.5.7. Acknowledgments :

The experiment reported in this section had been planned for years and postponed several times until it was eventually carried out in June 1977. The construction of a prototype sampler using an electronic speed regulator and the erection of the testing plant were sponsored and financed by Minemet-Industrie, 1 Avenue Albert Einst 78190 TRAPPES, FRANCE, manufacturers of sampling equipment and pioneers of the development of scientific sampling at industrial and laboratory scale.

General Superintendance Company (GSC), 1 place des Alpes, Geneva, Switzerland, was also interested in the same subject and had offered its assistance. Our tests were therefore carried out at Fos-sur-Mer (France) near the sampling and assaying facilities of the French Subsidiary of GSC, Société Générale de Surveillance.

We wish to thank the Managers of both Companies for the opportunity given to carry out this important experiment.

18.6. RECAPITULATION

The rules of correct extraction can be recapitulated as follows :

1) <u>Condition concerning the stream</u> : the stream should be sampled at a point where its trajectory is practically vertical,

2) <u>Condition concerning the cutter edges</u> : the cutter edges should be perpendicular to the stream, i.e. horizontal,

3) <u>Condition concerning the sampler lay-out</u> : the cutter should be set in such a way that the stream crosses the middle part of the area generated by the edges during their travel,

4) <u>Condition concerning the cutter width</u> : with a maximum particle size d_M larger than or equal to 3 mm, the cutter width should be larger than or equal to three times the diameter d_M. With a maximum particle size smaller than 3 mm, the cutter width should be larger than 10 mm, irrespective of the particle size,

5) <u>Condition concerning the cutter speed</u> : the cutter speed should not exceed $V_{on} = (1 + n)\ 0.3$ m/s if the actual cutter width $W = n\ W_o = 3\ n\ d_M$.

6) <u>Economical optimum</u> : it is defined by $W = W_o$ and $V_C = 0.6$ m/s.

7) <u>Condition concerning the cutter design</u> : the cutter should be so designed as to prevent any particle entering the cutter from bouncing out or overflowing.

18.7. COST OF EXTRACTION CORRECTNESS

Here again, we would like to point out the fact that sampling devices achieving a correct extraction are in no way more expensive than incorrect samplers but manufacturers and users are too often tempted to transgress the rules of extraction correctness recapitulated in the preceding section, at least at the primary sampling stage of facilities dealing with large tonnages of coarse materials, in order to reduce the primary sample weight and the cost of sample reduction. This amounts to putting bias on the same footing as random error which is the most dangerous mistake a designer of sampling equipment or sampling plants is likely to make.

When sampling a flowing stream, there are two ways of reducing the primary sample weight M_S which can be written :

$M_S = Q\, M_I$ with Q number of increments and M_I average increment weight.

- by reducing the number Q of increments,
- by reducing the average increment weight M_I.

1) <u>Reduction of the number Q of increments</u> : this can be achieved by increasing the free parameter T_{sy} or T_{st} of the selection scheme (see chapter 7). So doing, the variance $\sigma^2(CE)$ increases but the sampling remains unbiased (chapter 8).

2) <u>Reduction of the average increment weight M_I</u> : this can be achieved by increasing the cutter speed V_C and by reducing the cutter width W. But we have shown in this chapter that there are limits beyond which the extraction ceases to be correct. When these limits are transgressed, a bias is likely to take place. This leads to the notion of "<u>minimum increment weight</u>" M_{Io} with :

$M_{Io} = \mu_o W_o / V_{o1}$

This point will be further developed in section 33.5.

The solution to the designer's dilemma therefore consists in :
a) using a sampler corresponding to the economical optimum (see section 18.6.)
b) reducing the number Q of increments to the minimum Q_o compatible with the reproducibility requirements. Q_o can be calculated by means of the results of chapter 8 illustrated in chapter 15).

The minimum sample weight M_{So} is therefore :

$M_{So} = Q_o M_{Io} = Q_o \mu_o W_o / V_{o1}$

The cost of primary sample reduction is an increasing function of M_S but is not proportional to it. If for some reason it is necessary to reduce M_S below M_{So}, it is much less detrimental to reduce Q below Q_o than to reduce M_I below M_{Io}. Reducing Q increases the variance of the random errors QE_2, QE_3 and WE but does not affect the sampling accuracy. Reducing M_I below M_{Io} by reducing W below W_o and/or by increasing V above V_o introduces a bias that cannot be controlled and increases in the same time the sampling variance. Here again we must conclude that when they fail to respect rules as simple as those reviewed in this chapter, designers and manufacturers have no excuse but ignorance.

18.8. CHANCES OF OBTAINING UNBIASED SAMPLES THROUGH INCORRECT EXTRACTION

Reasoning as we did in section 17.8. we shall state that the correlation $Cor(P_i/a_i, M_i)$ of a fragment F_i can be regarded as the logical product of two elementary correlations :

1 - $Cor(P_i/d_i)$: between selection probability P_i and diameter d_i of fragment F_i.

2 - $Cor(d_i/a_i, M_i)$: between diameter d_i on the one hand, mass M_i and mineralogical composition a_i on the other. The correlation between d_i and M_i is a relationship $M_i = f_i \, d_i^3 \, \lambda_i$ with f_i a shape factor and λ_i the density of F_i.

Symbolically, we can write :

$$Cor(P_i/a_i, M_i) = Cor(P_i/d_i) \times Cor(d_i/a_i, M_i)$$

To cancel the product, it is necessary and sufficient to cancel at least one of the two factors. We can therefore think of two mathematical solutions :

1 - $Cor(P_i/d_i) = 0$: this equation is satisfied if and only if the extraction is correct. We know the rules to be respected to achieve extraction correctness.

2 - $Cor(d_i/a_i, M_i) = 0$: this correlation is partly a mathematical relationship and we are never entitled to assume that it is zero.

We therefore reach the conclusion that the only way to obtain unbiased samples is to achieve a correct increment extraction by respecting the rules reviewed in the present chapter.

With the delimitation error dealt with in chapter 17 it was always possible to give free rein to wishful thinking by assuming some form of distribution homogeneity throughout the lot or the stream cross-section, however dubious it may be, but with the extraction error, even wishful thinking is at a loss to provide an argument tending to prove that an incorrect extraction can deliver unbiased samples.

Of all sampling errors the increment extraction is in the same time the most dangerous and the most wide-spread.

FOURTH PART

DISCRETE MODEL OF THE INCREMENT SAMPLING PROCESS

The discrete model has been defined in section 5.4. According to this definition, the discrete model is applicable to any particulate material, irrespective of :

- its origin : mineral, vegetable, animal, synthetic or mixed materials such as town refuses,

- the interstitial fluid : air, water or any gas or liquid, this liquid being assumed to be a passive component of the material to be sampled,

- the critical component taken into consideration : mineralogical components, size fractions or moisture of a solid. The discrete model is not applicable, however, to the case when the critical content is the proportion of solids in a pulp.

We shall restrict our demonstrations to a simple, concrete problem : that of a particulate solid of mineral origin, the critical content taken into consideration remaining undefined.

When several components are regarded as critical, for instance with a crushed porphyry ore :

- copper content,
- molybdenum and gold contents,
- moisture content,
- percentage of + 10 mm,

the problem must be independently solved for each of these. If we are estimating the weight of a multipurpose sample, the most exacting of all solutions must be retained.

The discrete model has been specifically devised in order to analyse the phenomena taking place at the scale of individual particles or of small groups of particles.

Whereas the continuous model corresponds to the observation of the material to be sampled through a wide-angle lens, the discrete model corresponds to the observation through a magnifying lens bringing into light every detail of the particulate structure.

This fourth part is made of the following chapters :

Chapter 19 : Heterogeneity of a discrete set
Chapter 20 : Development of the discrete selection model
Chapter 21 : Linking up of the continuous and discrete models -
Fundamental error FE - Grouping and segregation error GE
Chapter 22 : Practical implementation of the theoretical results - correct selecti
Chapter 23 : Practical implementation of the theoretical results - incorrect selec

CHAPTER 19

HETEROGENEITY OF A DISCRETE SET

19.1. INTRODUCTION AND NOTATIONS

According to the definition given in section 5.4.1., the lot L is regarded in the discrete perspective as a set of discrete units U_m such as :

- the set L_F of fragments F_i ,
- the set L_G of groups G_n of fragments.

These groups G_n will be for instance the fragmental increments already defined (chapter 16). We shall assume that all groups G_n have about the same magnitude.
We shall use the following notations :

L   : lot of particulate material : M_L weight of L, A_L weight of critical component in L, a_L critical content of L.

U_m : a non-specified unit belonging to L : M_m weight of U_m, A_m weight of critical component in U_m, a_m critical content of U_m (m = 1, 2, ... N_U).

F_i : a fragment belonging to L. M_i weight of F_i, A_i weight of critical component in F_i, a_i critical content of F_i (i = 1, 2, ... N_F).

G_n : a group of fragments belonging to L. M_n weight of G_n, A_n weight of critical component in G_n, a_n critical content of G_n (n = 1, 2, ... N_G). We shall assume that the groups G_n are complementary.

F_{nj} : a fragment belonging to G_n. M_{nj} weight of F_{nj}, A_{nj} weight of critical component in F_{nj}, a_{nj} critical content of F_{nj} (j = 1, 2, ... N_n).

The following relationships are obvious :

$$M_L = \sum_n M_n = \sum_n \sum_j M_{nj} = \sum_i M_i \qquad A_L = \sum_n A_n = \sum_n \sum_j A_{nj} = \sum_i A_i$$

$$a_L = A_L/M_L = \sum_n A_n / \sum_n M_n = \sum_n \sum_j A_{nj} / \sum_n \sum_j M_{nj} = \sum_i A_i / \sum_i M_i$$

$$N_F = \sum_n N_n$$

We shall also define :

$F_{\bar{i}}$: average particle of L. $M_{\bar{i}}$ weight of $F_{\bar{i}}$, $A_{\bar{i}}$ weight of critical component in $F_{\bar{i}}$, $a_{\bar{i}}$ critical content of $F_{\bar{i}}$. By definition of the average particle :

$$M_{\bar{i}} = M_L/N_F \qquad A_{\bar{i}} = A_L/N_F \qquad a_{\bar{i}} = A_{\bar{i}}/M_{\bar{i}} = A_L/M_L = a_L$$

The average fragment of L has the same critical content as L.

$F_{n\bar{j}}$: average particle of G_n. $M_{n\bar{j}}$ weight of $F_{n\bar{j}}$, $A_{n\bar{j}}$ weight of critical component in $F_{n\bar{j}}$, $a_{n\bar{j}}$ critical content of $F_{n\bar{j}}$. By definition of the average particle :

$$M_{n\bar{j}} = M_n/N_n \qquad A_{n\bar{j}} = A_n/N_n \qquad a_{n\bar{j}} = A_{n\bar{j}}/M_{n\bar{j}} = A_n/M_n = a_n$$

The average fragment of G_n has the same critical content as G_n.

$G_{\bar{n}}$: average group G_n of L. $M_{\bar{n}}$ weight of $G_{\bar{n}}$, $A_{\bar{n}}$ weight of critical component in $G_{\bar{n}}$, $a_{\bar{n}}$ critical content of $G_{\bar{n}}$. By definition :

$$M_{\bar{n}} = M_L/N_G \qquad A_{\bar{n}} = A_L/N_G \qquad a_{\bar{n}} = A_{\bar{n}}/M_{\bar{n}} = A_L/M_L = a_L$$

The average group of L has the same critical content as the lot L.

As far as the sampling of L by selection of a certain number of units U_m (regarded as undissociable) is concerned, each unit U_m is described in a necessary and sufficient way by any two of the three parameters M_m, A_m and a_m. We shall retain M_m and a_m, knowing that $A_m = a_m M_m$.

The purpose of this chapter is to analyse the notion of heterogeneity of the sets L_F and L_G and to quantify this heterogeneity in terms of the parameters $a_{\bar{i}}$, $M_{\bar{i}}$ and a_n, M_n.

19.2. DEFINITION AND PROPERTIES OF HOMOGENEOUS AND HETEROGENEOUS DISCRETE MATERIAL

The notions of "homogeneity" and "heterogeneity" are necessarily associated to a given critical component. A lot of particulate material is said to be "homogeneo when the corresponding critical content is uniform for all units U_m making up the lot. It is said to be "heterogeneous" when this condition is not fulfilled. A gi material can be homogeneous as regards a certain component and heterogeneous as re gards another one. A pure mineral is homogeneous as regards its mineralogical composition but usually heterogeneous as regards its size distribution. On the contra ry, a complex sulphide ore ground to liberation size and closely calibrated between two consecutive screens of a same series is homogeneous as regards its size distribution and heterogeneous as regards its mineralogical composition.

When a material is homogeneous : $a_m \equiv a_L$.

Any sample made of whole units has a critical content a_S identical to a_L. The sampling of a homogeneous material is an exact process.

On the contrary, units extracted from a lot of heterogeneous material usually have different critical contents with the consequence that the critical content a_S of a sample S made of whole units is usually different from that of the lot L. The sampling of a heterogeneous material is therefore an error-generating process. Her the conclusion that sampling errors always result from one form or another of heterogeneity and the necessity of analysing this notion in a study dedicated to the theory of sampling.

19.3. CHARACTERIZING THE HETEROGENEITY OF A DISCRETE SET

19.3.1. Heterogeneity carried by a particle :

Our purpose in this section is to find out a mathematical definition of heterogeneity and our first step will be to see how we could define the heterogeneity h_i carried by a particle F_i.

Since a homogeneous material is defined by $a_i \equiv a_L$, irrespective of i, it seems natural to admit that h_i should be proportional to $(a_i - a_L)$. Now, at the scale of the lot L, the perturbation associated to the heterogeneity of F_i is obviously an increasing function of its weight M_i and we shall tentatively retain that h_i should be proportional to $(a_i - a_L) M_i$. This quantity is a weight of critical component. But we found it convenient, so far, to use relative, dimensionless characteristics (see variograms, errors, etc..) rather than absolute ones. This will be achieved if h_i is defined as the ratio of two weights of critical component and for instance the ratio of $(a_i - a_L) M_i$ to the weight $A_{\bar{i}}$ of critical component in the average particle $F_{\bar{i}}$ of L. We shall therefore define the "heterogeneity carried by the particle F_i " as the ratio :

$$h_i = \frac{(a_i - a_L) M_i}{a_{\bar{i}} M_{\bar{i}}} = \frac{N_F}{a_L M_L} (a_i - a_L) M_i$$

For the particle F_{nj} of G_n, we have in the same way :

$$h_{nj} = \frac{(a_{nj} - a_L) M_{nj}}{a_{\bar{i}} M_{\bar{i}}} = \frac{N_F}{a_L M_L} (a_{nj} - a_L) M_{nj}$$

In the same way again, we can define the heterogeneity carried by the average particle $F_{n\bar{j}}$ of the group G_n :

$$h_{n\bar{j}} = \frac{(a_{n\bar{j}} - a_L) M_{n\bar{j}}}{a_{\bar{i}} M_{\bar{i}}} = \frac{N_F}{a_L M_L} (a_{n\bar{j}} - a_L) M_{n\bar{j}} = \frac{N_F}{a_L M_L} (a_n - a_L) \frac{M_n}{N_n} = \frac{1}{N_n} \sum_j h_{nj}$$

According to its definition, the heterogeneity carried by a particle can be averaged, which is not true of contents for instance : the heterogeneity carried by the average particle of a group is the average of the heterogeneities carried by these particles.

The heterogeneity carried by the average particle $F_{\bar{i}}$ of L, proportional as it is to $(a_{\bar{i}} - a_L)$, is obviously zero.

These definitions, which may seem artificial, will be justified in section 20.5.

19.3.2. Heterogeneity carried by a group G_n of particles :

To be consistent with the definition of the heterogeneity carried by a single particle, the heterogeneity h_n carried by a group G_n of particles should be proportional to $(a_n - a_L) M_n$ and, in order to be dimensionless, inversely proportional to the weight of critical component in the average group $G_{\bar{n}}$. Hence the following definition :

$$h_n = \frac{(a_n - a_L) M_n}{a_{\bar{n}} M_{\bar{n}}} = \frac{N_G}{a_L M_L} (a_n - a_L) M_n$$

Obviously, the heterogeneity h_L carried by the lot L, which is proportional to $(a_L - a_L)$ is zero.

Now, we shall compare h_n with $h_{n\bar{j}}$, heterogeneities carried by the group G_n and by the average particle of this group respectively. If, as we have assumed in section 19.1. all groups have practically the same magnitude and for instance practically the same number N_n of particles, then :

$$N_G = \frac{N_F}{N_n} \quad \text{which involves :} \quad h_n = h_{n\bar{j}}$$

Inasmuch as all groups have about the same size, the heterogeneity carried by a group of particles is practically equal to the heterogeneity carried by its average particle.

19.3.3. Constitution heterogeneity and distribution heterogeneity of the lot L :

The constitution of L is characterized by the set L_F of N_F fragments F_i. We shall define the "<u>constitution heterogeneity</u> CH_L of the lot L" as the variance of the heterogeneities h_i carried by the N_F particles F_i :

$$CH_L = \sigma^2(h_i) = \frac{1}{N_F} \sum_i^{N_F} h_i^2 = \frac{1}{a_L^2 M_L^2} \sum_i^{N_F} (a_i - a_L)^2 M_i^2$$

CH_L is an intrinsic property of the set L_F, i.e. of the lot L in its present state of comminution. Blending and mixing have no effect on constitution heterogeneity.

The distribution of L is characterized by the set L_G of N_G groups G_n. We shall define the "<u>distribution heterogeneity</u> DH_L of the lot L" as the variance of the heterogeneities h_n carried by the N_G groups G_n :

$$DH_L = \sigma^2(h_n) = \frac{1}{N_G} \sum_n^{N_G} h_n^2 = \frac{1}{a_L^2 M_L^2} \sum_n^{N_G} (a_n - a_L)^2 M_n^2$$

These definitions will enable us to establish a relationship between the heterogeneities CH_L and DH_L and the sampling variance (section 20.5.).

19.3.4. Relationship between constitution and distribution heterogeneities :

We shall write :

$$h_i \equiv h_{nj} = (h_{nj} - h_n) + h_n$$

$$\sum_i h_i^2 = \sum_n \sum_j h_{nj}^2 = \sum_n \sum_j (h_{nj} - h_n)^2 + \sum_n N_n h_n^2$$

If, as previously assumed, all N_n are of the same order of magnitude, we obtain by dividing by N_F :

$$CH_L = \frac{1}{N_F} \sum_n \sum_j (h_{nj} - h_n)^2 + DH_L$$

It has been shown (Gy, 1975, chapter 11) that :

$$CH_{\bar{n}} = \frac{1}{N_F} \sum_n \sum_j (h_{nj} - h_n)^2$$

is nothing else than the average constitution heterogeneity of the groups G_n with the following consequences :

$$CH_L = CH_{\bar{n}} + DH_L \quad \text{and} \quad CH_L \geq DH_L \geq 0$$

19.3.5. Properties of the constitution and distribution heterogeneities :

19.3.5.1. Constitution homogeneity of the lot L :

The constitution of a lot L is said to be homogeneous when CH_L is zero :

<u>Constitution homogeneity</u> : $CH_L = 0$

CH_L being the variance of h_i is zero if and only if all values of h_i are equal to their mean $h_{\bar{i}}$ which is known to be nil. From the definition of h_i (assuming the values of M_i to be non-zero) this can be achieved if and only if all values of a_i are equal to a_L or in other words if the material under investigation is itself homogeneous. The apparently obvious statement that the constitution heterogeneity of a homogeneous material is zero confirms the consistency of our definitions. The property $CH_L = 0$ involves :

$CH_{\bar{n}} = 0$: which involves $CH_n = 0$ irrespective of n : all groups of particles that can be obtained from a homogeneous material are themselves homogeneous.

$DH_L = 0$: the distribution heterogeneity is nil.

19.3.5.2. Perfect distribution homogeneity of the lot L :

The distribution of a lot L is said to be "<u>perfectly homogeneous</u>" when the distribution heterogeneity DH_L is zero :

<u>Perfect distribution homogeneity</u> : $DH_L = 0$

DH_L being the variance of h_n is zero if and only if all values of h_n are equal to their mean $h_{\bar{n}}$ which is known to be nil. From the definition of h_n (assuming the values of M_n to be non-zero), this can be achieved if and only if all values of a_n are equal to a_L. This can be written as follows :

$$a_n = \frac{\sum_j a_{nj} M_{nj}}{M_n} = a_L \qquad \text{irrespective of n.}$$

The first obvious solution is :

$a_{nj} = a_L$ irrespective of n and j : the material is homogeneous and we know from the preceding section that the distribution heterogeneity is nil.

Now, the sets of values of a_{nj} and M_{nj} are data of the problem and the chances for the above equality to be satisfied, even after thorough mixing of the lot, are infinitely smaller than those of a foursome of card players to receive the thirteen cards of each colour respectively, which, according to our limited experience, is very small indeed. We shall therefore retain the conclusion that under natural conditions the distribution of L is never perfectly homogeneous. Then :

$DH_L > 0$

19.3.5.3. Natural distribution homogeneity of the lot L :

If DH_L cannot be zero under natural conditions, it is interesting to investigate whether and how it can be minimized and what is its minimal value.

Consider a Vee-mixer containing a layer of black magnetite and on top of it a layer of white silica sand. Such a distribution is perfectly heterogeneous. If we operate the mixer for a few minutes, we obtain a mix which seems, at first glance at least, to be uniformly grey : the material has been homogenized, which is what mixers are for. If however we split the lot between a certain number of groups and if we analyse their magnetite content by means of a magnet, we know that, irrespective of the mixing time, the magnetite content will slightly vary from one group to the next, the residual heterogeneity being an obvious consequence of the particulate structure of the material.

Now, if we analyse the mixing operation we observe that its theoretical model is to prepare a mix where the co-ordinates of the centre of gravity of any particle F_i are independent of its characteristics (size, density, shape etc...) i.e. of its weight M_i and of its critical content a_i. When this independence is achieved there is no point in further mixing. This state is what we shall call the "natural distribution homogeneity" of the lot L. All particles of L are distributed at random throughout the domain (D_L) occupied by the lot L.

In order to make easier the statistical interpretation of this random distribution, we shall consider the lot L as a sample of N_F particles extracted at random from an infinite lot L_∞ having the same average properties as L. We shall call σ^2

the variance of h_i in this infinite population.

The hypothesis of random distribution is equivalent to the hypothesis according to which all groups G_n of neighbouring particles are samples of N_n particles extracted one by one and at random from the lot L_∞. It is a known result of mathematical statistics (analysis of variance) that in such conditions :

- the quantity $\sum_n \sum_j h_{nj}^2 / \sigma^2 = N_F \, CH_L / \sigma^2$ follows a chi-squared distribution with $N_F - 1$ degrees of freedom,

- the quantity $\sum_n \sum_j (h_{nj} - h_n)^2 / \sigma^2 = N_F \, CH_{\bar{n}} / \sigma^2$ follows a chi-squared distribution with $N_F - N_G$ degrees of freedom,

- the quantity $\sum_n N_n h_n^2 / \sigma^2 = N_F \, DH_L / \sigma^2$ follows a chi-squared distribution with $N_G - 1$ degrees of freedom.

Expressed in other words, we can state that we have three independent unbiased estimators of σ^2. These estimators are :

- the "total variance" $\dfrac{N_F}{N_F - 1} CH_L$

- the "residual variance" $\dfrac{N_F}{N_F - N_G} CH_{\bar{n}}$

- the "variance between groups" $\dfrac{N_F}{N_G - 1} DH_L$

CH_L is an intrinsic property of L with a fixed value. On the other hand, DH_L is a random variable. We can therefore state that the natural distribution homogeneity is characterized by a minimal value of DH_L the mean of which is :

$$m(DH_L)_{min} = \frac{N_G - 1}{N_F - 1} CH_L$$

This minimum depends on the number N_G of groups or in other words on the average size of the groups or on the observation scale. We shall introduce the "grouping factor γ" the physical meaning of which will be precised in due course and defined as follows :

$$\gamma = \frac{N_F - N_G}{N_G - 1} \quad \text{or} \quad \frac{N_G - 1}{N_F - 1} = \frac{1}{1 + \gamma}$$

Now, as a general rule, we may admit that : $1 \ll N_G \ll N_F$ (the meaning of $\ll$ being "very small in comparison with"). Then, for all practical purposes :

$$\gamma = \frac{N_F}{N_G} = N_{\bar{n}}$$

The order of magnitude of γ is the average number of particles in the groups G_n. We shall now retain :

$$m(DH_L)_{min} = \frac{1}{1 + \gamma} CH_L$$

19.3.5.4. Various forms of distribution homogeneity :

The words "homogeneous" and "heterogeneous" are very often misleading as they can be used with different meanings. Since perfect distribution homogeneity can be achieved only through artificial means, we shall from now on drop the qualificative "natural". Distribution homogeneity can take the following forms :

1) <u>Three-dimensional distribution homogeneity</u> : this is achieved if the condition of homogeneity defined in the preceding section is satisfied when the fragmental function $a_F(X)$ defined in section 5.3.4.3. is practically uniform throughout the domain (D_L) occupied by the lot L in the geometrical three-dimensional space.

Vee-mixers and the like are built according to a model tending to achieve a three-dimensional distribution homogeneity. This is the only true distribution homogeneity. The degenerate forms that will be described now are anisotropic hybrids between homogeneous and heterogeneous distributions.

2) <u>Two-dimensional distribution homogeneity</u> : this is achieved in two-dimensional lots (section 5.3.2.1.) if the condition of homogeneity is satisfied when the fragmental function $a_F(X)$ is practically uniform throughout the domain (D_L') occupied by the projection of L on the extension plane.

Such a state can be observed when gravity segregation takes place in a lot previously characterized by a three-dimensional homogeneity. We would like to seize this opportunity to point out the fact that with multi-size materials containing a large proportion of coarse particles or with pulps containing minerals of very different densities, the state of three-dimensional homogeneity, if ever achieved is definitely unstable and tends to become a two-dimensional homogeneity. For those familiar with mineral processing techniques, a discontinuous laboratory jig achieves a near perfect vertical segregation and a good two-dimensional horizontal distribution homogeneity.

3) <u>One-dimensional distribution homogeneity</u> : this is achieved in one-dimensional lots such as elongated piles if the condition of homogeneity is satisfied when the fragmental function $a_F(X)$ is practically uniform throughout the domain (D_L'') occupied by the projection of L on the extension axis.

This is the kind of homogeneity which is achieved by the bed-blending systems where the pile is made of a very large number (several hundreds to several thousands) of layers deposited by a stacker moving at a uniform speed along the pile. The user should not forget that homogeneity is attained in only one dimension, the material remaining very heterogeneous in a transversal plane such as the plane or surface generated by the reclaiming device.

4) <u>Revolution distribution homogeneity or symmetry towards a vertical axis</u> : this is achieved in cylindrical or conical piles if the condition of homogeneity is satisfied with groups of particles delimited by two vertical planes containing the revolution axis.

This is the kind of homogeneity which is tentatively attained when mixing the material prior to splitting by means of the coning and quartering method (which will be discussed in chapter 24). Apart from this symmetry around the vertical axis, the distribution is perfectly heterogeneous.

5) <u>Three-dimensional distribution heterogeneity</u> : such a distribution is observed when none of the homogeneities described in the preceding paragraphs is achieved. Our world being as it is governed by an omnipresent gravity (the author has not yet been invited to conduct sampling experiments aboard a spaceship), three-dimensional heterogeneity is the rule, distribution homogeneity, even of a degenerate kind, the exception.

Since the state of distribution homogeneity is characterized by a minimal value of DH_L, we can state that three-dimensional distribution heterogeneity is characterized by :

$$DH_L > m(DH_L)_{min} = \frac{1}{1+\gamma} CH_L$$

6) <u>Experimental check</u> : it is theoretically possible to collect series of groups G_n, to estimate a_n and M_n, to calculate h_n and its variance $\sigma^2(h_n) = DH_L$. On the other hand it is also possible to estimate $(DH_L)_{min}$ (chapter 22). But it is not so easy to carry out a statistically valid comparison between DH_L and $(DH_L)_{min}$ nor to decide whether DH_L is significantly larger than its minimum. This would require repeating the same operation a large number of times, which is seldom economically acceptable.

It is always safe to assume that the distribution is heterogeneous and dangerous to speculate upon the existence of one form or another of homogeneity.

<u>19.3.5.5. Maximal distribution heterogeneity</u> :

We already know that : $DH_L \leq CH_L$. We can therefore define the maximum of DH_L as :

$(DH_L)_{max} = CH_L$ which involves $CH_n = 0$ irrespective of n.

Maximal distribution heterogeneity is achieved when the composition of all groups is homogeneous. This can be observed under various circumstances :

1) Each group G_n is made of only one particle. Then, by definition, $DH_L = CH_L$.

2) Each group G_n is made of several particles but all these particles have the same critical content : $a_{nj} = a_n$ irrespective of j. In practice, this is achieved when simultaneously :

- the material under investigation is completely liberated (each particle is made of a pure mineral),
- the various minerals are perfectly segregated,
- the particles are grouped according to their mineral composition.

At the limit, such distributions can be observed in the bed of batch laboratory jigs concentrating liberated ores.

19.3.5.6. Segregation factor :

From the results obtained in the preceding sections, we can precise the range of the distribution heterogeneity DH_L :

1) <u>Natural range</u> :

$$\frac{1}{1+\gamma} CH_L \leq DH_L \leq CH_L \quad \text{which can be written :} \quad \frac{1}{1+\gamma} CH_L \leq DH_L \leq \frac{1+\gamma}{1+\gamma} CH_L$$

We shall now introduce the "segregation factor" ξ defined by :

$$DH_L = \frac{1+\gamma\xi}{1+\gamma} CH_L = (1+\gamma\xi) \frac{N_G - 1}{N_F - 1} CH_L$$

The natural range of ξ is obviously : $0 \leq \xi \leq 1$ with :

$\xi = 1$: when the distribution heterogeneity is maximal : state of perfect segrega-
$0 < \xi < 1$: natural range of distribution heterogeneity,
$\xi = 0$: when the distribution heterogeneity is minimal : state of natural distribution homogeneity.

2) <u>artificial range</u> :

$$0 \leq DH_L \leq \frac{1}{1+\gamma} CH_L$$

This range corresponds to negative values of ξ: $-\frac{1}{\gamma} \leq \xi < 0$ with :

$-\frac{1}{\gamma} < \xi < 0$: artificial range of distribution homogeneity,

$\xi = -\frac{1}{\gamma}$: artificial state of perfect homogeneity

19.3.5.7. Influence of the observation scale on the value of DH_L :

The observation scale is characterized by the grouping factor γ and it is interesting to investigate how DH_L varies when γ varies. For a given lot L and a given natural distribution characterized by a certain value of ξ the following table summarizes the ranges of N_G, γ and DH_L (table 19.1. page 225). DH_L is a decreasing function of the grouping factor γ and an increasing function of the number N_G of groups.

19.3.5.8. Influence of the comminution state of the material on the value of CH_L :

The set of particles L_F can be adulterated either by agglomeration or by comminution :

Table 19.1. *Influence of the group size or of the observation scale.*

$N_n = 1$ irrespective of n (N_F groups of one fragment each)		$1 < N_n < N_F$		$N_n = N_F$ (One group of N_F fragments)
N_F	>	N_G	>	1
0	<	γ	<	$+\infty$
CH_L	>	DH_L	>	0

1) <u>agglomeration</u> : consider a lot L made of N_F particles (set L_F) and of N_G groups G_n of particles (set L_G) and assume that particles are agglomerated in such a way that each group G_n becomes a new particle (set L'_F). By definition :

$$CH_{L'} = DH_L \leq CH_L$$

Agglomeration can only reduce the constitution heterogeneity of the material under investigation.

2) <u>comminution</u> : by inverse reasoning we easily deduce that comminution can only increase the constitution heterogeneity of the material.

19.4. GENERAL EXPRESSION OF THE DISTRIBUTION HETEROGENEITY DH_L

DH_L can be written under the developed form :

$$DH_L = (1 + \gamma\xi)(N_G - 1) \frac{N_L}{N_L - 1} \sum_i \frac{(a_i - a_L)^2 M_i^2}{a_L^2 M_L^2} = \frac{1 + \gamma\xi}{1 + \gamma} CH_L$$

which precises the part played by all relevant data of the problem and which will be useful when trying to establish a relationship between the sampling variance and the distribution heterogeneity of the material to be sampled. This point will be dealt with in section 20.5.2.

The distribution heterogeneity depends on three factors or groups of factors :

CH_L : <u>constitution heterogeneity</u> of the material to be sampled characterizing the comminution state of the material (see section 19.3.5.8.),

γ : <u>grouping factor</u> characterizing the size of the groups taken into consideration or in other words the observation scale,

ξ : <u>segregation factor</u> characterizing the degree of heterogeneity of the distribution.

CHAPTER 20

DEVELOPMENT OF THE DISCRETE SELECTION MODEL

20.1. INTRODUCTION AND NOTATIONS

According to the definition given in section 5.4. the lot L of material to be sampled is, in the discrete perspective, regarded as the set L_U of N_U discrete units U_m and its sampling consists of a selection according to which each unit U_m is individually and independently submitted to the selecting process with a given probability P_m of being selected. These units U_m can be :

- either individual fragments F_i with i = 1, 2, .. N_F
- or groups of fragments G_n with i = 1, 2, ... N_G.

We shall develop the model for unspecified units U_m, assuming that as far as the present sampling stage of L is concerned, each of these units makes an undissociable whole.

We shall also assume in the following sections that :
- the same selection process is repeated K times,
- the sample S_k obtained as the result of the k th trial is analysed in order to take a census of the selected units,
- the lot L is reconstituted prior to the next trial,
- K is a large number tending to become infinite.

These assumptions obviously correspond to a virtual, not an actual process, with the purpose of expressing the moments of the various random characteristics of the sample S_k.

The following notations will be retained :

L : lot of particulate material, M_L weight of L, A_L weight of critical component in L, a_L critical content of L, N_L number of units in L (corresponding to the notation N_U of chapter 19, changed here for the sake of convenience).

U_m : unit belonging to L (m = 1, 2, ... N_L), M_m weight of U_m, A_m weight of critical component, a_m critical content, P_m selection probability. We shall not make any hypothesis on the uniformity of P_m until section 20.5.

S_k : sample obtained as the result of the k th selection trial (k = 1, 2, ... K), M_{Sk} weight of S_k, A_{Sk} weight of critical component, a_{Sk} critical content,

N_{Sk} number of units in S_k. All characteristics of S_k are random variables. The purpose of the next sections is to express their moments.

Z : virtual set of units obtained by reunion of the K samples S_k. M_Z weight of Z, A_Z weight of critical component in Z, a_Z critical content of Z, N_Z number of units in Z.

N_{Zm} : frequency of unit U_m in the set Z. It is a random variable.

We shall recall the following property which will be used in the next sections. Consider N_L random variables x_m, independent in probability of one another and x a linear function of the x_m defined by :

$$x = \sum_m \Omega_m x_m \quad \text{with } \Omega_m \text{ a parameter associated with unit } U_m.$$

Then, the mean and variance of x can be expressed as follows :

$$m(x) = \sum_m \Omega_m m(x_m) \quad \text{and} \quad \sigma^2(x) = \sum_m \Omega_m^2 \sigma^2(x_m)$$

20.2. DISTRIBUTIONS OF N_{Zm}, N_{Sk}, M_{Sk} and A_{Sk}

20.2.1. Distribution of N_{Zm} frequency of unit U_m in the set Z :

By definition, the unit U_m has been submitted to K independent selection trials with a uniform probability P_m of being selected. N_{Zm}, which is the number of successful trials, follows a binomial distribution characterized by the moments :

$$m(N_{Zm}) = K P_m \quad \text{and} \quad \sigma^2(N_{Zm}) = K P_m (1 - P_m)$$

20.2.2. Distribution of N_{Sk} number of units in S_k :

The number N_Z of units in the set Z can be expressed in two different ways :

1) N_Z is the sum of the K numbers of units N_{Sk} in the K samples S_k : then :

$$N_Z = \sum_k N_{Sk} \quad \text{with} \quad k = 1, 2, \ldots K \quad \text{which entails :}$$

$$m(N_Z) = K m(N_{Sk}) \quad \text{and} \quad \sigma^2(N_Z) = K \sigma^2(N_{Sk})$$

2) N_Z is the sum of the N_L frequencies N_{Zm} of the units U_m : then :

$$N_Z = \sum_m N_{Zm} \quad \text{with} \quad m = 1, 2, \ldots N_L \quad \text{which entails :}$$

$$m(N_Z) = \sum_m m(N_{Zm}) = K \sum_m P_m$$

$$\sigma^2(N_Z) = \sum_m \sigma^2(N_{Zm}) = K \sum_m P_m (1 - P_m)$$

These results are the consequence of the property recalled overleaf. By matching both expressions of $m(N_Z)$ and $\sigma^2(N_Z)$ we obtain :

$$m(N_{Sk}) = \sum_m P_m$$

$$\sigma^2(N_{Sk}) = \sum_m P_m (1 - P_m)$$

20.2.3. Distribution of M_{Sk}, weight of S_k :

The weight M_Z of the set Z can be expressed in two different ways :

1) M_Z is the sum of the weights M_{Sk} of the K samples S_k , then :

$$M_Z = \sum_k M_{Sk} \quad \text{with} \quad k = 1, 2, \ldots K \quad \text{which involves :}$$

$$m(M_Z) = K\, m(M_{Sk}) \quad \text{and} \quad \sigma^2(M_Z) = K\, \sigma^2(M_{Sk})$$

2) M_Z is the sum of the weights of all units making up the set Z. As unit U_m is present in Z with a frequency N_{Zm}, it readily follows that :

$$M_Z = \sum_m M_m\, N_{Zm} \quad \text{with } m = 1, 2, \ldots N_L$$

In this expression, M_m is a numerical parameter and N_{Zm} is a random variable. Then, according to the property recalled overleaf :

$$m(M_m\, N_{Zm}) = K\, M_m\, P_m \quad \text{and} \quad \sigma^2(M_m\, N_{Zm}) = K\, M_m^2\, P_m\, (1 - P_m)$$

which involves :

$$m(M_Z) = K \sum_m M_m\, P_m \quad \text{and} \quad \sigma^2(M_Z) = K \sum_m M_m^2\, P_m\, (1 - P_m)$$

By matching both expressions of $m(M_Z)$ and $\sigma^2(M_Z)$ we obtain :

$$m(M_{Sk}) = \sum_m M_m\, P_m$$

$$\sigma^2(M_{Sk}) = \sum_m M_m^2\, P_m\, (1 - P_m)$$

20.2.4. Distribution of A_{Sk}, weight of critical component in S_k :

As the properties of A_{Sk} are identical with those of M_{Sk} we can directly write :

$$m(A_{Sk}) = \sum_m A_m\, P_m = \sum_m a_m\, M_m\, P_m$$

$$\sigma^2(A_{Sk}) = \sum_m A_m^2\, P_m\, (1 - P_m) = \sum_m a_m^2\, M_m^2\, P_m\, (1 - P_m)$$

20.3. DISTRIBUTION OF a_S, CRITICAL CONTENT OF THE SAMPLE S - INCORRECT SELECTION

This section deals with the most general problem : no assumption is made as regards the selection probabilities P_m. From now on and in order to lighten the notations, the subscript k of S_k, M_{Sk}, A_{Sk} and a_{Sk} which is no more relevant will be omitted. The results of this section are valid irrespective of the selecting conditions, provided however that they remain probabilistic.

20.3.1. Distribution law of the critical content a_S :

By definition, this critical content can be written :

$$a_S = \frac{A_S}{M_S}$$

We know from the preceding sections that both A_S and M_S are random variables and it is a known fact that the quotient of two random variables follows in the most general case a complex distribution law of an unspecified kind. It has been shown however (Geary, 1930 and Bastien, 1960) that when both the numerator and the denominator of this quotient follow a normal law and when furthermore the coefficient of variation of the denominator remains small as compared to unity (say smaller than 0.03), the quotient can be regarded as following a practically normal law. Now, what are the chances for A_S and M_S of being normally distributed ?

When the number of units in the sample S is large enough (say larger than 30), the Central Limit Theorem of Laplace and Liapounoff may be invoked and we may admit that both A_S and M_S are likely to be approximately normal, as a consequence of which a_S follows a normal distribution too. There is however one exception to this rule : that of very low grade materials such as the ores of precious metals and minerals for which it is no more the total number of units but the number of units with non-zero values of A that must be taken into account. With all usual materials, the normal approximation is acceptable.

The second condition concerns the coefficient of variation (or relative standard deviation) of M_S or the relative variance $u^2(M_S)$ which can be written :

$$u^2(M_S) = \frac{\sigma^2(M_S)}{m^2(M_S)} = \frac{\sum_m M_m^2 P_m (1 - P_m)}{(\sum_m M_m P_m)^2}$$

If a certain variance u_o^2 that lies somewhere about 10^{-3} is regarded as acceptable and if we choose for instance to carry out a correct selection with $P_m = P$ irrespective of m, then, the inequality : $u^2(M_S) \leq u_o^2$ can be solved for P :

$$P \geq P_o = \frac{\sum_m M_m^2}{\sum_m M_m^2 + M_L^2 u_o^2}$$

We shall assume in the next sections that A_S and M_S are both normally distributed and that P is larger than P_o. If one of these assumptions is not founded, our conclusions must be accepted with a certain degree of uncertainty.

20.3.2. Moments of a_S - Introduction :

In order to lighten our formulas, we shall use the following parameters :

$N_o = m(N_S) = \sum_m P_m$ N_o is usually different from N_L

$M_o = m(M_S) = \sum_m M_m P_m$ M_o is usually different from M_L

$A_o = m(A_S) = \sum_m A_m P_m = \sum_m a_m M_m P_m$ A_o is usually different from A_L

$a_o = \dfrac{A_o}{M_o}$ a_o is usually different from a_L

and the following auxiliary variables :

ε : relative deviation of A_S from its mean A_o :

$$\varepsilon = \frac{A_S - A_o}{A_o} \quad \text{which can be written :} \quad A_S = (1 + \varepsilon) A_o$$

η : relative deviation of M_S from its mean M_o :

$$\eta = \frac{M_S - M_o}{M_o} \quad \text{which can be written :} \quad M_S = (1 + \eta) M_o$$

ψ : relative deviation of a_S from its mean a_o (which will be shown later on) :

$$\psi = \frac{a_S - a_o}{a_o} \quad \text{which can be written :} \quad a_S = (1 + \psi) a_o$$

The critical content a_S of the sample S can also be expressed as :

$$a_S = \frac{A_S}{M_S} = \frac{(1 + \varepsilon)}{(1 + \eta)} a_o = (1 + \psi) a_o \quad \text{which involves} \quad 1 + \psi = \frac{(1 + \varepsilon)}{(1 + \eta)}$$

As already pointed out in section 8.7.2. the difficulty arises from the fact that the denominator $(1 + \eta)$ is a random variable. For this reason, it is convenient to develop the quotient $1/(1 + \eta)$:

$$1 + \psi = (1 + \varepsilon)(1 - \eta + \eta^2 - \eta^3 + \ldots \text{etc}) \quad \text{or} \quad \psi = (\varepsilon - \eta) + (\eta^2 - \varepsilon\eta) + \ldots \text{etc.}$$

This development is absolutely convergent when $|\eta| < 1$. When furthermore, as previously assumed, $u(M_S) = \sigma(\eta)$ is very small, then each term of the development can be regarded as negligible as compared to the preceding term. We shall retain two approximations :

1) <u>First approximation (subscript 1)</u> :

$\psi_1 = \varepsilon - \eta \quad$ which involves $\quad (a_S)_1 = (1 + \psi_1) a_o = (1 + \varepsilon - \eta) a_o$

2) <u>Second approximation (subscript 2)</u> :

$\psi_2 = \varepsilon - \eta + \eta^2 - \varepsilon\eta \quad$ which involves $\quad (a_S)_2 = (1 + \psi_2) a_o = (1 + \varepsilon - \eta + \eta^2 - \varepsilon\eta) a_o$

We shall calculate the moments of a_S for both approximations.

20.3.3. Moments of a_S :

20.3.3.1. First approximation :

We feel inappropriate to reproduce in this book long and tedious demonstrations which have been already published in a previous work (Gy, 1975, section 20.14) where the reader can check their accuracy. We shall therefore directly give the following results :

$$m(a_S)_1 = a_o = \frac{\sum_m a_m M_m P_m}{\sum_m M_m P_m} \qquad \text{usually different from } a_L$$

$$\sigma^2(a_S)_1 = \frac{\sum_m (a_m - a_o)^2 M_m^2 P_m (1 - P_m)}{(\sum_m M_m P_m)^2}$$

Or, in terms of the sampling error SE :

$$(SE)_1 = \frac{(a_S)_1 - a_L}{a_L}$$

$$m(SE)_1 = B(a_S)_1 = \frac{a_o - a_L}{a_L} \qquad \text{usually different from zero.}$$

The bias $B(a_S)_1$ is called the "<u>incorrectness bias</u>" for the simple reason that it cancels when the selection is correct, which will be shown in section 20.5.

$$\sigma^2(SE)_1 = \frac{\sum_m (a_m - a_o)^2 M_m^2 P_m (1 - P_m)}{a_L^2 (\sum_m M_m P_m)^2} = \frac{\sigma^2(a_S)_1}{a_L^2} = \frac{a_o^2}{a_L^2} u^2(a_S)_1$$

With the exception of low grade materials such as ores of precious metals and minerals, this first approximation is sufficient. We shall however mention the results of the second approximation.

20.3.3.2. <u>Second approximation</u> :

$$m(a_S)_2 = m(a_S)_1 - \frac{\sum_m (a_m - a_o) M_m^2 P_m (1 - P_m)}{(\sum_m M_m P_m)^2}$$

$$\sigma^2(a_S)_2 = \sigma^2(a_S)_1 - \frac{2 \sum_m (a_m - a_o)^2 M_m^3 P_m (1 - P_m)(1 - 2P_m)}{(\sum_m M_m P_m)^3}$$

Or, in terms of the sampling error SE :

$$(SE)_2 = \frac{(a_S)_2 - a_L}{a_L}$$

$$m(SE)_2 = B(a_S)_2 = \frac{m(a_S)_2 - a_L}{a_L}$$

and :

$$\sigma^2(SE)_2 = \frac{\sigma^2(a_S)_2}{a_L^2} = \frac{a_o^2}{a_L^2} u^2(a_S)_2$$

233

20.4. PRACTICAL IMPLEMENTATION OF FORMULAS INVOLVING SUMS $\sum_i$

20.4.1. Introduction :

When the units U_m are individual fragments F_i, the practical implementation of the formulas obtained in the preceding sections involves the calculation or estimation of the following quantities :

$$M_o = \sum_i M_i P_i \quad \text{with } i = 1, 2, \ldots N_L$$

$$A_o = \sum_i a_i M_i P_i$$

$$m(a_S)_1 = a_o = A_o/M_o$$

$$m(a_S)_2 = a_o - \frac{1}{M_o^2} \sum_i (a_i - a_o) M_i^2 P_i (1 - P_i)$$

$$\sigma^2(a_S)_1 = \frac{1}{M_o^2} \sum_i (a_i - a_o)^2 M_i^2 P_i (1 - P_i)$$

All these quantities involve sums such as $\sum_i$ extended to the N_L particles of the lot L while N_L is practically always much too large to be enumerated or even roughly estimated. Furthermore, individual values of a_i, M_i, P_i are never known. Then, the practical question arises of how to estimate these sums, failing which the theoretical results would be definitely useless. These difficulties can be overcome by observing that :

1) <u>The weight M_i</u> of F_i can be expressed in terms of its density λ_i and of its volume v_i, itself proportional to the cube of its "diameter" d_i that can be objectively defined and estimated,

2) <u>The critical content a_i</u> is nearly always directly correlated to the density λ_i and often indirectly correlated to the diameter d_i,

3) <u>The selection probability P_i</u> deviates from a uniform value as a result of delimitation and extraction errors (chapters 17 and 18) depending mainly on the particle diameter d_i, and to a lesser extent on the density λ_i.

4) <u>The "personality" of a given fragment F_i</u> characterized by the parameters M_i, a_i, P_i, is therefore dominated by its diameter d_i and to a lesser extent by its density λ_i.

20.4.2. Practical estimation of the sums extended to the N_L particles of the lot L :

We shall assume now that the lot L has been virtually broken up into a certain number of size-density fractions $L_{\alpha\beta}$. This virtual operation can to a certain extent be actually carried out on the lot or on a representative sample of it, by series of screenings and dense-liquids or dense-media separations. Such operations are commonplace in mineral processing laboratories, despite the fact that they are usually regarded as too costly for the sole purpose of sampling errors estimation.

We shall use the following notations:

L : lot of particulate material (M_L, A_L, a_L, N_L).

L_α : a size fraction of L characterized by an average particle diameter d_α $(M_{L\alpha}, A_{L\alpha}, a_{L\alpha}, N_{L\alpha})$.

$L_{\alpha\beta}$: a density fraction of L_α characterized by a density λ_β $(M_{L\alpha\beta}, A_{L\alpha\beta}, a_{L\alpha\beta}, N_{L\alpha\beta}$

S : a probabilistic sample extracted from L (M_S, A_S, a_S, N_S),

S_α : a size fraction of S characterized by the same diameter d_α as above $(M_{S\alpha}, A_{S\alpha}, a_{S\alpha}, N_{S\alpha})$.

$S_{\alpha\beta}$: a density fraction of S_α characterized by the same density λ_β as above $(M_{S\alpha\beta}, A_{S\alpha\beta}, a_{S\alpha\beta}, N_{S\alpha\beta})$.

$\tau_{\alpha\beta}$: numerical sampling ratio of the fraction $L_{\alpha\beta}$ defined by : $\tau_{\alpha\beta} = \dfrac{N_{S\alpha\beta}}{N_{L\alpha\beta}}$

We shall now assume that:

1) all particles belonging to the same size-density fraction $L_{\alpha\beta}$ can be assimilated to the average particle $F_{\alpha\beta}$ defined by:
- its weight $M_{F\alpha\beta} = M_{L\alpha\beta}/N_{L\alpha\beta}$
- its critical content $a_{F\alpha\beta} = a_{L\alpha\beta}$

This hypothesis amounts to assimilating the lot L with a virtual lot L' with:

$$L \equiv \sum_i F_i \quad \text{and} \quad L' \equiv \sum_\alpha \sum_\beta N_{L\alpha\beta} F_{\alpha\beta}$$

2) all particles belonging to the same size-density fraction $L_{\alpha\beta}$ are submitted to the selection process with the same probability $P_{\alpha\beta}$ of being selected. This hypothesis is well supported by the observation that the most dangerous deviations from selection correctness depend almost solely on particle size and density.

All sums involved in this chapter can be put under the general form:

$$\sum_i a_i^x M_i^y \Phi(P_i) \quad \text{with } x = 0 \text{ or } 1 \text{ or } 2 \quad \text{and} \quad y = 1 \text{ or } 2.$$

$\Phi(P_i) = P_i$ or $P_i(1 - P_i)$ or $P_i(1 - P_i)(1 - 2P_i)$

It is easy to show that $\sum_i$ can be replaced by the estimator $\sum_\alpha \sum_\beta$:

$$\sum_i a_i^x M_i^y \Phi(P_i) = \sum_\alpha \sum_\beta N_{L\alpha\beta} a_{L\alpha\beta}^x M_{F\alpha\beta}^y \Phi(P_{\alpha\beta}) = \sum_\alpha \sum_\beta a_{L\alpha\beta}^x M_{F\alpha\beta}^{y-1} M_{L\alpha\beta} \Phi(P_{\alpha\beta})$$

from which it readily follows:

$$m(M_S) = M_0 = \sum_i M_i P_i = \sum_\alpha \sum_\beta M_{L\alpha\beta} P_{\alpha\beta}$$

$$m(A_S) = A_0 = \sum_i a_i M_i P_i = \sum_\alpha \sum_\beta a_{L\alpha\beta} M_{L\alpha\beta} P_{\alpha\beta}$$

$$m(a_S)_1 = a_o = A_o/M_o$$

$$m(a_S)_2 = m(a_S)_1 - \frac{1}{M_o^2} \sum_\alpha \sum_\beta (a_{L\alpha\beta} - a_o) M_{L\alpha\beta} M_{F\alpha\beta} P_{\alpha\beta} (1 - P_{\alpha\beta})$$

$$\sigma^2(a_S)_1 = \frac{1}{M_o^2} \sum_\alpha \sum_\beta (a_{L\alpha\beta} - a_o)^2 M_{L\alpha\beta} M_{F\alpha\beta} P_{\alpha\beta} (1 - P_{\alpha\beta})$$

The parameters involved in these formulas can be practically estimated in the following way from the results of an actual size-density analysis :

$M_{L\alpha\beta}$: by direct weighing,

$M_{F\alpha\beta}$: either by weighing an enumerable number of particles extracted at random from $L_{\alpha\beta}$ or in terms of the characteristics of the average particle $F_{\alpha\beta}$:

$$M_{F\alpha\beta} = f\, d_\alpha^3\, \lambda_\beta = v_\alpha\, \lambda_\beta \quad \text{with}$$

f : an average shape factor always near 0.5 (dimensionless)
d_α : average diameter of the size fraction L_α from which $L_{\alpha\beta}$ has been obtained,
λ_β : average density of the density fraction $L_{\alpha\beta}$,
v_α : average volume of $F_{\alpha\beta}$.

$a_{L\alpha\beta}$: by assaying a representative sample of $L_{\alpha\beta}$.

$P_{\alpha\beta}$: by estimating its a posteriori equivalent, the sampling ratio $\tau_{\alpha\beta}$ (see for instance the experiment reported in section 18.5.).

20.4.3. Properties of the selection probability $P_{\alpha\beta}$ and of the sampling ratio $\tau_{\alpha\beta}$:

According to the hypothesis made in section 20.4.2. and to the results of section 20.2.2., all particles of $L_{\alpha\beta}$ are submitted to the selection scheme with a uniform probability $P_{\alpha\beta}$ of being selected and $N_{S\alpha\beta}$ is a random variable with mean :

$$m(N_{S\alpha\beta}) = N_{L\alpha\beta}\, P_{\alpha\beta}$$

and the sampling ratio $\tau_{\alpha\beta}$ is a random variable with mean :

$$m(\tau_{\alpha\beta}) = \frac{m(N_{S\alpha\beta})}{N_{L\alpha\beta}} = P_{\alpha\beta}$$

Let's observe now that :

$$N_S = \sum_\alpha \sum_\beta N_{S\alpha\beta} \quad : N_S \text{ is a random variable with mean :}$$

$$m(N_S) = \sum_\alpha \sum_\beta m(N_{S\alpha\beta}) = \sum_\alpha \sum_\beta N_{L\alpha\beta}\, P_{\alpha\beta}$$

Practically :

$$m\left\{\frac{N_{S\alpha\beta}}{N_S}\right\} = \frac{m(N_{S\alpha\beta})}{m(N_S)} = \frac{N_{L\alpha\beta}\, P_{\alpha\beta}}{\sum_\alpha \sum_\beta N_{L\alpha\beta}\, P_{\alpha\beta}}$$

Let's now define the average selection probability $\bar{P}$:

$$\bar{P} = \sum_\alpha \sum_\beta \frac{N_{L\alpha\beta} P_{\alpha\beta}}{N_L} \quad \text{it readily follows}: \quad m\left\{\frac{N_{S\alpha\beta}}{N_S}\right\} = \frac{P_{\alpha\beta}}{\bar{P}} \frac{N_{L\alpha\beta}}{N_L}$$

The quantities $N_{L\alpha\beta}/N_L$ and $N_{S\alpha\beta}/N_S$ are the proportions of the size-density fraction $\alpha\beta$ in the lot L and the sample S respectively. They are tied by the above relationship. We shall distinguish two cases:

1) <u>the selection is correct</u> and characterized by: $P_{\alpha\beta} = \bar{P} = P = $ constant: then

$$m\left\{\frac{N_{S\alpha\beta}}{N_S}\right\} = \frac{N_{L\alpha\beta}}{N_L} \quad : \text{a correct selection respects the size-density analysis.}$$

2) <u>the selection is incorrect</u>: then:

$$m\left\{\frac{N_{S\alpha\beta}}{N_S}\right\} \neq \frac{N_{L\alpha\beta}}{N_L} \quad : \text{an incorrect selection does not respect the size-density analysis}$$

20.5. MOMENTS OF N_S, M_S, A_S and a_S - CORRECT SELECTION

We shall now assume that the selection is correct or in other words that the values of P_m, P_i or $P_{\alpha\beta}$ as the case may be are uniformly equal to a constant P.

The following results are readily obtained, they are expressed for an unspecified unit U_m:

20.5.1. Number of units N_S in the sample S:

$m(N_S) = P\, N_L$

$\sigma^2(N_S) = P(1-P)\, N_L$

$u^2(N_S) = \dfrac{1-P}{P\, N_L}$

20.5.2. Weight M_S of the sample S:

$m(M_S) = P\, M_L$

$\sigma^2(M_S) = P(1-P) \sum_m M_m^2$

$u^2(M_S) = \left\{\dfrac{1-P}{P}\right\} \dfrac{\sum_m M_m^2}{M_L^2}$

20.5.3. Weight A_S of critical component in the sample S:

$m(A_S) = P\, a_L\, M_L$

$\sigma^2(A_S) = P(1-P) \sum_m a_m^2 M_m^2$

$u^2(A_S) = \left\{\dfrac{1-P}{P}\right\} \dfrac{\sum_m a_m^2 M_m^2}{a_L^2 M_L^2}$

20.5.4. Critical content a_S of the sample S :

20.5.4.1. Mean - First approximation :

$$m(a_S)_1 = a_o = a_L \quad \text{or} \quad B(a_S)_1 = 0 \quad \text{and} \quad m(SE)_1 = 0$$

This result means that at least in first approximation, a correct selection provides an unbiased sample which was not true of an incorrect selection.

$$m(a_S)_2 = a_L - \left\{\frac{1-P}{P}\right\} \sum_m (a_m - a_L) \frac{M_m^2}{M_L^2} \quad \text{with } m = 1, 2, \ldots N_U$$

20.5.4.2. Mean - Second approximation :

$$m(SE)_2 = B(a_S)_2 = -\left\{\frac{1-P}{P}\right\} \sum_m \frac{(a_m - a_L) M_m^2}{a_L \; M_L^2}$$

$B(a_S)_2$ is what we shall call the "__correctness bias__". It is the unavoidable bias occurring even when the selection is correct, which is the best we can do. According to our experience, with the exception of very low grade materials such as ores of precious metals and minerals, the correctness bias is always negligible.

If U_m is a single particle F_i, we have the following formula :

$$m(a_S)_2 = a_L - \left\{\frac{1-P}{P}\right\} \sum_i (a_i - a_L) \frac{M_i^2}{M_L^2} \quad \text{with } i = 1, 2, \ldots N_F$$

$$m(SE)_2 = -\left\{\frac{1-P}{P}\right\} \sum_i \frac{(a_i - a_L) M_i^2}{a_L \; M_L^2}$$

If U_m is a group G_n of particles :

$$m(a_S)_2 = a_L - \left\{\frac{1-P}{P}\right\} \sum_n (a_n - a_L) \frac{M_n^2}{M_L^2} \quad \text{with } n = 1, 2, \ldots N_G$$

$$m(SE)_2 = -\left\{\frac{1-P}{P}\right\} \sum_n \frac{(a_n - a_L) M_n^2}{a_L \; M_L^2}$$

20.5.4.3. Relative variance - First approximation :

$$\sigma^2(SE)_1 = u^2(a_S)_1 = \left\{\frac{1-P}{P}\right\} \sum_m \frac{(a_m - a_L)^2 M_m^2}{a_L^2 \; M_L^2}$$

If U_m is a single particle F_i :

$$\sigma^2(SE)_1 = u^2(a_S)_1 = \left\{\frac{1-P}{P}\right\} \sum_i \frac{(a_i - a_L)^2 \; M_i^2}{a_L^2 \; M_L^2} = \frac{1-P}{P \; N_F^2} \sum_i h_i^2 = \frac{1-P}{P \; N_F} CH_L$$

In these expressions, $i = 1, 2, \ldots N_F$ and CH_L is the constitution heterogeneity of the lot L introduced in chapter 19.

If U_m is a group G_n of particles, an increment extracted from the lot for instance,

$$\sigma^2(SE)_1 = u^2(a_S)_1 = \frac{1-P}{P} \sum_n \frac{(a_n - a_L)^2 M_n^2}{a_L^2 \; M_L^2} = \frac{1-P}{P \, N_G^2} \sum_n h_n^2 = \frac{1-P}{P \, N_G} DH_L$$

In these expressions, $n = 1, 2, \ldots N_G$ and DH_L is the distribution heterogeneity of the lot L introduced in chapter 19.

These properties link up the theory of heterogeneity presented in the preceding chapter and the theory of the discrete selection process. They justify a posteriori the definitions retained for h_i and h_n, for the constitution heterogeneity CH_L and the distribution heterogeneity DH_L. This important point will be developed in chapter 21.

20.5.5. Expression of the sums $\sum_i$ extended to the N_L particles of the lot :

Using the same notations as in section 20.4.2. we obtain :

$$m(a_S)_2 = a_L - \frac{1-P}{P \, M_L^2} \sum_\alpha \sum_\beta (a_{L\alpha\beta} - a_L) M_{L\alpha\beta} M_{F\alpha\beta}$$

$$\sigma^2(a_S)_1 = \frac{1-P}{P \, M_L^2} \sum_\alpha \sum_\beta (a_{L\alpha\beta} - a_L)^2 M_{L\alpha\beta} M_{F\alpha\beta}$$

These formulas will be used in chapter 22 for the estimation of the moments of the sampling error.

20.6. MOMENTS OF N_S, M_S, A_S and a_S - CORRECT SELECTION - UNIFORM WEIGHTING

The weight M_m of unit U_m plays the part of a weighting factor. When it is reduced to a constant $M = M_L/N_U$, the moments of N_S, M_S, A_S and a_S can be simplified as follows :

20.6.1. Moments of N_S :

$m(N_S) = P \, N_U$

$\sigma^2(N_S) = P \, (1 - P) \, N_U$

$u^2(N_S) = \dfrac{1-P}{P \, N_U}$

20.6.2. Moments of M_S :

$m(M_S) = P \, M_L$

$\sigma^2(M_S) = P \, (1 - P) \, \dfrac{M_L^2}{N_U}$

$u^2(M_S) = \dfrac{1-P}{P \, N_U} = u^2(N_S)$

20.6.3. Moments of A_S :

$$m(A_S) = P A_L = P a_L M_L$$

$$\sigma^2(A_S) = P(1-P)\frac{M_L^2}{N_U^2} \sum_m a_m^2$$

$$u^2(A_S) = \frac{1-P}{P a_L^2 N_U^2} \sum_m a_m^2$$

20.6.4. Moments of a_S :

$$m(a_S)_1 = m(a_S)_2 = a_L$$

$$\sigma^2(a_S) = \frac{1-P}{P N_U^2} \sum_m (a_m - a_L)^2 - \frac{2(1-P)(1-2P)}{P^2 N_U^3} \sum_m (a_m - a_L)^2$$

$$u^2(a_S) = \frac{1-P}{P a_L^2 N_U^2} \sum_m (a_m - a_L)^2 - \frac{2(1-P)(1-2P)}{P^2 a_L^2 N_U^3} \sum_m (a_m - a_L)^2$$

20.6.5. Moments of the sampling error SE :

By definition :

$$m(SE) = B(a_S) = 0$$

The sampling bias is identically zero.

$$\sigma^2(SE)_1 = u^2(a_S)_1 = \frac{1-P}{P a_L^2 N_U^2} \sum_m (a_m - a_L)^2$$

This first approximation is sufficient for all practical purposes.

20.6.6. Remark :

Although the unit weight M_m is a constant, the sample weight M_S remains a random variable with a non-zero variance. This is due to the fact that the number of units N_S in the sample S is itself a random variable. The relative variance of M_S is equal to that of N_S.

20.7. PRACTICAL IMPLEMENTATION OF THE RESULTS OF THIS CHAPTER

Two chapters will be dedicated to the practical implementation of the results of the discrete selection model :

- Chapter 22 will deal with the estimation of the sampling error when the conditions of correctness are fulfilled,

- Chapter 23 will present examples of sampling error and more specifically of sampling bias when these conditions of correctness are not fulfilled.

CHAPTER 21

LINKING UP THE CONTINUOUS AND DISCRETE MODELS

FUNDAMENTAL ERROR FE - GROUPING AND SEGREGATION ERROR GE

21.1. INTRODUCTION AND NOTATIONS

We shall recall the following notations that belong to the continuous model :

L : a lot of particulate material flowing from time $t = 0$ to time $t = T_L$.

$a(t)$: critical content of the material flowing at time t,

$\mu(t)$: rate of flow at time t,

$v_a(\theta)$: variogram of the function $a(t)$,

T_I : time during which the stream is diverted in order to extract an increment from the lot. T_I can be easily calculated from the features of the sampler. We shall assume T_I to be a constant from one increment to the next.

I_q : increment extracted from L at time t_q,

Q : number of increments extracted from L, $(q = 1, 2, \ldots Q)$,

S : sample obtained by reunion of the Q increments I_q

We repeatedly pointed out in the second part of this book that all properties tied to the particulate structure of the material to be sampled were taken into account by the "short-range quality fluctuation component" $a_1(t)$ of the critical content function $a(t)$. This component is specific to particulate materials and for instance cancels with liquids. The components $a_2(t)$ and $a_3(t)$ describe the long-range non-periodic and periodic continuous quality fluctuations where the particulate structure of the material is irrelevant. They would describe the properties of a liquid or of a compact solid as well.

The rate of flow or weighting function $\mu(t)$ describe the quantity fluctuations. In these fluctuations, the particulate structure of the material is relevant only to a minor extent : it is responsible for slight residual fluctuations that prevent the hypothesis $\mu(t) = \mu_0$ = constant from being strictly observed in actual practice. This fact will be interpreted in section 21.2.3.

The function $a_1(t)$ is objectively defined (chapters 6 and 10) as the component of $a(t)$ represented in the variogram $v_a(\theta)$ by the constant term v_{a1} which can be easily estimated from the results of a variographic experiment (chapter 14).

The short-range quality fluctuation error QE_1 is objectively defined as the component of the quality fluctuation error QE that contains the constant term v_{a1} of the variogram (chapter 10). This error accounts for all "perturbations" introduced in the continuous model by the discrete structure of the material to be sampled. When sampling a liquid or a compact solid for instance, QE_1 is likely to be zero or at least negligible. It is therefore at the level of this error QE_1 that the continuous and the discrete models can be linked up.

In order to isolate QE_1 and to analyse its moments both in the continuous and in the discrete perspectives, we shall consider the sampling of a flowing stream described by the following functions :

$a(t) = a_L + a_1(t)$ which involves $QE_2 = QE_3 = 0$ and $QE = QE_1$

$\mu(t) = \mu_0 =$ constant   which involves $WE = 0$ and $CE = QE = QE_1$.

We shall furthermore assume that both increment delimitation and extraction are correctly carried out and that the increment delimitation and extraction error are identically nil, which involves $DE = EE = 0$ and $SE = CE = QE_1$. The total sampling error SE is therefore reduced to its component QE_1.

This chapter serves two purposes :

1) We shall express the moments of the sampling error SE in the light of both the continuous and the discrete models and match the corresponding expressions.

2) We shall analyse this error and break it up into a sum of two independent components.

21.2. MOMENTS OF SE = QE_1 ACCORDING TO THE CONTINUOUS AND DISCRETE MODELS

21.2.1. Continuous perspective :

The lot L is regarded as a continuous set L_C (C for continuous) from which a certain number Q_C of increments are extracted. According to the results of chapter 10, the moments of the sampling error SE can be expressed as follows, irrespective of the selection scheme :

$m_C(SE) = m(QE_1) = 0$

$\sigma_C^2(SE) = \sigma^2(QE_1) = \dfrac{v_{a1}}{Q_C}$

21.2.2. Discrete perspective :

The same lot L is regarded as a discrete set of N_U units U_m of the same average bulk as the increments I_q taken into consideration in the continuous model (those of the variographic experiment) and playing here the part of "potential increments". By definition : $N_U = T_L / T_I$

Two models have been developed in order to study the sampling of this set of N_U units :

1) <u>the discrete selection model</u> devised in chapter 20 of this work. According to this model, each unit U_m is submitted to the selecting process with a certain probability P_m of being selected. We shall assume here that this selection is correct and characterized by a uniform probability P. If we chose $P = Q_C/N_U$, then the number Q_D (subscript D for discrete) of units actually extracted from the lot L is, as already pointed out in chapter 20, a random variable with mean :

$$m(Q_D) = P\, N_U = Q_C$$

In the same chapter 20 we expressed the moments of the sampling error SE (section 20.5.).

$$m_D(SE)_1 = 0 \qquad \text{(first approximation)}$$

$$m_D(SE)_2 = -\frac{1-P}{P\, a_L\, M_L^2} \sum_m (a_m - a_L)\, M_m^2 \qquad \text{(second approximation)}$$

$$\sigma_D^2(SE)_1 = \frac{1-P}{P\, a_L^2\, M_L^2} \sum_m (a_m - a_L)^2\, M_m^2 \qquad \text{(first approximation)}$$

2) <u>the "equiprobable model"</u> presented in previous works (for instance Gy, 1953). According to this model (the interest of which is purely historical), the set of N_U units is assumed to be sampled in such a way that a definite number Q_E (subscript E for equiprobable) of units are extracted from the lot, independently, one by one and at random, with an equal probability (hence its name) for all combinations of Q_E units of being obtained. If we introduce the sampling ratio :

$$\tau = \frac{Q_E}{N_U}$$

the moments of the sampling error can be written, with the notations of the present chapter :

$$m_E(SE)_1 = 0 \qquad \text{(first approximation)}$$

$$m_E(SE)_2 = -\frac{1-\tau}{\tau\, a_L\, M_L^2} \sum_m (a_m - a_L)\, M_m^2 \qquad \text{(second approximation)}$$

$$\sigma_E^2(SE)_1 = \frac{1-\tau}{\tau\, a_L^2\, M_L^2} \sum_m (a_m - a_L)^2 M_m^2 \qquad \text{(first approximation)}$$

Although the approaches are different and independent (Q_E is a definite number whereas Q_D is a random variable), the moments are expressed by similar formulas, with the discrete selection and equiprobable models, the only difference being that the "a priori" notion of selecting probability P at the basis of the discrete model is replaced in the equiprobable model by the "a posteriori" notion of sampling ratio τ. In the discrete model, the sampling ratio τ is a random variable with mean :

$m(\tau) = P$

This quick recall being made, the equiprobable model will not be mentioned any more.

21.2.3. Comparison of the results of the continuous and discrete models :

The purpose of this section is to show that both models are consistent within the limits of the hypotheses assumed in their respective definitions.

21.2.3.1. Expression of the mean m(SE) :

$$m_C(SE) = m_D(SE)_1 = 0$$

Both expressions of the mean are nil, at least, as far as $m_D(SE)$ is concerned in first approximation. There remains to interpret the fact that in second approximation $m_D(SE)$ is not zero. We assumed in section 21.1. that the rate of flow $\mu(t)$ was a constant, recalling that due to the particulate structure of the material to be sampled this condition could not be strictly respected in practice. The hypothesis $\mu(t) = \mu_o$ = constant assumed in the continuous definition of L would, if assumed also in the discrete perspective, involve :

$$M_m = \mu_o T_I = M_L T_I / T_L = M_L / N_U = \text{constant}$$

and would obviously result (section 20.6) in :

$$m_D(SE)_2 = 0$$

Both models are therefore consistent. From a practical standpoint, even assuming the best regulation available, the function $\mu(t)$ remains slightly modulated about its mean μ_o with the following consequences :

1) $M_m = \mu(t) T_I \neq \mu_o T_I$ is not a constant but a time variable,

2) $m_D(SE)_2 \neq 0$ can accordingly be interpreted as a residual weighting bias m(W) (see chapter 13) which is more readily apprehended by the discrete than by the continuous model.

21.2.3.2. Expression of the variance $\sigma^2(SE)$:

$$\sigma^2_C(SE) = \frac{v_{a1}}{Q_C}$$

$$\sigma^2_D(SE) = \frac{1-P}{P \, a_L^2 \, M_L^2} \sum_m (a_m - a_L)^2 M_m^2$$

The hypotheses made in section 21.1. involve that v_{a1} is equal to the relative variance $u^2(a_q)$ of the critical contents of the increments extracted during the variographic experiment (chapter 14), itself an estimate of the relative variance $u^2(a_m)$ of the whole set. Then, for all practical purposes, we can write :

$$v_{a1} = u^2(a_q) = u^2(a_m) = \frac{1}{N_U \, a_L^2} \sum_m (a_m - a_L)^2$$

$$\sigma_C^2(SE) = \frac{1}{N_U \, Q_C \, a_L^2} \sum_m (a_m - a_L)^2$$

Now, if we assume in the discrete model that :

$$M_m = M_L / N_U$$

which is the result of the constant rate of flow hypothesis retained with the continuous model in section 21.1., and if we replace the selection probability P by its a posteriori equivalent the sampling ratio $\tau = Q_D / N_U$, we can write :

$$\sigma_D^2(SE) = \left\{\frac{1}{Q_D} - \frac{1}{N_U}\right\} \frac{1}{N_U \, a_L^2} \sum_m (a_m - a_L)^2$$

But, if Q_D is negligible in comparison with N_U, which is always true with the increment process, then $1/N_U$ is negligible in comparison with $1/Q_D$ and both formulas become equivalent :

$$\sigma_D^2(SE) = \sigma_C^2(SE) = \frac{1}{Q} u^2(a_m)$$

Both models are therefore consistent and lead to the same results if the same hypotheses are assumed in both cases. But as far as the error QE_1 is concerned, tied as it is to the particulate structure of the material to be sampled, the discrete model is obviously closer to reality and definitely more efficient than the continuous model, which is indeed the reason for which it has been developed.

21.3. ANALYSIS OF THE SHORT-RANGE QUALITY FLUCTUATION ERROR QE_1 – DEFINITION OF THE FUNDAMENTAL ERROR FE AND OF THE GROUPING AND SEGREGATION ERROR GE

21.3.1. Incidence of the increment size on the variance $\sigma^2(QE_1)$:

Consider a lot L of particulate material and assume it to be correctly sampled with uniform selection probability P according to the two following schemes :

1) L is regarded as the set L_F of N_F individual fragments F_i and each of these is submitted to the selecting process with a selection probability P,

2) L is regarded as the set L_G of N_G groups G_n of fragments, the potential increments introduced in section 21.2.2. assumed to be undissociable, and each of these is submitted to the selecting process with the same selecting probability P as above.

The variance $\sigma^2(QE_1)$ is, according to the conclusions of the preceding section, equal to the variance $\sigma^2(SE)$ which has been expressed for these two schemes in section 20.5.4.3. We shall use the subscripts F and G to distinguish the two sampling schemes :

$$\sigma_F^2(QE_1) = \sigma_F^2(SE) = \frac{1-P}{P \, N_F} CH_L \quad \text{and} \quad \sigma_G^2(QE_1) = \sigma_G^2(SE) = \frac{1-P}{P \, N_G} DH_L$$

Now, according to the results of section 19.3.5.6. the distribution heterogeneity DH_L is related to the constitution heterogeneity CH_L in the following way :

$$DH_L = \frac{1 + \gamma\xi}{1 + \gamma} CH_L = (1 + \gamma\xi) \frac{N_G - 1}{N_F - 1} CH_L$$

Then, it readily follows :

$$\sigma_G^2(QE_1) = (1 + \gamma\xi) \frac{N_F(N_G - 1)}{(N_F - 1)N_G} \sigma_F^2(QE_1)$$

or, more simply, as N_G and N_F are always large numbers :

$$\sigma_G^2(QE_1) = (1 + \gamma\xi) \sigma_F^2(QE_1)$$

Now, as γ and ξ are non-negative quantities (at least in the natural range of the variance $\sigma_F^2(QE_1)$ appears in this equality as the minimum of $\sigma_G^2(QE_1)$. In plain words, this property means that for a given lot L and for a given selection probability P, the variance of QE_1 cannot be smaller than the variance attained when the selection process is applied to the set of individual fragments. This minimum is reached when the product $\gamma\xi$ is zero which is achieved when either γ or ξ is nil. This point will be developed in section 21.3.3.

21.3.2. Definition of the fundamental error FE :

We call "<u>fundamental error</u>" FE the error arising when implementing a correct selection scheme, with uniform selection probability P, on the set L_F of individual fragments F_i. The moments of FE have been expressed in sections 20.5.1. and 20.5.2. :

$$m(FE) = -\frac{1 - P}{P\, a_L\, M_L^2} \sum_i (a_i - a_L)\, M_i^2 \qquad \text{with } i = 1, 2, \ldots N_F$$

$$\sigma^2(FE) = \frac{1 - P}{P\, a_L^2\, M_L^2} \sum_i (a_i - a_L)^2\, M_i^2 = \frac{1 - P}{P\, N_F} CH_L$$

$m(FE)$ and $\sigma^2(FE)$ will be referred to as "<u>fundamental bias</u> and <u>fundamental variance</u>". The qualificative "fundamental" underlines the fact that for a given lot and a given selection probability, the error QE_1 as well as the total sampling error S can never be smaller than FE. All errors can be cancelled, except the fundamental error FE which is to be regarded as an incompressible minimum.

In terms of fundamental error, the variance $\sigma_G^2(QE_1)$ can be written :

$$\sigma_G^2(QE_1) = (1 + \gamma\xi)\, \sigma^2(FE)$$

This simple expression justifies the definition given to the grouping factor γ (section 19.3.5.3.) and that given to the segregation factor ξ (section 19.3.5.6.)

Let's now precise our notations and call :

N_{FL} : number of fragments in the lot L (formerly N_F),

N_{FS} : number of fragments in the sample S (formerly N_S).

According to the results of section 20.5.1. :

$$m(N_{FS}) = P \, N_{FL}$$

Then, the variance of FE writes :

$$\sigma^2(FE) = \left(\frac{1}{m(N_{FS})} - \frac{1}{N_{FL}}\right) CH_L$$

or, when estimating the fundamental variance of a sample made of N_{FS} fragments :

$$\sigma^2(FE) = \left(\frac{1}{N_{FS}} - \frac{1}{N_{FL}}\right) CH_L$$

As the sample is very often very small in comparison with the lot, then $1/N_{FL}$ is negligible in comparison with $1/N_{FS}$, this already simple expression reduces itself to a still simpler one :

$$\sigma^2(FE) = \frac{CH_L}{N_{FS}}$$

In plain words, the fundamental variance is proportional to the constitution heterogeneity CH_L of the lot and inversely proportional to the number of fragments in the sample.

21.3.3. Definition of the grouping and segregation error GE :

We call "grouping and segregation error" GE (and sometimes more simply segregation error), the error defined by the following moments :

$$m(GE) = m_G(QE_1) - m_F(QE_1) = m_G(QE_1) - m(FE)$$

$$\sigma^2(GE) = \sigma_G^2(QE_1) - \sigma_F^2(QE_1) = \sigma_G^2(QE_1) - \sigma^2(FE) = \gamma\xi\sigma^2(FE) = \frac{1-P}{P \, N_{FL}} \gamma\xi \, CH_L$$

The name of this error is obviously justified by the fact that its variance is proportional to the grouping factor γ and to the segregation factor ξ. Its moments can be written as follows :

$$m(GE) = -\frac{1-P}{P \, a_L \, M_L^2} \left[\sum_n (a_n - a_L) M_n^2 - \sum_i (a_i - a_L) M_i^2 \right]$$

$$\sigma^2(GE) = \frac{1-P}{P \, a_L^2 \, M_L^2} \left[\sum_n (a_n - a_L)^2 M_n^2 - \sum_i (a_i - a_L)^2 M_i^2 \right] = \left(\frac{1}{N_{FS}} - \frac{1}{N_{FL}}\right) \gamma\xi CH_L = \frac{\gamma\xi CH_L}{N_{FS}}$$

21.3.4. Breaking up of QE_1 :

From the definitions of FE and GE it follows that :

$$m_G(QE_1) = m(FE) + m(GE)$$

$$\sigma_G^2(QE_1) = \sigma^2(FE) + \sigma^2(GE)$$

These properties will be symbolically summarized by :

$$QE_1 = FE + GE$$

The short-range quality fluctuation error QE_1 defined in the development of the continuous selection model can be regarded as the sum of two independent errors :

- the fundamental error FE
- the grouping and segregation error GE

21.4. CANCELLING AND MINIMIZING OF THE FUNDAMENTAL ERROR FE

21.4.1. Cancelling of FE :

The fundamental error would be cancelled if both its mean and variance could be cancelled. Consider the variance as expressed in section 21.3.2. :

$$\sigma^2(FE) = \frac{1-P}{P\, N_{FL}} CH_L$$

From a mathematical standpoint, the equation : $\sigma^2(FE) = 0$ admits two solutions

1) $P = 1$: The selecting probability is equal to unity. Then all fragments F_i are necessarily recovered in the sample. It is definitely reassuring for the theoretician to find out that the sampling error vanishes when the sample is identical with the lot. The contrary would be worrying.

2) $CH_L = 0$: The constitution heterogeneity of the lot is zero. This property has already been reviewed in section 19.3.5.1. We know that CH_L cancels if and only if the material to be sampled is perfectly homogeneous, all fragments having the same critical content a_L as the lot itself.

The same solutions obviously cancel the fundamental bias too.

From a practical standpoint, we can state that the fundamental error never cancels as perfect constitution homogeneity is never actually observed. For a given material, assumed to be heterogeneous, we can, theoretically at least, define conditions that would cancel all sampling errors, except the fundamental error FE. This error does not depend on our awkwardness nor does it depend on the amount of money we accept to spend : it solely depends on the mineralogical constitution of the material to be sampled which is one of the intangible data of the problem.

21.4.2. Minimizing of FE :

According to their definitions, both the mean and variance of FE depend on the same factors : the set of values of a_i and M_i and the selecting probability P. Depending on the problem to be solved, we can attempt to minimize :

- the bias $m(FE)$
- the variance $\sigma^2(FE)$
- the mean square $r^2(FE) = m^2(FE) + \sigma^2(FE) = \{1 + b^2(FE)\} \sigma^2(FE)$

with $b(FE)$ coefficient of bias (defined in section 1.3.).

Now, with the possible exception of very low grade ores such as ores of precious metals or minerals, the fundamental bias is always negligible, as well in absolute as in relative value. Then, for all practical purposes, minimizing FE amounts to minimizing its variance. This can be achieved in two different ways :

1) By reducing the factor $(1 - P)/P$, i.e. by increasing the selecting probability P. In plain words, by increasing the sample weight. But the heavier the sample, the higher the processing cost, which means that the economical optimum must result from a compromise between reproducibility and cost. There are no general rules to solve this problem but according to our practical experience, the usual tendency is to put the emphasis on economy rather than on sample representativeness, which can be dangerous. We shall come back to this point in section 29.2.

2) By reducing the factor CH_L/N_{FL}. We know from section 19.3.5.8. that comminution increases the constitution heterogeneity CH_L, but we shall see in chapter 22 that it reduces the quotient CH_L/N_{FL}. We shall therefore minimize the fundamental error FE by crushing or grinding the material prior to its sampling whenever possible. But comminution is not always possible and when it is, it costs money. Here again, the economical optimum lies in a compromise between reproducibility and cost.

21.5. CANCELLING AND MINIMIZING OF THE GROUPING AND SEGREGATION ERROR GE

21.5.1. Cancelling GE :

Consider the variance as expressed in section 21.3.3. :

$\sigma^2(GE) = \gamma \xi \, \sigma^2(FE)$

From a mathematical standpoint, we can think of three solutions :

1) $\sigma^2(FE) = 0$: this point has already been dealt with in section 21.4.1.

2) $\gamma = 0$: according to its definition (section 19.3.5.3.) γ cancels out when the groups (in the present case the increments) are made of one and only one particle.

3) $\xi = 0$: according to its definition (section 19.3.5.6.) ξ cancels out when the distribution of the fragments throughout the domain occupied by the lot is "naturally homogeneous" or in other words, random.

From a practical standpoint, we already know that the first solution is ruled out. The second one can practically never be achieved. The third one corresponds to a thorough mixing of the material prior to its sampling (see remarks of section 19.3.5.4.).

21.5.2. Minimizing GE :

We can act on the same three factors in order to minimize the variance $\sigma^2(GE)$:

1) Minimizing $\sigma^2(FE)$: see section 21.4.2.

2) Minimizing the grouping factor γ : this is achieved by extracting increments as small as possible. From a practical standpoint, the limit is dictated by the absolute necessity of avoiding the increment extraction error EE. This point has been dealt with in section 18.7. where the notion of "minimum correct increment weight" has been introduced. This reinforces the importance of the economical optimum defined in section 18.4.5.3.

3) Minimizing the segregation factor ξ : when sampling flowing streams of particulate materials, we have practically no means of acting on ξ : we must take it as it is.

Mixing, which would reduce ξ, is limited to lots small enough to fall within the range of three-dimensional mechanical mixers such as the Vee-mixer. Such devices are very efficient but can handle no more than a few hundred kilograms.

Blending and more specifically bed-blending is well adapted to flowing streams but it must be clearly understood that it does not achieve a three-dimensional homogenizing which alone would reduce ξ. A recent theory of bed-blending, derived from the present theory of sampling and yet unpublished, shows that bed-blending does not practically affect the segregation factor ξ.

CHAPTER 22

PRACTICAL IMPLEMENTATION OF THE THEORETICAL RESULTS - CORRECT SELECTION

22.1. INTRODUCTION

It is always possible to implement a correct sampling by carrying out :
- a correct selection (chapters 7 and 8),
- a correct increment delimitation (chapter 17),
- a correct increment extraction (chapter 18).

We know that when all these conditions of correctness are respected the sampling operation may be assimilated to the model developed in chapter 20, the total sampling error being reduced to the short-range quality fluctuation error QE_1. We also know that this error QE_1 may be regarded as the sum of two independent errors the fundamental error FE and the grouping and segregation error GE.

From a practical standpoint, the first problem to be solved is to estimate the moments of the fundamental error FE which alone can never be cancelled. The present chapter deals with the practical solution of this problem, assuming the selection to be correct and characterized by a uniform selecting probability P. Examples of how to calculate the fundamental bias and variance by various methods will provide a practical illustration of the theoretical results.

The practical estimation of the moments of the sampling error involved when implementing an incorrect sampling scheme characterized by a non-uniform selecting probability will be dealt with in chapter 23.

22.2. ESTIMATION OF THE MOMENTS OF THE FUNDAMENTAL ERROR FE - INTRODUCTION OF Y AND Z

Combining the results of section 20.5.3 with the definition of the fundamental error given in section 21.3.2. we obtain, remembering that the second equality is only approximative :

$$m(FE) = - \frac{1 - P}{P \, a_L \, M_L^2} \sum_i (a_i - a_L) \, M_i^2 = - \frac{1 - P}{P \, a_L \, M_L^2} \sum_\alpha \sum_\beta (a_{\alpha\beta} - a_L) \, M_{\alpha\beta} \, M_{F\alpha\beta}$$

$$\sigma^2(FE) = \frac{1 - P}{P \, a_L^2 \, M_L^2} \sum_i (a_i - a_L)^2 \, M_i^2 = \frac{1 - P}{P \, a_L^2 \, M_L^2} \sum_\alpha \sum_\beta (a_{\alpha\beta} - a_L)^2 \, M_{\alpha\beta} \, M_{F\alpha\beta}$$

with the following notations already introduced in section 20.4.2. but slightly lightened by omission of the subscript L wherever it is not absolutely necessary :

L_α : a size fraction of L. It is characterized by d_α average particle diameter, M_α weight of L_α, a_α critical content of L_α, N_α number of fragments in L_α, F_α average fragment of L_α, v_α volume of F_α.

$L_{\alpha\beta}$: a density fraction of L_α. It is characterized by an average specific gravity λ_β, a weight $M_{\alpha\beta}$, a critical content $a_{\alpha\beta}$, a number of fragments $N_{\alpha\beta}$, an average fragment $F_{\alpha\beta}$, the weight $M_{F\alpha\beta}$ of $F_{\alpha\beta}$ which is defined as follows :

$$M_{F\alpha\beta} = \frac{M_{\alpha\beta}}{N_{\alpha\beta}} = v_\alpha \lambda_\beta$$

L_β : a density fraction of L. It is characterized by a density λ_β, a weight M_β, a critical content a_β. By definition :

$$L \equiv \sum_\alpha L_\alpha \equiv \sum_\beta L_\beta \equiv \sum_\alpha \sum_\beta L_{\alpha\beta}$$

We shall now introduce the parameters Y and Z defined as follows :

$$Y = \frac{1}{a_L M_L} \sum_i (a_i - a_L) M_i^2 = \frac{1}{a_L M_L} \sum_\alpha v_\alpha \sum_\beta \lambda_\beta (a_{\alpha\beta} - a_L) M_{\alpha\beta} \quad \text{(dimension of a weight)}$$

$$Z = \frac{1}{a_L^2 M_L} \sum_i (a_i - a_L)^2 M_i^2 = \frac{1}{a_L^2 M_L} \sum_\alpha v_\alpha \sum_\beta \lambda_\beta (a_{\alpha\beta} - a_L)^2 M_{\alpha\beta} \quad \text{(dimension of a weight)}$$

Then, M_S being the sample weight assimilated to its mean $M_S = P M_L$:

$$m(FE) = -\frac{1-P}{P M_L} Y = -\left[\frac{1}{M_S} - \frac{1}{M_L}\right] Y$$

$$\sigma^2(FE) = \frac{1-P}{P M_L} Z = \left[\frac{1}{M_S} - \frac{1}{M_L}\right] Z$$

Y and Z are intrinsic properties of the material to be sampled, irrespective of the bulk of the lot or of the representative sample on which they are estimated. The purpose of the next sections is to present three methods for the estimation of Y and Z.

22.3. ESTIMATION OF Y AND Z - METHOD No. 1

22.3.1. The critical component is a mineral :

In other words, we are sampling for assaying. This first method consists of a direct exploitation of the results of a complete size-density analysis. From a practical standpoint, density analysis can be achieved by means of heavy liquids for specific gravities up to 3.33 and heavy suspensions for higher specific gravities. Desnoes (1965) successfully developed a simple method using suspensions of mercury droplets in bromoform for specific gravities up to 5.0.

Such size-density analyses are sometimes available and for instance in connection with a mineral processing investigation. In some cases, the cost of a size-density analysis is found to be acceptable for the sole purpose of estimating the moments of the fundamental error. We shall assume in the next paragraphs that this analysis has been carried out on a representative sample of the material, the "working batch" for which the notation L will be retained.

The method consists of the following operations :

1) Selection of a series of screen apertures and of a series of densities of heavy liquids or suspensions. The larger the number of size and density fractions, the closer the approximation but also the higher the cost. The finest screen of the series should not be finer than 0.2 mm (about 80 mesh) because finer materials cannot be accurately separated in heavy suspensions.

2) Wet scrubbing of the batch with the purpose of liberating the agglomerated particles (of clay for instance), decanting on the finest screen, washing and filtering, drying of the oversize and undersize separately.

3) Dry screening of the dry oversize. The wet and dry undersizes obtained below the finest screen are gathered (fraction L_u), weighed (weight M_u) and assayed (critical content a_u).

4) The other size fractions are then separately fractionned in heavy media. All size-density fractions are dried, weighed and assayed and the values of $M_{\alpha\beta}$ and $a_{\alpha\beta}$ are tabulated.

5) The volume v_α of the average particle F_α of the size fraction L_α can be directly measured if the size is coarse enough (say coarser than a few mm). For finer fractions it is easier and accurate enough to estimate :

$$v_\alpha = f_\alpha \frac{d^3_{\alpha 1} + d^3_{\alpha 2}}{2} = f_\alpha \, d^3_\alpha$$

where f_α is a shape factor (dimensionless) usually near 0.5, d_α the diameter of F_α, $d_{\alpha 1}$ and $d_{\alpha 2}$ the openings of the sieves used as upper and lower limits of the size fraction L_α.

6) The density λ_β is usually estimated as the average :

$$\lambda_\beta = \frac{\lambda_{\beta 1} + \lambda_{\beta 2}}{2}$$

where $\lambda_{\beta 1}$ is the density of the liquid or suspension on the surface of which $L_{\alpha\beta}$ floats and $\lambda_{\beta 2}$ that of the liquid or suspension in which $L_{\alpha\beta}$ sinks.

7) The weight M_L of the lot is calculated as :

$$M_L = \sum_\alpha \sum_\beta M_{\alpha\beta} + M_u$$

and the critical content a_L as :

$$a_L = \frac{1}{M_L} (\sum_\alpha \sum_\beta a_{\alpha\beta} M_{\alpha\beta} + a_u M_u)$$

8) the values of Y and Z are calculated according to the formulas given in section 22.2.

9) <u>remark</u> : this method is certainly costly, time consuming but it is also very accurate. An example of its application is given in section 22.8.1.

22.3.2. <u>The critical component is a size fraction</u> :

In other words, we are sampling for a size analysis. The expressions of Y and Z are valid, irrespective of the critical component taken into consideration. In the present case, their expressions can be simplified as follows. We shall call :

$L_{\alpha c}$: the critical size fraction, weight $M_{\alpha c}$, average fragment $F_{\alpha c}$ with a volume $v_{\alpha c}$, average density $\lambda_{\alpha c}$ usually not very different from the average density λ_L of the lot L.

We are interested in the proportion $M_{\lambda c}/M_L$ which is by definition the critical content a_L to be estimated :

$$a_L = \frac{M_{\alpha c}}{M_L}$$

The content $a_{\alpha\beta}$ is the proportion of particles belonging to $L_{\alpha c}$ in the fraction $L_{\alpha\beta}$. Obviously :

- if $L_{\alpha\beta} = L_{\alpha c}$: then $a_{\alpha\beta} = 1$ irrespective of β.
- if $L_{\alpha\beta} \neq L_{\alpha c}$: then $a_{\alpha\beta} = 0$ irrespective of β.

which involves :

$$Y = v_{\alpha c} \lambda_{\alpha c} - \sum_\alpha v_\alpha \sum_\beta \lambda_\beta \frac{M_{\alpha\beta}}{M_L}$$

$$Z = \left[\frac{1}{a_L} - 2 \right] v_{\alpha c} \lambda_{\alpha c} + \sum_\alpha v_\alpha \sum_\beta \lambda_\beta \frac{M_{\alpha\beta}}{M_L}$$

An example of application of these results will be presented in section 22.8.1.

22.4. ESTIMATION OF Y AND Z - METHOD No.2

The first method is seldom applicable : it requires a lot of experimental work and costs a lot of money which Superintendents are reluctant to spend. It was therefore necessary to develop a much simpler, quicker and cheaper method. The method which is described in this section was developed and presented 25 years ago (Gy, 1953). It involves new approximations but the loss in accuracy is largely compensated by a huge gain in time and money. This method is specific of the estimation of Z when the critical component is a well defined mineral.

22.4.1. Principle of the method :

Let's recall the definition of Y and Z given in section 22.2. :

$$Y = \sum_\alpha v_\alpha \sum_\beta \lambda_\beta \frac{(a_{\alpha\beta} - a_L) M_{\alpha\beta}}{a_L \; M_L}$$

$$Z = \sum_\alpha v_\alpha \sum_\beta \lambda_\beta \frac{(a_{\alpha\beta} - a_L)^2 M_{\alpha\beta}}{a_L^2 \; M_L}$$

The purpose of the second method is to simplify these formulas, knowing that the simplified formula of Y will be used in the third method. We shall assume two hypotheses defining new approximations :

1) <u>First hypothesis</u> : experience teaches that the critical content $a_{\alpha\beta}$ usually varies much more from one density fraction to the next than from one size fraction to the next. We shall therefore assume that all values of $a_{\alpha\beta}$ may be replaced by the average critical content a_β of the density fraction L_β : then we shall retain:
$a_{\alpha\beta} = a_\beta$ irrespective of α.

2) <u>Second hypothesis</u> : the study of a number of practical examples shows that the proportions $M_{\alpha\beta}/M_\beta$ usually vary little from one density fraction to the next. We shall therefore assume that all values of $M_{\alpha\beta}/M_\beta$ can be replaced by their average M_α/M_L. Then, we shall retain :
$$\frac{M_{\alpha\beta}}{M_\beta} = \frac{M_\alpha}{M_L} \text{ irrespective of } \beta.$$

Thanks to these hypotheses, actually supported by experience, Y and Z which were double sums can be regarded now as the products of two single sums :

$$Y = \left[\sum_\alpha v_\alpha \frac{M_\alpha}{M_L}\right] \times \left[\sum_\beta \lambda_\beta \frac{(a_\beta - a_L) M_\beta}{a_L \; M_L}\right] = Y_1 \; Y_2$$

$$Z = \left[\sum_\alpha v_\alpha \frac{M_\alpha}{M_L}\right] \times \left[\sum_\beta \lambda_\beta \frac{(a_\beta - a_L)^2 M_\beta}{a_L^2 \; M_L}\right] = Z_1 \; Z_2$$

with :

$$Y_1 = Z_1 = \sum_\alpha v_\alpha \frac{M_\alpha}{M_L} \qquad Y_2 = \sum_\beta \lambda_\beta \frac{(a_\beta - a_L) M_\beta}{a_L \; M_L} \quad \text{and} \quad Z_2 = \sum_\beta \lambda_\beta \frac{(a_\beta - a_L)^2 M_\beta}{a_L^2 \; M_L}$$

Our purpose is now to allow a quick, easy and cheap estimation of $Z = Z_1 Z_2$, knowing that the sampling is correct (which we know how to achieve) and practically unbiased with Y negligible. The breaking up of Y into a product of two terms will be used in method No.3.

22.4.2. Estimation of $Y_1 = Z_1$:

According to its definition :

$$Z_1 = \sum_\alpha v_\alpha \frac{M_\alpha}{M_L}$$

We have previously defined (section 22.3.1.) the volume v_α :

$$v_\alpha = f_\alpha d_\alpha^3 = f_\alpha \frac{d_{\alpha 1}^3 + d_{\alpha 2}^3}{2} \qquad \text{hence :} \qquad Z_1 = \sum_\alpha f_\alpha d_\alpha^3 \frac{M_\alpha}{M_L}$$

This expression calls for the following remarks :

1) The "shape factor" f can be defined as a coefficient of cubicity. Accurate estimations carried out on quite a large variety of pure minerals and natural ore and for a given material on a number of size classes showed (Gy, 1967, section 5.) that, irrespective of the material (with a few duly mentioned exceptions) and of the size class, the shape factor f_α deviates very little from an average value :

f = 0.5 (dimensionless)

For a sphere f = 0.524. For a cube f = 1.0. From a practical standpoint we sha therefore admit that :
- for a given material, f_α can be regarded as a constant f, independent of α.
- whenever the particles are neither flaky nor elongated but are roughly "spheroïdal", f takes the average value : f = 0.5 (dimensionless).

2) We shall characterize an aggregate by means of two parameters :

d : the "nominal particle diameter" or "maximum particle size" or "top particl size", defined as the opening of the square mesh retaining a 5 % oversize,

g : the "size range factor" defined as follows :

$$g = \sum_\alpha \frac{M_\alpha}{M_L} \frac{d_\alpha^3}{d^3} \qquad \text{that can also be written :} \qquad \sum_\alpha \frac{M_\alpha}{M_L} d_\alpha^3 = g\, d^3$$

It so happens that, with the definition retained for d, the factor g (dimensio less) usually deviates very little from the average value : g = 0.25. This point has been extensively studied in a previous work (Gy, 1967, section 5.4.).

According to these definitions the sum Z_1 can be expressed as the product :

$Z_1 = f\, g\, d^3$

In this expression, f, g and d can be readily estimated without any experimen tation (see section 22.6.3. to 22.6.5. for more details).

22.4.3. Estimation of Z_2 :

By definition :

$$Z_2 = \sum_\beta \lambda_\beta \frac{(a_\beta - a_L)^2 M_\beta}{a_L^2 M_L}$$

Here again, the general idea is to transform Z_2 into a product of simple facto easy to estimate, characterizing all relevant properties of the material to be sampled. We shall first observe that two extreme values of Z_2 can be defined :

1) <u>Minimum of Z_2</u> : if the constitution of the material is assumed to be homogeneous, then (see section 19.2.) :

$a_\beta = a_L$ irrespective of β. Hence : $(Z_2)_{min} = 0$

2) <u>Maximum of Z_2</u> : if the mineralogical components are now assumed to be completely liberated, each density fraction L_β is made of a pure mineral. If we call $L_{\beta c}$ the density class containing the critical component (density λ_c) we have :

- if $L_\beta \equiv L_{\beta c}$ then $a_\beta = 1$ and $\dfrac{M_\beta}{M_L} = a_L$

$$\lambda_\beta \frac{(a_\beta - a_L)^2}{a_L^2} \frac{M_\beta}{M_L} = \lambda_c \frac{(1 - a_L)^2}{a_L}$$

- if $L_\beta \neq L_{\beta c}$ then $a_\beta = 0$ and $\lambda_\beta \dfrac{(a_\beta - a_L)^2}{a_L^2} \dfrac{M_\beta}{M_L} = \lambda_\beta \dfrac{M_\beta}{M_L}$

We shall call "<u>mineralogical composition factor</u>" c this maximum of Z_2 :

$$c = (Z_2)_{max} = \lambda_c \frac{(1 - a_L)^2}{a_L} + \sum_\beta \lambda_\beta \frac{M_\beta}{M_L} \quad \text{with} \quad \lambda_\beta \neq \lambda_c$$

With a number of ores and concentrates it is very often possible to admit that the material is made of two density fractions :

- the critical (often valuable) component (density λ_c),
- the non-critical (often valueless) components (density λ_g) usually referred to as "<u>gangue minerals</u>". Then :

$$c = \frac{1 - a_L}{a_L} \{(1 - a_L) \lambda_c + a_L \lambda_g\}$$

The value of c can be easily estimated as soon as the mineralogical composition of the material is known, at least approximately. This point will be developed in section 22.6.1. The dimension of c is that of a specific gravity. Since these are usually expressed in grams per cubic centimeter (or tons per cubic meters), it is convenient to express c in g/cm^3.

From the preceding remarks, it follows that :

$0 \leq Z_2 \leq c$

In order to recall this property, we shall write :

$Z_2 = c \ell$ with $0 \leq \ell \leq 1$

This expression defines the dimensionless "<u>liberation factor</u>" ℓ. We shall describe (section 22.6.2.) how this factor can be estimated but if no better estimate is available, it is always safe to assume the maximal value : $\ell = 1$.

22.4.4. Simplified expressions of Z and $\sigma^2(FE)$:

From the results and definitions of the preceding sections, it readily follows

$$Z = Z_1 Z_2 = c \ell f g d^3 = C d^3 \quad \text{and}$$

$$\sigma^2(FE) = \left[\frac{1}{M_S} - \frac{1}{M_L}\right] C d^3$$

In this expression, C is the "sampling constant", with :

$$C = c \ell f g$$

Units : when using this formula, it is imperative to use a consistent system of units. Since c and C have the dimension of a specific gravity which is usually expressed in g/cm^3, the maximum particle diameter d must be expressed in centimete (cm) and the weights in grams (g). The factors ℓ, f , g are dimensionless.

Section 22.6. will be dedicated to a more detailed study of the factors c , ℓ , f , g , sections 22.7. and 22.8. to numerical examples.

22.4.5. Conclusions concerning the second method :

At the price of a few minor approximations, we have been able to express the fundamental variance as the product of several factors, each taking into account one and only one relevant property of the material to be sampled. These factors can be easily estimated in each particular case according to rules that will be presented in section 22.6. and illustrated in section 22.7.

This second method is quick and cheap as it requires no experimentation. It is the choice method when sampling ores and minerals. We shall now describe a third method, more accurate than the second one but much cheaper than the first.

22.5. ESTIMATION OF Y AND Z - METHOD No.3

This method can be used either when difficulties arise in estimating the facto c and ℓ (which may happen with metallurgical products such as slags or mattes) or when a higher degree of accuracy is required in the estimation of $m(FE)$ and $\sigma^2(FE)$

This third method is not applicable when the critical component is either moisture or a size fraction. Let's recall :

$$Y = Y_1 Y_2 \quad \text{and} \quad Z = Z_1 Z_2 \quad \text{with :}$$

$$Y_1 = Z_1 = \sum_\alpha v_\alpha \frac{M_\alpha}{M_L} = f g d^3$$

which can be easily estimated and with a good accuracy,

$$Y_2 = \sum_\beta \lambda_\beta \frac{(a_\beta - a_L) M_\beta}{a_L \quad M_L} \quad \text{and} \quad Z_2 = \sum_\beta \lambda_\beta \frac{(a_\beta - a_L)^2 M_\beta}{a_L^2 \quad M_L}$$

The third method is based on the observation that $a_{\alpha\beta}$ and $M_{\alpha\beta}/M_L$ vary relatively little from one size fraction to the next and that we can estimate Y_2 and Z_2 on a representative random sample of the coarsest size fraction L_1. It consists in :

1) Extracting Q fragments F_q (q = 1, 2, ... Q with Q larger than or equal to 50) one by one and at random from the coarsest size fraction L_1. This fraction does not have to be isolated : the fragments F_q can be selected among the coarsest fragments of the lot, fragments that can be easily located by the sampling operator.

2) Measuring the volume of the Q fragments F_q (volume v_q)

3) Drying the Q fragments

4) Weighing the Q dry fragments (weight M_q)

5) Assaying the Q fragments individually (critical content a_q)

6) Calculating the weight $M_{L1} = \sum_q M_q$ of the batch L_1 submitted to the test,

7) Calculating the critical content $a_{L1} = \sum_q a_q \dfrac{M_q}{M_{L1}}$

8) Calculating the estimates Y_2' and Z_2' of Y_2 and Z_2 with :

$$Y_2' = \sum_q \frac{(a_q - a_{L1}) M_q^2}{v_q \, a_{L1} \, M_{L1}} \quad \text{and} \quad Z_2' = \sum_q \frac{(a_q - a_{L1})^2 M_q^2}{v_q \, a_{L1}^2 \, M_{L1}}$$

So described, the method is applicable to maximum particle diameters larger than 2 cm (or 20 mm). A more detailed modus operandi is available in Gy, 1975, section 22.5.4.2.

It should be observed that the average fragment volume :

$$\bar{v} = \frac{1}{Q} \sum_q v_q$$

makes a good estimator of the product fd^3 and the product $g\bar{v}$ a good estimator of Y_1 and Z_1 :

$$Y_1' = Z_1' = g\bar{v}$$

This method is simple and does not require special equipment. It can be implemented practically everywhere and is regarded by the users as cheap enough and reliable. We shall present examples of its application in section 22.8.

22.6. PROPERTIES AND PRACTICAL ESTIMATION OF c , ℓ , f , g and d

The subject has been treated in great detail in Gy, 1967, chapter 5.

22.6.1. Mineralogical composition factor c (g/cm^3) :

It is defined without ambiguity whenever there is a constant proportionality factor between the content determined by chemical assay (e.g. Sn content of a cassiterite ore) and the corresponding physical content expressed in mineralogical component which alone must be taken into consideration in the present theory. This point has been emphasized in section 1.1.2. For instance, with a cassiterite ore :

a_L = cassiterite content = 1.27 tin content.

Then, the mineralogical factor c can be calculated from its definition :

$$c = \frac{1 - a_L}{a_L} \{(1 - a_L) \lambda_c + a_L \lambda_g\}$$

where λ_c is the density of the critical component and λ_g that of the gangue. Both should be expressed in g/cm^3 and a_L, a visual estimate of the critical content, is expressed in decimal value, not in percents : e.g. a 10 % content means 10/100 and must be written 0.1, which means 0.1 gram of critical component per gram of solids.

Particular case No.1 :

But there are cases when the assumed critical content cannot be expressed with precision :

1) the critical metal actually assayed is present in several minerals with different compositions and different densities : e.g. lead present in galena and cerussite :
  - galena content = 1.31 x Pb content and density = 7.5
  - cerussite content = 1.29 x Pb content and density = 6.5

Whenever the differences are as small as that, the obvious solution is to use average values such as :
  - critical content a_L = 1.3 x Pb content and density = 7

2) the critical metal actually assayed is present in a single mineral but the composition of this mineral and its density are liable to shift from one deposit to the next or from one side of a given deposit to the other side, or from top to bottom, etc.. : e.g. Mn in psilomelanes, W in wolframite, Ni in garnierite, etc. Here again, the solution is to use average compositions and densities, never forgetting that we are dealing with orders of magnitudes. The problem is never to decide whether we must take 38 or 42 kg but whether we should take a sample of 40 kg, 400 kg or 4 tons.

3) the critical metal actually assayed is present in a metallurgical product (e.g. Ni in a matte, Cu in a slag, etc..) the composition of which is not defined. In such a case, the third method is definitely advisable (example of a Ni-matte in section 22.8.3.).

Particular case No.2 :

The critical component and the gangue materials have approximately the same densities, or else we are looking only for the order of magnitude of c. Then, if λ is the average density of the material to be sampled :

$$c = \frac{1 - a_L}{a_L} \lambda$$

Example : a fluorspar ore with about 50 % CaF_2 and an average density of 3.0 g/cm^3
$c = 3.0$ g/cm^3

Particular case No.3 :

The material to be sampled is very rich, say richer than 90 or 95 % such as most concentrates, we can use the simplified formula :

$$c = (1 - a_L) \lambda_g \quad \text{with} \quad \lambda_g \text{ density of the gangue.}$$

Example : a barite concentrate with calcite gangue and 98 % $BaSO_4$: $a_L = 0.98$
$\lambda_c = 4.5$ g/cm^3 and $\lambda_g = 2.7$ g/cm^3.
$c = 0.054$ g/cm^3 (simplified formula)
$c = 0.053$ g/cm^3 (complete formula)

Particular case No.4 :

The material is very low grade, say lower than 0.05 such as most run-of-mine ores of the metal mining industry, processing tailings, or when the critical component is an impurity, we can use the following formula :

$$c = \frac{\lambda_c}{a_L}$$

Example 1 : galena flotation tailing with pyrite gangue : $a_L = 0.0023$ (0.23 % Pb)
$\lambda_c = 7.5$ g/cm^3 and $\lambda_g = 5.0$ g/cm^3 :
$c = 3260$ g/cm3 (simplified formula)
$c = 3251$ g/cm^3 (complete formula)

Example 2 : porphyry copper ore with 0.8 % Cu as chalcopyrite in a silica gangue :
$a_L = 0.0231$ (or 2.31 % $CuFe S_2$), $\lambda_c = 4.2$ g/cm^3 and $\lambda_g = 2.65$ g/cm^3 :
$c = 181$ g/cm^3 (simplified formula)
$c = 176$ g/cm^3 (complete formula)

Example 3 : alluvial gold ore with a content of 1 g/t (one gram per ton : 10^{-6}).
$a_L = 10^{-6}$ and $\lambda_c = 19$ g/cm^3 :
$c = 19 \times 10^6$ g/cm^3 (simplified as well as complete formula).

Range of the mineralogical composition factor c :

Of all parameters involved in the sampling constant C, the factor c is the one which can take the most extreme values, from 0.05 g/cm^3 for a very rich concentrate to 20 millions g/cm^3 for a gold ore, i.e. a ratio of 1 to 4 x 10^8. Fortunately, thanks to its definition, it is always possible to estimate its value with a sufficient accuracy.

22.6.2. Liberation factor ℓ (dimensionless):

We showed in a previous work (Gy, 1967, section 5.3.1.) that ℓ has actually the physical meaning of a liberation factor of the critical component : comminution can but increase the value of ℓ until it reaches its maximal value $\ell_{max} = 1$ which is achieved when the critical component is completely liberated, i.e. when it is present only as fragments of pure mineral.

From its definition, we know that ℓ can vary between 0 and 1 but for all practical purposes we shall never use values of ℓ smaller than 0.03 (dimensionless) which leaves a variation range from 1 to 33, much smaller than the one of c.

The liberation factor is seldom experimentally estimated. It is nearly always evaluated by analogy according to the following formula involving the "liberation diameter d_ℓ", defined as the maximum particle diameter ensuring the complete (or practically complete) liberation of the critical component, a parameter that can be estimated, roughly at least, by mineralogical examination. The notion of liberation size is familiar to mineral processing engineers. We shall therefore estimate ℓ by means of the formulas :

- if $d \leq d_\ell$: $\ell = 1$
- if $d > d_\ell$: $\ell = \sqrt{\dfrac{d_\ell}{d}}$

with d actual maximum fragment diameter. This rule of thumb is not based upon scientific considerations but upon practical experimentation carried out on all kinds of ore. Table 22.1. allows a quick estimation of ℓ in terms of either d/d_ℓ or d_ℓ/d.

Table 22.1. Estimation of ℓ as a function of d/d_ℓ and of d_ℓ/d :

d/d_ℓ	d_ℓ/d	$\ell = \sqrt{d_\ell/d}$
1	1	1.00
2	0.5	0.71
5	0.2	0.45
10	0.1	0.32
20	0.05	0.22
50	0.02	0.14
100	0.01	0.10
200	0.005	0.07
500	0.002	0.05
1000	0.001	0.03

22.6.3. Particle shape factor f (dimensionless) :

A very accurate method for estimating the shape factor of a given material has been described in an earlier work (Gy, 1967, section 5.1.1.). Its application to a large number of materials showed that for all practical purposes f can be regarded as a constant :

f = 0.5 (dimensionless)

The only important exception is that of alluvial gold ores for which it is the shape factor of the gold particles that must be taken into consideration. This factor can vary between 0.5 (when the particle shape is spheroidal) and 0.2 (when the particles are flat or elongated).

22.6.4. Size range factor g (dimensionless) :

An extensive study involving some 114 size analyses covering all kinds of materials and all size ranges has been related in Gy (1967, section 5.4.).

If we call "size range" the ratio d/d' of the upper size limit d (about 5 % oversize) to the lower size limit d' (about 5 % undersize), then for all practical purposes the value of g (dimensionless) can be estimated as follows :

Large size range (d/d' > 4) :   g = 0.25
Medium size range (4 to 2) :   g = 0.50
Small size range (d/d' < 2) :   g = 0.75
Uniform size (d/d' = 1) :   g = 1.00

Fluctuations of g : the study of 114 size analyses with large size ranges, including the most usual models of size distribution such as the Rosin-Rammler-Bennet model, showed the value of g to range from 0.17 to 0.40 with an average g = 0.25.

22.6.5. Maximum particle diameter d (centimeters cm) :

It is also referred to as maximum particle size, nominal particle size or upper size limit. It is defined in the present theory as the opening of the square mesh that retains a 5 % oversize. This definition is not arbitrary : it is the definition that reduces the range of fluctuation of the factor g to a minimum, allowing the most precise estimation of the product $g\,d^3$. Furthermore, it corresponds to the habits of the mineral processing industry.

When no screen analysis is available, d must be visually estimated as accurately as possible. When a detailed analysis is available, it is always possible and advisable to calculate the product $g\,d^3$ according to its definition given in section 22.4.2.

When estimating the fundamental variance, d must always be expressed in cm.

22.6.6. Sampling slide rule (see fig. 22.1. page 280) :

In order to help the user to calculate the sampling constant C and the fundamental standard deviation $\sigma(FE)$ we have in turns developed :

- charts (1955), in French and German ,
- circular nomograms (1956), in French, English and German,
- slide rules (1965), in French and English.

These slide rules implement the simplified formula :

$$\sigma^2(FE) = \frac{C\,d^3}{M_S}$$

where C is expressed in terms of the critical content a_L. Three scales of a_L are available :

- all materials except gold and coal (with average values of λ_c and λ_g),
- gold, with a_L expressed in grams per ton (g/t),
- coal, with a_L expressed in ash content.

Both the French and the English slide rules implement metric units. They are delivered with detailed instructions for use, illustrated by various examples. They can be obtained through the author as well as his previous publications.

22.7. RESOLUTION OF SAMPLING PROBLEMS INVOLVING THE FUNDAMENTAL VARIANCE

The formula recalled in the preceding section links up three variables : $\sigma^2(FE)$, d and M_S , and one parameter C (that can be assumed, in first approximation, to be a constant characterizing the properties of the material to be sampled with the exception of its maximum particle size d). Given the material and its sampling constant C, we can fix the value of two of the three variables and solve the equation for the third one, which makes three different problems :

22.7.1. Estimation of the fundamental variance :

Our purpose is to estimate $\sigma^2(FE)$ when extracting a sample of weight M_S from a lot characterized by a sampling constant C and a maximum particle size d :

$$\sigma^2(FE) = \left[\frac{1}{M_S} - \frac{1}{M_L}\right] C\,d^3 \quad \text{or, when } M_L \text{ is very large :} \quad \sigma^2(FE) = \frac{C\,d^3}{M_S}$$

Example 1 : Estimation of the fundamental variance incurred when sampling a 0.8 % Cu chalcopyrite ore crushed at 12.5 mm (d = 1.25 cm) and when a sample S of weight M_S = 50 kg (5 x 10^4 g) has been obtained :

d = 1.25 cm

d_ℓ = 0.01 cm (mineralogical examination) : d/d_ℓ = 125

a_L = 0.0231 g $CuFeS_2$ / g solids

From these data we can calculate or estimate :

$c = 176 \text{ g/cm}^3$
$\ell = 0.09$
$f = 0.5$
$g = 0.25$
$C = 2.0 \text{ g/cm}^3$
$\sigma(FE) = 0.88 \times 10^{-2}$ (calculation)
$\sigma(FE) = 1.00 \times 10^{-2}$ (direct reading from the slide rule)

The 95 % probability confidence interval $\pm 2 \sigma(FE)$ is $\pm 2 \times 10^{-2}$. The same confidence interval is for the copper content :

$a_S (1 \pm 2\sigma) = 0.8 \% \text{ Cu} \pm 0.016 \% \text{ Cu}$

22.7.2. Estimation of the minimum sample weight :

What is the minimum sample weight M_{So} necessary to represent a lot characterized by a constant C and a maximum particle diameter d when a fundamental variance σ_o^2 is regarded as the maximum acceptable. The equation to be solved is :

$$M_{So} = \frac{C d^3}{\sigma_o^2}$$

When the tolerated fundamental error is expressed as $\pm a_o$, the value of σ_o to use in the above formula is :

$$\sigma_o = \frac{a_o}{2 a_L} \quad \text{which involves :} \quad M_{So} = \frac{4 C d^3 a_L^2}{a_o^2}$$

In this expression, a_o and a_L must be expressed in the same unit, irrespective of the unit. Then, there is a 95 % probability for the fundamental error to fall within the limits of $\pm a_o$.

Example 2 : Same material as in example 1 (overleaf). Minimum sample weight M_{So} if a confidence interval $\pm a_o = 0.05 \%$ Cu is tolerated :

$C = 2.0 \text{ g/cm}^3$
$d = 1.25 \text{ cm}$
$a_o = 0.05 \% \text{ Cu}$
$a_L = 0.80 \% \text{ Cu}$
$M_{So} = 4000 \text{ grams}$ (calculation)
$M_{So} = 5 \text{ kg}$ (direct reading from the slide rule)

The difference between 4 and 5 kg is unimportant, the point being to estimate the order of magnitude of the minimum sample weight

22.7.3. Estimation of the maximum particle size :

It may happen that both the sample weight M_S and the tolerated variance σ_o^2 are fixed and the problem is to determine at what particle size d_o the material must be crushed or pulverized prior to its sampling. The equation to be solved is :

$$d_o^3 = \frac{M_S \sigma_o^2}{C}$$

- Then, if $d \leq d_o$: no comminution is required prior to any sampling, the fundamental variance will in any case be smaller than σ_o^2.

- If $d > d_o$: the whole lot must be comminuted to d_o or to a smaller size prior to sampling.

If, for some reason or other, comminution of the whole lot is to be regarded as impossible, the data of the problem are incompatible and the terms of the problem must be reconsidered, by increasing either M_S or σ_o^2 or both.

Remark : the so-called "sampling constant" C actually varies as a function of d through the liberation factor ℓ (see section 22.6.2.). When solving a problem such as this, it is advisable to calculate the value of d_o corresponding to various values of ℓ, and retain as solution to the problem the one which is compatible with the value of ℓ expressed in table 22.1.(see example 4 below).

Example 3 : Same material as in examples 1 and 2. At what size should the laboratory sample be pulverized if a 1 gram assay portion is to be analysed and if a 1 % relative confidence interval is tolerated. We know that the liberation of chalcopyrite is likely to be complete at that size.

$a_o = 0.01 \times 0.8 \% \text{ Cu} = 0.008 \% \text{ Cu} \qquad \text{or } \sigma_o = 5 \times 10^{-3}$
$C = 2.0 \div 0.09 = 22$
$d_o = 0.0104 \text{ cm}$ (calculation)
$d_o = 100$ microns (direct reading from the slide rule)

Example 4 : At what size d_o should one crush a lot of run-of-mine cassiterite ore containing about 0.2 % SnO_2 in order to be able to extract a 50 kg sample with a tolerated variance $\sigma_o^2 = 10^{-4}$.

$a_L = 0.002$ (gram of SnO_2 per gram of solids)
$c = \dfrac{\lambda_c}{a_L} = \dfrac{7.0}{0.002} = 3500$
$f = 0.5$
$g = 0.25$
$C = 438 \ell$

The liberation size is assumed (examination) to be $d_\ell = 0.2$ cm

We shall solve the equation for d_o, retaining the following values of ℓ : 0.8 - 0.4 - 0.2 - 0.1 - 0.05. Table 22.2. gives, in front of each assumed value of ℓ :

- the value of $C = 438 \, \ell$
- the calculated value $d_o = \sqrt[3]{\dfrac{50,000 \times 10^{-4}}{C}} = \sqrt[3]{\dfrac{5}{C}}$
- the value of the ratio d_o/d_ℓ
- the corresponding value of ℓ read by interpolation in table 22.1.

Table 22.2. Estimation of the maximum particle size - Example of a tin ore.

Assumed value of ℓ in the calculation of C	Corresponding value of C $C = 438 \, \ell$	Corresponding solution d_o (cm)	Corresponding value $\dfrac{d_o}{d_\ell}$	Corresponding value of ℓ (table 22.1.)
0.80	350	0.24	12	0.30
0.40	175	0.31	16	0.26
0.20	88	0.38	19	0.23
0.10	44	0.48	24	0.20
0.05	22	0.61	31	0.18

By comparing the first and the last columns it is easy to see that for all values of d_o smaller than 0.35 cm (or 3.5 mm or about 6 mesh) the assumed value of ℓ (first column) is larger than the actual value (last column). Then, we are on the safe side, the actual variance will be smaller than 10^{-4}. For $d_o = 0.35$ cm, the assumed value is practically equal to the actual value : this is the right solution. For d_o larger than 0.35 cm, the actual value is larger than the assumed value : the actual variance is then larger than 10^{-4}.

The conclusion is that the whole lot of ore must be crushed and ground down to about 3.5 mm prior to its sampling.

22.8. PRACTICAL APPLICATION OF METHODS No. 1, 2 AND 3

22.8.1. Example 5 : application of methods No. 1 and 2 :

A magnetite ore from equatorial Africa was crushed to 20 mm in closed circuit with a screen. Its assumed iron content was 55 % Fe or 76 % Fe_3O_4 or $a_L = 0.76$ gram of magnetite per gram of solids. The specific gravity of magnetite is 5.0 g/cm^3 and that of the siliceous gangue was 2.65 g/cm^3. We shall show how to estimate the values of Y and Z by the methods described in sections 22.3. and 22.4. This experiment was carried out in Minemet laboratories, in Paris, in 1961.

22.8.1.1. Method No.1 :

A 100 kg representative sample was washed, dried and screened on the following series of screens : 10 mm - 4 mm - 2 mm - 1 mm - 0.5 mm - 0.25 mm - 0.12 mm .

The wet and dry undersizes of the finest screen (0.12 mm) representing a negligible weight were discarded. The remaining seven size fractions were separated into five density fractions by means of two heavy liquids (densities 2.96 and 3.31 g/cm^3) and two heavy suspensions (densities 4.00 and 4.50 g/cm^3) of mercury droplets in bromoform. The fractions obtained were assayed for Fe after weighing. We obtained or calculated the following data :

$M_{\alpha\beta}$: by weighing

$a_{\alpha\beta}$: by assaying

v_α : by calculating $v_\alpha = (d_{\alpha 1}^3 + d_{\alpha 2}^3)/4$

λ_β : by calculating $\lambda_\beta = (\lambda_{\beta 1} + \lambda_{\beta 2})/2$

$M_{F\alpha\beta}$: by calculating $M_{F\alpha\beta} = v_\alpha \lambda_\beta$

It is interesting to point out the fact that in this method the critical content may be expressed in any unit and for instance :

in % metal : 55 % Fe
in % mineral : 76 % Fe_3O_4
in gram of metal per gram of solids : 0.55 (dimensionless)
in gram of mineral per gram of solids : 0.76 (dimensionless)

provided that all contents be expressed in the same way.

Table 22.3 gives all relevant data of the problem. In each box, we find three lines :

- first line : $M_{\alpha\beta}/M_L$ (dimensionless)
- second line : $a_{\alpha\beta}$ (Fe %)
- third line : $M_{F\alpha\beta}$ (grams)

From these results, we calculate , for a 100 kg batch :

$\sum_\alpha \sum_\beta M_{\alpha\beta} / M_{F\alpha\beta} = \sum_i 1 = N_L = 161{,}206{,}000$ (rounded off)

$\sum_\alpha \sum_\beta a_{\alpha\beta} M_{\alpha\beta} / \sum_\alpha \sum_\beta M_{\alpha\beta} = \sum_i a_i M_i / \sum_i M_i = 0.5513$ (gram Fe per gram solids)

$\sum_\alpha \sum_\beta M_{F\alpha\beta} M_{\alpha\beta} = \sum_i M_i^2 = 623{,}700$ (grams)^2

$\sum_\alpha \sum_\beta M_{F\alpha\beta} M_{\alpha\beta} a_{\alpha\beta} = \sum_i M_i^2 a_i = 342{,}200$ (grams)^2

$\sum_\alpha \sum_\beta M_{F\alpha\beta} M_{\alpha\beta} a_{\alpha\beta}^2 = \sum_i M_i^2 a_i^2 = 189{,}500$ (grams)^2

$\sum_\alpha \sum_\beta M_{F\alpha\beta}^2 M_{\alpha\beta} = \sum_i M_i^3 = 6{,}051{,}000$ (grams)^3

$\sum_\alpha \sum_\beta M_{F\alpha\beta}^2 M_{\alpha\beta} a_{\alpha\beta} = \sum_i M_i^3 a_i = 3{,}341{,}000$ (grams)^3

$\sum_\alpha \sum_\beta M_{F\alpha\beta}^2 M_{\alpha\beta} a_{\alpha\beta}^2 = \sum_i M_i^3 a_i^2 = 1{,}858{,}000$ (grams)^3

Table 22.3. Size-density analysis of a magnetite ore - First line : $M_{\alpha\beta}/M_L$
Second line : $a_{\alpha\beta}$ = Fe % - Third line : $M_{F\alpha\beta} = \nu_\alpha \lambda_\beta$

$L_{\alpha\beta}$	$L_{\alpha 1}$	$L_{\alpha 2}$	$L_{\alpha 3}$	$L_{\alpha 4}$	$L_{\alpha 5}$	$L_{\alpha\Sigma}$
Density → Size ↓	5.00-4.50 g/cm3 λ_β = 4.75	4.50-4.00 g/cm3 λ_β = 4.25	4.00-3.31 g/cm3 λ_β = 3.655	3.31-2.96 g/cm3 λ_β = 3.135	- 2.96 g/cm3 λ_β = 2.805	Sums or means
$L_{1\beta}$ 20-10 mm ν = 2.25	0.25225 58.05 10.6875	0.30583 54.25 9.5625	0.04512 42.45 8.2237	0.00981 22.75 7.0537		0.61300 54.44
$L_{2\beta}$ 10- 4 mm ν = 0.266	0.06130 59.90 1.2635	0.06903 53.70 1.1305	0.01177 42.65 0.9722	0.00250 26.30 0.8339		0.14460 54.95
$L_{3\beta}$ 4 - 2 mm ν = 0.018	0.05085 62.10 0.0855	0.02730 53.40 0.0765	0.00726 41.95 0.0658	0.00198 24.40 0.0564		0.08739 56.85
$L_{4\beta}$ 2 - 1 mm ν = 2.25 10^{-3}	0.03306 65.55 0.0107	0.01710 54.90 0.0096	0.00521 37.20 0.0082	0.00198 28.90 0.0071	0.00085 16.35 0.0063	0.05820 58.27
$L_{5\beta}$ 1 - 0.5 mm ν = 281 10^{-6}	0.02608 66.55 1334×10^{-6}	0.00684 55.00 1194×10^{-6}	0.00337 34.15 1027×10^{-6}	0.00418 21.55 881×10^{-6}	0.00113 16.37 788×10^{-6}	0.04160 56.14
$L_{6\beta}$ 0.5-0.25 mm ν = 35 10^{-6}	0.02142 67.70 167×10^{-6}	0.00405 56.55 149×10^{-6}	0.00202 31.50 128×10^{-6}	0.00145 23.65 110×10^{-6}	0.00126 17.70 99×10^{-6}	0.03020 59.64
$L_{7\beta}$ 0.25-0.12mm ν = 4 10^{-6}	0.01541 67.86 21×10^{-6}	0.00330 63.30 19×10^{-6}	0.00106 33.50 16×10^{-6}	0.00359 17.70 14×10^{-6}	0.00164 17.05 12×10^{-6}	0.02500 55.26
$L_{\Sigma\beta}$	0.46037 60.44	0.43345 54.24	0.07581 41.29	0.02549 22.85	0.00488 16.94	1.00000 55.13

Y = 0.028 grams
Z = 0.054 grams

$$m(FE) = -0.028 \left[\frac{1}{M_S} - \frac{1}{M_L} \right]$$

$$\sigma^2(FE) = 0.054 \left[\frac{1}{M_S} - \frac{1}{M_L} \right]$$

As an illustration of the formulas of section 20.5, we have calculated the moments of N_S, M_S, A_S and a_S (both in first and second approximation) for the following values of the selecting probability : P = 0.5 - 0.1 - 0.01 - 0.001. These results are given in tables 22.4. to 22.7.

Table 22.4. Magnetite ore - Distribution of N_S number of fragments in sample S.

Selection Probability	0.5	0.1	0.01	0.001
$m(N_S)$ (units)	80.6×10^6	16.1×10^6	1.61×10^6	0.16×10^6
$\sigma^2(N_S)$	40.3×10^6	14.5×10^6	1.60×10^6	0.16×10^6
$\sigma(N_S)$ (units)	6.35×10^3	3.81×10^3	1.26×10^3	401
$u(N_S)$	8×10^{-5}	2.4×10^{-4}	7.8×10^{-4}	2.5×10^{-3}

Table 22.5. Magnetite ore - Distribution of M_S weight of the sample S.

Selection Probability	0.5	0.1	0.01	0.001
$m(M_S)$ (grams)	50,000	10,000	1000	100
$\sigma^2(M_S)$	16×10^4	5.6×10^4	6175	623
$\sigma(M_S)$ (grams)	395	237	79	25
$u(M_S)$	7.9×10^{-3}	2.4×10^{-2}	7.9×10^{-2}	25×10^{-2}

Table 22.6. Magnetite ore - Distribution of A_S weight of critical component in the sample S.

Selection Probability	0.5	0.1	0.01	0.001
$m(A_S)$ (grams)	27,500	5500	550	55
$\sigma^2(A_S)$	8.5×10^4	3.1×10^4	3388	342
$\sigma(A_S)$ (grams)	292	175	58	19
$u(A_S)$	1.1×10^{-2}	3.2×10^{-2}	10.5×10^{-2}	33.6×10^{-2}

Table 22.7. Magnetite ore - Distribution of a_S, critical content of the sample S.

Selection Probability	0.5	0.1	0.01	0.001
Mean $m(a_S)$				
1st approximation	0.5513	0.5513	0.5513	0.5513
Corrective term	-8×10^{-8}	-8×10^{-7}	-8×10^{-6}	-8×10^{-5}
Bias $B(a_S)$	-1.6×10^{-7}	-1.6×10^{-6}	-1.6×10^{-5}	-1.6×10^{-4}
Variance $\sigma^2(a_S)$				
1st approximation	16.5×10^{-8}	1.48×10^{-6}	16.3×10^{-6}	1.65×10^{-4}
Corrective term	0	-18×10^{-10}	-25×10^{-8}	-25×10^{-6}
Ratio Cor.T./1st ap.	0	-1.2×10^{-3}	-1.5×10^{-2}	-15×10^{-2}
Standard-deviation $\sigma(a_S)$	4.06×10^{-4}	1.22×10^{-3}	4.0×10^{-3}	1.18×10^{-2}
Relative standard deviation $u(a_S)$	74×10^{-5}	22×10^{-4}	73×10^{-4}	2.14×10^{-2}
Confidence interval $m(a_S) \pm 2 \sigma(a_S)$ % Fe	55.13 ± 0.08	55.13 ± 0.24	55.13 ± 0.80	55.13 ± 2.36
Square of the coefficient of bias $b(a_S)$	0	50×10^{-8}	5×10^{-6}	50×10^{-6}

These tables call for the following remarks:

1) For a selection probability $P = 0.1$ (sample weight of the order of 10 kg), the relative standard deviation of M_S reaches 2.4×10^{-2} which is near the limit 3×10^{-2} mentioned insection 20.3.1. This means that in this particular case, for values of P smaller than 0.1 or for sample weights smaller than 10 kg the distribution of a_S is liable to deviate from a normal distribution.

2) The bias $B(a_S)$ is negligible, even with a selection probability as small as 0.001 corresponding to a sample weight of only 100 grams which is much smaller than the weight usually acceptable for a 20 mm material. Hence the conclusion which is consistently supported by a number of experiments of the same kind, that for all practical purposes and with the sole exception of very low grade materials, a correct sampling may be regarded as unbiased.

3) The relative importance of the corrective term of the variance $\sigma^2(a_S)$ varies from 0 for $P = 0.5$ (a consequence of the fact that it is proportional to $1 - 2P$) to 15 % for $P = 0.001$. The first approximation is therefore sufficient throughout the useful domain of P.

4) The square $b^2(a_S)$ of the coefficient of bias $b(a_S) = |B(a_S)|/u(a_S)$, which is involved in the definition of the degree of representativeness $r^2(FE)$:

$r^2(FE) = m^2(FE) + \sigma^2(FE) = \{1 + b^2(a_S)\} \sigma^2(FE)$

is perfectly negligible throughout the useful domain of P (50×10^{-6} to be added to 1 for P = 0.001). Then, for all practical purposes and <u>as long as the sampling remains correct</u>, the notion of representativeness, a property of the mean square $r^2(FE)$ can be assimilated to that of reproducibility, a property of the variance $\sigma^2(FE)$. This ceases to be true when the sampling is incorrect and when the bias ceases to be negligible.

5) The confidence interval of the sample critical content a_S corresponding to a 95 % probability is :

± 0.08 % Fe for P = 0.5 and a sample weight M_S = 50 kg, and

± 0.24 % Fe for P = 0.1 and a sample weight M_S = 10 kg which is somewhat larger than the usually acceptable error for a single sampling stage. This means that the sample weight should not be smaller than say 20 kg to 30 kg for such an iron ore crushed at 20 mm.

6) Table 22.8. recapitulates the values taken by the sums :

$$\sum_{\beta} M_{F\alpha\beta} M_{\alpha\beta} \qquad \sum_{\beta} M_{F\alpha\beta} M_{\alpha\beta} a_{\alpha\beta} \qquad \sum_{\beta} M_{F\alpha\beta} M_{\alpha\beta} a_{\alpha\beta}^2$$

involved in the calculation of m(FE) and $\sigma^2(FE)$, for each of the seven size fractions.

Table 22.8. *Importance of the various size fractions in the calculation of m(FE) and $\sigma^2(FE)$.*

L_α	$\sum_{\beta} M_{F\alpha\beta} M_{\alpha\beta}$	$\sum_{\beta} M_{F\alpha\beta} M_{\alpha\beta} a_{\alpha\beta}$	$\sum_{\beta} M_{F\alpha\beta} M_{\alpha\beta} a_{\alpha\beta}^2$
L_1 = 20 - 10 mm	606,058.4	332,473.7	183,959.7
L_2 = 10 - 4 mm	16,901.7	9,372.8	5,251.8
L_3 = 4 - 2 mm	702.7	404.2	236.4
L_4 = 2 - 1 mm	57.9	34.2	20.8
L_5 = 1 - 0.5 mm	5.1	3.0	1.8
L_6 = 0.5 - 0.25 mm	0.4	0.3	0.2
L_7 = 0.25 - 0.12 mm	0.05	0.03	0.02
$\sum_\alpha$	623,726.3	342,288.2	189,470.7

The first term of the sums, corresponding to the coarsest size fraction represents more than 97 % of the sum, whilst the sum of the first two terms represents approximately 99.85 % of the total. This conclusion is quite general. The first size fraction covering the range from d to d/2 fixes the order of magnitude of the sum and the fraction ranging from d to d/4 is a sufficient approximation of the sum extended to the whole size analysis. Then, and this is consistently confirmed by experience, we can for all practical purposes concentrate our attention on the coarsest size fractions and forget the other ones, which simplifies the experimentation considerably. The third method is indeed based on this observation.

22.8.1.2. Method No.1 - the critical component is a size fraction :

In other words, we are sampling for a size analysis. This section presents an application of the results of section 22.3.2. Table 22.9. illustrates the calculation of Y and Z when each size fraction is regarded in turn as critical.

Table 22.9. Sampling for a size analysis - Estimation of Y and Z.

Critical size fraction $L_{\alpha c}$ mm	$v_{\alpha c}$ cm^3	$\lambda_{\alpha c}$ g/cm^3	Y grams	Z grams
L_1 = 20 - 10 mm	2.25	4.394	+ 3.65	2.58
L_2 = 10 - 4 mm	0.266	4.394	- 5.07	11.98
L_3 = 4 - 2 mm	0.018	4.466	- 6.16	7.15
L_4 = 2 - 1 mm	2.25×10^{-3}	4.422	- 6.23	6.38
L_5 = 1 - 0.5 mm	0.28×10^{-3}	4.364	- 6.24	6.27
L_6 = 0.5 - 0.25 mm	35.0×10^{-6}	4.451	- 6.24	6.24
L_7 = 0.25 - 0.12 mm	4.40×10^{-6}	4.278	- 6.24	6.24

From a practical standpoint, the problem is usually put as follows : how heavy must be the sample if the proportion of a given critical size fraction is to be known within the limits of a confidence interval $\pm a_0$ %. **Example** : we want to know the proportion of each size fraction with a relative precision of ± 5 % : we calculate : $2\sigma_0 = 0.05$ and $\sigma_0^2 = 625 \times 10^{-6}$. Assuming the weight M_L to be very large, the equation to be solved is :

<u>Minimum sample weight</u> $M_{So} = Z/\sigma_0^2 = 1600 Z$ (grams)

Table 22.10 gives the corresponding value of M_{So} for each size fraction :

Table 22.10. Sampling for a size analysis - Minimum sample weight.

Critical size fraction $L_{\alpha c}$ mm	Minimum sample weight M_{So}
L_1 = 20 - 10 mm	4,100 g or 4.1 kg
L_2 = 10 - 4 mm	19,200 g or 19.2 kg
L_3 = 4 - 2 mm	11,400 g or 11.4 kg
L_4 = 2 - 1 mm	10,200 g or 10.2 kg
L_5 = 1 - 0.5 mm	10,000 g or 10.0 kg
L_6 = 0.5 - 0.25 mm	10,000 g or 10.0 kg
L_7 = 0.25 - 0.12 mm	10,000 g or 10.0 kg

The required precision will be achieved for all size fractions if the sample weight is larger than 19.2 kg or in round figures 20 kg.

If a sample weight of 20 kg is retained, the bias $m(FE) = -Y/M_S$ ranges between $+1.8 \times 10^{-4}$ and -3.1×10^{-4}. This confirms the general conclusion that when the sampling is correct, the bias is negligible, on the size analysis as well as on the metal content.

22.8.1.3. Method No.2 :

This method is applicable only when the critical component is a mineral and when the variance alone is to be estimated :

$$\sigma^2(FE) = \left[\frac{1}{M_S} - \frac{1}{M_L}\right] Z \quad \text{with} \quad Z = c\,\ell\,f\,g\,d^3 = C\,d^3$$

1) <u>estimation of c</u> : we need the following mineralogical data :
- specific gravity of magnetite : $\lambda_c = 5.0$ g/cm^3
- specific gravity of the gangue: $\lambda_g = 2.65$ g/cm^3
- critical content a_L : we know from the results of the test that $a_L = 55.13$ % Fe or 0.76 g Fe$_3$O$_4$ per g solids but as a general rule, one never knows the value of a_L before the sampling which is precisely meant to provide a sample to the assaying laboratory. But the geologist, the miner, the metallurgist always know the order of magnitude of the metal or mineral content and the nature of the metal-bearing mineral(s). Let's assume in this case that a_L, that must imperatively be expressed in gram of critical mineral per gram of solids, lies somewhere between 0.7 and 0.8 g Fe$_3$O$_4$/g solids. According to the definition of c we easily calculate :

$0.79 < c < 1.46$ g/cm^3 or $c = 1.12 \pm 0.34$ g/cm^3

2) <u>estimation of ℓ</u> : we need an estimate of the liberation size of the magnetite crystals. A quick estimation by means of a magnifying glass shows this liberation size to be of the order of 0.1 mm. Then, the maximum particle size being 20 mm, we can easily calculate $d/d_\ell = 200$ and read from table 22.1. (p 262) :

$$\ell = 0.07 \quad \text{(dimensionless)}$$

3) <u>estimation of f</u> : a simple examination of the ore shows the general particle shape to be spheroidal. We may therefore take the usual value :

$$f = 0.5 \quad \text{(dimensionless)}$$

4) <u>estimation of g</u> : according to the definition given in section 22.6.4., we may take the usual value :

$$g = 0.25 \quad \text{(dimensionless)}$$

5) <u>calculation of C</u> : we easily calculate :

$C = c\,\ell\,f\,g = 1.0 \times 10^{-2} \pm 0.3 \times 10^{-2}$ g/cm^3 or $C_{max} = 1.3$ and $C_{min} = 0.7 \times 10^{-2}$ g/

6) <u>Nominal particle diameter d</u> : we can select from the batch a few fragments obviously belonging to the coarsest size fraction and measure or estimate (it is imperative to express d in cm even though it is more usual to speak of mm) :

$$d = 2.0 \text{ cm}$$

7) <u>Calculation of $Z = C d^3$</u> : we immediately obtain :

$$Z = 8 \times 10^{-2} \pm 2.4 \times 10^{-2} \text{ grams or } Z_{max} = 0.104 \text{ and } Z_{min} = 0.056 \text{ gram}.$$

We should observe that the costly experiment presented in section 22.8.1.1. had provided the estimate :

$$Z = 0.054 \text{ gram}$$

which is practically equal to the lower limit estimated here.

8) <u>calculation of $\sigma^2(FE)$</u> : with M_S and M_L imperatively expressed in grams :

$$\sigma^2(FE) = \left[\frac{1}{M_S} - \frac{1}{M_L}\right] Z = \left[\frac{1}{M_S} - \frac{1}{M_L}\right](0.08 \pm 0.024) \quad \text{(dimensionless)}$$

or, whenever M_S is small in comparison with M_L :

$$\sigma^2(FE) = \frac{0.080 \pm 0.024}{M_S}$$

If we assume for instance $M_S = 10^5$ grams (100 kg) and $M_L = 10^6$ grams (1 ton) :

$$\sigma^2(FE) = (10^{-5} - 10^{-6}) \times (0.08 \pm 0.024) = (72 \pm 22) \times 10^{-8}$$

$$\sigma(FE) = (8.4 \pm 1.3) \times 10^{-4}$$

If we assume now $M_S = 10^5$ grams and $M_L = 10^7$ grams (10 tons), $1/M_L$ is negligible :

$$\sigma^2(FE) = (80 \pm 24) \times 10^{-8}$$

$$\sigma(FE) = (8.8 \pm 1.3) \times 10^{-4}$$

9) <u>estimation of the minimum sample weight M_{So}</u> : let's fix for instance $\sigma_o^2 = 10^{-6}$. The lot weight M_L is 10^6 grams (1 ton). We must solve for M_{So} the equality :

$$\sigma_o^2 = \left[\frac{1}{M_{So}} - \frac{1}{M_L}\right] Z \quad \text{or} \quad M_{So} = \frac{1}{\frac{\sigma_o^2}{Z} + \frac{1}{M_L}} = \frac{M_L Z}{M_L \sigma_o^2 + Z}$$

$Z_{max} = 0.102$ gram gives the solution : $M_{So} = 92,600$ grams
$Z_{min} = 0.056$ gram gives the solution : $M_{So} = 53,000$ grams
$\quad$ or $M_{So} = 73 \pm 20$ kg

In such a case we would take a 100 kg sample.

10) <u>estimation of the maximum particle size d_o</u> : if we fix for instance $M_S = 10^3$ g and $M_L = 5 \times 10^5$ grams, $1/M_L$ can be neglected and we solve the equation :

$$d_o^3 = \frac{M_S \sigma_o^2}{C}$$

If we fix now $\sigma_o^2 = 10^{-6}$ we easily calculate :

with $C_{max} = 0.013$ g/cm^3 : $d_o^3 = 0.0769$ cm^3 and $d_o = 0.42$ cm (or 4.2 mm)
with $C_{min} = 0.007$ g/cm^3 : $d_o^3 = 0.1429$ cm^3 and $d_o = 0.52$ cm (or 5.2 mm)

In such a case, we would crush down to 4 to 5 mm, according to the available equipment.

11) <u>use of the slide rule</u> : we can solve the same problems by means of the slide rule in a matter of seconds :

a) <u>estimation of $\sigma^2(FE)$</u> : for a sample weight of 10^5 g (read 100 kg on the rule)
setting the rule to $a_L = 0.7$ we obtain in an instant : $\sigma(FE) = 10^{-3}$
$\sigma^2(FE) = 10^{-6}$
setting the rule to $a_L = 0.8$ we obtain   : $\sigma(FE) = 7 \times 10^{-4}$
$\sigma^2(FE) = 50 \times 10^{-8}$
which can be written : $\sigma(FE) = (8.5 \pm 1.5) \times 10^{-4}$
$\sigma^2(FE) = (75 \pm 25) \times 10^{-8}$ (compare with (8) above)

b) <u>estimation of M_{So}</u> : with $\sigma_o^2 = 10^{-6}$ or $\sigma_o = 10^{-3}$ we directly read :
for $a_L = 0.7$: $M_{So} = 100$ kg
for $a_L = 0.8$: $M_{So} = 50$ kg   (compare with (9) above)

c) <u>estimation of d_o</u> : for a sample weight $M_S = 1$ kg and $\sigma_o = 10^{-3}$ we read :
for $a_L = 0.7$: $d_o = 4.4$ mm
for $a_L = 0.8$: $d_o = 5.3$ mm   (compare with (10) above)

d) <u>estimation of d_o</u> : if we fix now $M_S = 100$ kg and $\sigma_o = 10^{-3}$ we read :
for $a_L = 0.7$: $d_o = 20$ mm
for $a_L = 0.8$: $d_o = 25$ mm

As the material passes a 20 mm screen, no comminution is required prior to sampling.

22.8.2. Example No.6 - Method No.3 :

This example involves another iron ore coming from another African mine. Contrary to the ore illustrating our first example, the present material contains a variety of iron-bearing minerals, making it rather difficult to estimate with precision the mineralogical factor c of the second method. It was then decided to constitute a working batch of 46 fragments extracted one by one and at random among the coarsest fragments of the material. Following the method described in section 22.5. we estimated for each fragment :

  - its weight M_q (grams)
  - its volume v_q (cm^3)
  - its iron content a_q (Fe %)

We recall that with the third method, as well as with the first one, the critical content can be expressed in any unit provided all contents be expressed with the same unit.

The results of this experiment are given in table 22.11.

Table 22.11. Estimation of Y and Z - Illustration of Method No.3 - Iron ore.

q	M_q grams	v_q cm^3	a_q % Fe	q	M_q grams	v_q cm^3	a_q % Fe
1	163	42	55.31	25	189	49	53.83
2	121	44	21.33	26	157	48	43.91
3	116	39	37.86	27	116	42	38.66
4	127	39	45.62	28	66	24	38.09
5	78	29	13.91	29	141	53	6.01
6	102	26	50.86	30	128	39	39.91
7	60	23	42.65	31	138	50	21.36
8	142	48	44.13	32	89	31	22.80
9	105	34	26.91	33	150	42	45.73
10	119	45	36.84	34	100	41	4.79
11	61	19	43.22	35	208	60	47.55
12	149	57	4.11	36	127	39	41.39
13	95	37	6.04	37	146	49	48.58
14	95	39	7.33	38	117	30	59.07
15	198	77	43.91	39	102	38	20.53
16	107	27	53.14	40	71	29	10.38
17	172	43	61.58	41	81	29	44.82
18	120	29	62.72	42	82	36	3.40
19	119	33	38.20	43	105	31	41.28
20	104	28	44.48	44	116	33	48.13
21	121	33	44.93	45	110	31	46.83
22	91	32	47.56	46	67	26	7.64
23	148	40	63.52				
24	52	20	7.30				
				Σ	5371	1733	37.91

From these results, we can compute :

$$M_Q = \sum_q M_q = 5371 \text{ grams}$$

$$v_Q = \sum_q v_q = 1733 \text{ cm}^3 \quad \text{and} \quad \bar{v} = \frac{v_Q}{46} = 37.67 \text{ cm}^3$$

$$a_Q = \frac{\sum_q a_q M_q}{\sum_q M_q} = 37.91 \text{ \% Fe} \quad \text{and} \quad a_Q^2 = 1437.17 \text{ (\% Fe)}^2$$

$$\Sigma_1 = \sum_q (a_q - a_Q) M_q \lambda_q = \sum_q (a_q - a_Q) M_q^2 / v_q = +386.7 \quad (g^2/cm^3)$$

$$\Sigma_2 = \sum_q (a_q - a_Q)^2 M_q \lambda_q = \sum_q (a_q - a_Q)^2 M_q^2 / v_q = +498.5 \quad (g^2/cm^3)$$

$$Y_1 = Z_1 = 0.25 \bar{v} = 9.4 \text{ cm}^3$$

$$Y_2 = \Sigma_1 / a_Q M_Q = +0.19 \text{ g/cm}^3 \quad \text{and} \quad Y = Y_1 Y_2 = 1.8 \text{ grams}$$

$$Z_2 = \Sigma_2 / a_Q^2 M_Q = +0.63 \text{ g/cm}^3 \quad \text{and} \quad Z = Z_1 Z_2 = 5.9 \text{ grams}$$

$$m(FE) = -1.8 \left[\frac{1}{M_S} - \frac{1}{M_L}\right] \quad \text{and} \quad \sigma^2(FE) = 5.9 \left[\frac{1}{M_S} - \frac{1}{M_L}\right]$$

with M_S and M_L imperatively expressed in grams.

Now, the average volume $\bar{v}$ is a good estimator of the product $f \, d^3$. If we retain for f the usual value 0.5 which is applicable here, we can estimate :

$$d^3 = \frac{\bar{v}}{f} = 75.34 \text{ cm}^3 \quad \text{or} \quad d = 4.22 \text{ cm}$$

We can also estimate :

$$C = Z/d^3 = 0.078 \text{ g/cm}^3$$

22.8.3. Example No.7 - Method No.3 :

The present example deals with a nickel and cobalt matte produced by Société Le Nickel in New-Caledonia. This matte was crushed to about 60 mm prior to sampling and shipping. We conducted the same experiment as above, with exactly 50 fragments. On every fragment we estimated weight, volume and three critical contents : Ni %, Co % and Fe %. We obtained the following results :

M_Q = 42,961 grams

$\bar{v}$ = 137.34 cm^3

	Ni	Co	Fe	Unit
a_Q	74.08	1.92	2.38	%
Z_2	256 x 10^{-5}	246 x 10^{-3}	825 x 10^{-3}	g/cm^3
Z	0.0879	8.45	28.3	g

Table 22.12. gives the variance $\sigma^2(FE)$ and the confidence interval $\pm \, 2 \, \sigma(a_S)$ for various values of the sample weight and for the three metals.

Table 22.12. Example No.3 - Method No.3 - Nickel-cobalt matte.

Sample weight M_S	Nickel		Cobalt		Iron	
	$\sigma^2(FE)$	$\pm \, 2\sigma(a_S)$	$\sigma^2(FE)$	$\pm \, 2\sigma(a_S)$	$\sigma^2(FE)$	$\pm \, 2\sigma(a_S)$
10 tons	9 x 10^{-9}	0.014 %	8.5 x 10^{-7}	0.003 %	2.83 x 10^{-6}	0.008 %
1 ton	9 x 10^{-8}	0.044 %	8.5 x 10^{-6}	0.011 %	2.83 x 10^{-5}	0.025 %
100 kg	9 x 10^{-7}	0.14 %	8.5 x 10^{-5}	0.035 %	2.83 x 10^{-4}	0.08 %
10 kg	9 x 10^{-6}	0.44 %	8.5 x 10^{-4}	0.11 %	2.83 x 10^{-3}	0.25 %
1 kg	9 x 10^{-5}	1.4 %	8.5 x 10^{-3}	0.35 %	2.83 x 10^{-2}	0.8 %

If we suppose now that the tolerated variance is $\sigma_o^2 = 10^{-6}$ for Ni, and 10^{-4} for Co and Fe, and that the problem is to determine the minimum sample weight ensuring that these three conditions are simultaneously fulfilled, we can compute :

Minimum sample weights :

Ni : $M_{So} = \dfrac{Z(Ni)}{\sigma_o^2(Ni)} = \dfrac{0.0879}{10^{-6}}$ = 87,900 grams or 88 kg

Co : $M_{So} = \dfrac{Z(Co)}{\sigma_o^2(Co)} = \dfrac{8.45}{10^{-4}}$ = 84,500 grams or 84 kg

Fe : $M_{So} = \dfrac{Z(Fe)}{\sigma_o^2(Fe)} = \dfrac{28.3}{10^{-4}}$ = 283,000 grams or 283 kg.

If the same sample is to be used for the three metals, we must take the largest of the three figures, i.e. 283 kg. In practice, we shall retain the order of magnitude 250 - 300 kg.

22.9. RECAPITULATION AND CONCLUSIONS

The purpose of this chapter was to illustrate the theoretical chapters and more specifically chapters 20 and 21 and to show the reader how to estimate the sampling constants C and Z. We presented three methods :

1) <u>Method No.1</u> : Precise but time consuming and costly : the reader will have practically no opportunity to use it. It was developed as a first step towards a more practical method. One of the advantages of this method was to allow the calculation of the fundamental bias m(FE) in a number of practical cases, which led to the conclusion that with the possible exception of very low grade materials such as the ores of precious metals and minerals, the fundamental bias was always negligible. We unfortunately lack examples to illustrate the sampling of gold or diamond ores.

2) <u>Method No.2</u> : Easy, quick and cheap : it is the choice method for the geologist, miner or metallurgist. It makes it possible to fix the order of magnitude of the sampling constant C and to solve one of the three classical sampling problems in a matter of minutes or even of seconds when using the slide rule devised for this purpose.

3) <u>Method No.3</u> : Reliable and reasonably cheap : this method is useful when a difficulty arises in the implementation of the second method or when a higher degree of precision is required in estimating the fundamental bias and variance. It requires a minimum of experimental work that can be carried out in practically all laboratories of the mining and metallurgical industries.

But we would like to point out the fact that the second method which was developed 25 years ago, breaking up as it does the fundamental variance into a product of simple factors precises remarkably well the influence of the various characteristics of the material to be sampled :

$$\sigma^2(FE) = \left[\frac{1}{M_S} - \frac{1}{M_L}\right] c\, \ell\, f\, g\, d^3$$

M_S : <u>sample weight</u> (in grams)
M_L : <u>lot weight</u> (in grams)
c : <u>mineralogical composition factor</u>, depending mainly on the critical content a_L of the material to be sampled and to a lesser extent on the specific gravities of the constituting minerals (in grams per cm^3),
ℓ : <u>liberation factor</u> characterizing the degree of liberation of the critical component (dimensionless),
f : <u>shape factor</u> characterizing the shape of the particles (dimensionless),
g : <u>size range factor</u> characterizing the dispersion of the particle size distribution (dimensionless),
d : <u>nominal or maximum particle diameter</u> characterizing the size of the coarsest particles (in cm).

This formula underlines the parts played by the sample weight and the lot weight respectively. More specifically, it disposes of the deep-rooted idea that, to achieve a given reproducibility standard, the sample weight should be proportional to the lot weight. On the contrary, whenever the sample weight is small in comparison with the lot weight (say less than 10 %), the fundamental variance is practically independent of the lot weight, depending solely upon the <u>absolute</u> value of the sample weight.

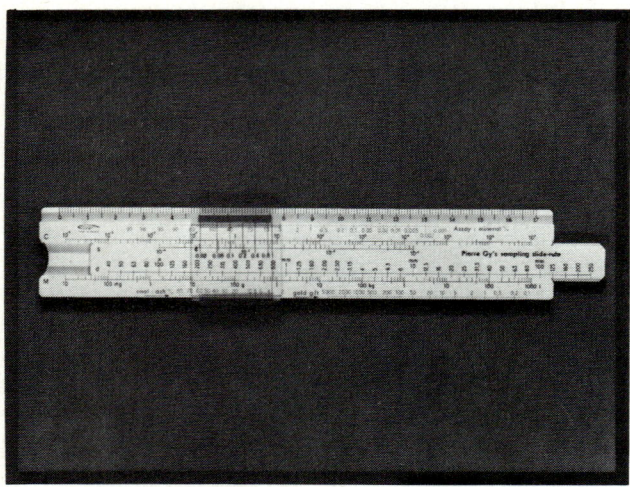

Fig. 22.1. Sampling slide rule.

CHAPTER 23

PRACTICAL IMPLEMENTATION OF THE THEORETICAL RESULTS

INCORRECT SELECTION

23.1. INTRODUCTION

The purpose of this chapter is to show how the theoretical formulas obtained in section 20.4. can be practically implemented and more specifically how the incorrectness bias can be estimated. We shall first observe that it is not at the level of the point selection step (covered by the second part of this book) that the sampling process is likely to depart from correctness but at the level of the increment delimitation and extraction errors (covered by the third part of this work). We pointed out in chapters 17 and 18 that these deviations from correctness affect individual particles F_i, alloting to each of these a certain non-uniform probability P_i of being selected. Now, if we can, to some extent at least, estimate the relevant characteristics of individual particles after a size-density fractioning of the lot for instance (as shown in section 22.3.), the question arises of how to estimate the selecting probability P_i involved in the formulas of the moments of the sampling error.

A theoretical analysis shows and experience confirms that, as far as deviation from correctness is concerned, the most relevant characteristic of a given particle F_i is its diameter d_i and we may, in order to simplify the formulas involving P_i assume that :

1) All particles of the size fraction L_α of the lot L can be assimilated to the average particle F_α defined by the following characteristics :

- diameter d_α
- specific gravity λ_α
- shape factor f_α
- volume $v_\alpha = f_\alpha d_\alpha^3$
- weight $M_{F\alpha} = M_{L\alpha}/N_{L\alpha} = f_\alpha d_\alpha^3 \lambda_\alpha$
- critical content $a_{F\alpha} = a_{L\alpha}$

with $M_{L\alpha}$, $N_{L\alpha}$ and $a_{L\alpha}$ weight, number of particles and critical content of L_α .

2) These particles are selected with a uniform selecting probability P_α.

These hypotheses amount to assimilating the lot L strictly defined as

$$L \equiv \sum_i F_i$$

with an imaginary lot L' defined as

$$L' \equiv \sum_\alpha N_{L\alpha} F_\alpha$$

Reasoning as we did in section 20.4.2. we can put the moments of the sample characteristics M_S, A_S and a_S under the form of sums of terms such as :

$$\sum_i a_i^x M_i^y \Phi(P_i)$$ that can be replaced by the estimator :

$$\sum_\alpha N_{L\alpha} a_{L\alpha}^x M_{F\alpha}^y \Phi(P_\alpha) = \sum_\alpha a_{L\alpha}^x M_{F\alpha}^{y-1} M_{L\alpha} \Phi(P_\alpha)$$

with : x = 0 or 1 or 2 , y = 1 or 2 and $\Phi(P) = P$ or $P(1 - P)$ or $P(1 - P)(1 - 2P)$

From these expressions it readily follows that (with M_o, A_o and a_o defined in section 20.3.2.) :

$$m(M_S) = M_o = \sum_i M_i P_i = \sum_\alpha M_{L\alpha} P_\alpha$$

$$m(A_S) = A_o = \sum_i a_i M_i P_i = \sum_\alpha a_{L\alpha} M_{L\alpha} P_\alpha$$

$$m(a_S)_1 = a_o = A_o / M_o$$

$$m(a_S)_2 = a_o - \frac{1}{M_o^2} \sum_\alpha (a_{L\alpha} - a_o) M_{L\alpha} M_{F\alpha} P_\alpha (1 - P_\alpha)$$

$$\sigma^2(a_S)_1 = \frac{1}{M_o^2} \sum_\alpha (a_{L\alpha} - a_o)^2 M_{L\alpha} M_{F\alpha} P_\alpha (1 - P_\alpha)$$

The factors involved in these formulas can be practically estimated when a size analysis is available :

$M_{L\alpha}$: by direct weighing,

$M_{F\alpha}$: either by weighing a countable number of particles extracted at random from L_α (when the particles are coarse enough) or by estimating the characteristics of the average particle F_α (when the particles are too fine) :

$$M_{F\alpha} = \frac{M_{L\alpha}}{N_{L\alpha}} \quad \text{or} \quad M_{F\alpha} = f_\alpha d_\alpha^3 \lambda_\alpha$$

For all practical purposes, f_α and λ_α can usually be replaced by constants f and

P_α : by experimentally estimating its a posteriori equivalent, the sampling ratio τ_α, a random variable with mean :

$$m(\tau_\alpha) = P_\alpha$$

Any estimate of τ_α is therefore an unbiased estimator of P_α.

23.2. INCORRECT EXTRACTION CURVE

From a practical standpoint, the rules of increment delimitation correctness are widely accepted and increment delimitation errors (chapter 17) tend to become rarer and rarer, at least in Countries where the present theory has been known for some time. Now, due to a passive resistance offered by a number of equipment manufacturers whose interests are directly at stake, the rules of increment extraction correctness (chapter 18) remain too often ignored with the consequence that extraction probabilities, which should be uniformly equal to unity for all particles, do actually vary to a large extent from one size fraction to the next.

We call "incorrectness curve" or "incorrect extraction curve" the curve representing the function :

$$P_\alpha = \Omega(d_\alpha) \quad \text{or} \quad \tau_\alpha = \Omega(d_\alpha)$$

Figures 23.1., 23.2. and 23.3. show three classical examples of incorrect extraction curves.

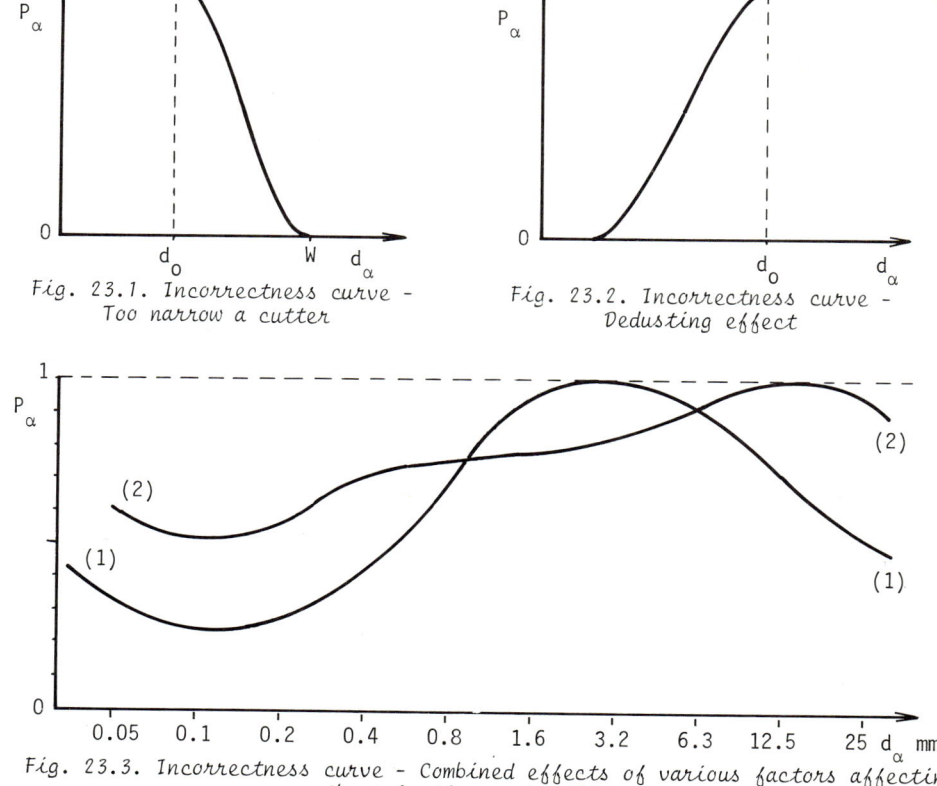

Fig. 23.1. Incorrectness curve - Too narrow a cutter

Fig. 23.2. Incorrectness curve - Dedusting effect

Fig. 23.3. Incorrectness curve - Combined effects of various factors affecting the selecting probability

Fig. 23.1. represents the incorrectness curve of a sampler equiped with too narrow a cutter, this curve decreasing and tending towards zero as soon as the particle diameter d_α exceeds a certain "critical diameter" d_o defined in section 18.5.2. tied to the cutter width W. Obviously, such a sampler should not be used for sampling materials containing fragments coarser than d_o. Samplers equiped with too fast a cutter are characterized by similar curves tending asymptotically towards zero. For a sampler equipped with a cutter of width W, the extraction probability of a particle coarser than W is obviously zero.

Fig. 23.2. illustrates the incorrectness of a sampler where an upward air draught achieves a partial dedusting of the sample. Such a sampler should not be used with materials containing particles finer than d_o. As already mentioned in chapter 18 the solution is to protect the cutter from air draughts (for instance by means of vane feeders whenever possible).

Fig. 23.3. illustrates the complex behaviour of a once popular rotary sampler cumulating various defects. Curves No.1 and 2 were actually obtained by testing the same sampler with two different ores under different conditions and call for the following remarks :

1) an upward air draught removes a large proportion of the finer size fractions. Curve No.1 shows that the dedusting effect can reach 77 % for d_α = 0.1 mm (P_α = 0.2

2) both curves show a minimum in the neighbourhood of 0.1 mm which is due to the fact that finer particles are partly stuck on the surface of coarser fragments, either mechanically or electrically (depending on whether the material is moist or dry), thus being protected to some extent against the dedusting effect. These particles recover their individuality in the size analysis when this is carried out with efficiency.

3) the cutter openings were designed in such a way that coarse particles rebounded in an erratic but selective way, with a tendency for the coarsest particles to jump over the openings.

23.3. PRACTICAL DETERMINATION OF THE CURVE OF INCORRECT EXTRACTION

23.3.1. Method No.1 : This very accurate method has already been presented in detail in section 18.5. It should be used by sampling equipment manufacturers to check all existing or new devices.

23.3.2. Method No.2 : This method consists in :

1) preparing a working batch L by reunion of known weights $M_{L\alpha}$ of carefully screened size fractions L_α.

2) carefully mixing the lot.

3) sampling the lot by means of the device under investigation.

4) repeating steps 1) to 3) as many times as necessary to carry out an efficient statistical analysis (say 30 to 50 times but for the statistical part of this method, the reader will usefully refer to chapter 31).

5) carrying out an accurate size analysis of all samples collected, using the same series of sieves as for the composition of the batch L.

6) weighing the fractions S_α : weight $M_{S\alpha}$.

7) calculating for each sample the ratio :

$$\tau_\alpha = \frac{M_{S\alpha} \, M_L}{M_{L\alpha} \, M_S}$$

8) checking by the method described in chapter 31 whether τ_α is significantly different from unity. If τ_α is not significantly different from 1, retain $P_\alpha = 1$. If it is, retain $P_\alpha = \tau_\alpha$.

9) plotting P_α against d_α.

The material and statistical parts of such an experiment are developed in chapters 32 and 31, respectively.

23.3.3. Method No.3 : This method consists in :

1) preparing a lot of unknown size analysis,

2) alternating the taking of increments of comparable weights by means of the sampler under investigation and by means of a reference sampling method, <u>assumed to be correct</u> (e.g. stopped belt sampling). The number of couples of increments should be large enough to allow a statistical analysis.

3) carrying out an accurate size analysis of both the actual samples (weights $M_{S\alpha}$) and the reference samples (weights $M_{R\alpha}$).

4) calculating the average sampling ratio τ of the reference sampling method in terms of its geometrical or mechanical characteristics.

5) estimating the weights $M_{L\alpha} = M_{R\alpha}/\tau$, estimators of the actual proportions of the fractions L_α in the lot L.

6) calculating for each couple of increments the sampling ratio τ_α :

$$\tau_\alpha = \frac{M_{S\alpha} \, M_L}{M_{L\alpha} \, M_S} = \tau \, \frac{M_{S\alpha} \, M_L}{M_{R\alpha} \, M_S}$$

7) see step 8) of method No. 2

8) see step 9) of method No. 2

Here again, the reader will usefully refer to chapters 31 and 32 for the statistical and material aspects of such an experiment.

23.3.4. Method No. 4 : This cheap method consists in :

1) preparing a lot of unknown size analysis.

2) taking increments by means of the sampler under investigation and gathering these to form a single sample.

3) carrying out an accurate size analysis of both the sample S (weights $M_{S\alpha}$) and the sampling reject R (weights $M_{R\alpha}$).

4) calculating $M_{L\alpha} = M_{S\alpha} + M_{R\alpha}$

5) calculating the sampling ratio τ_α of each size fraction L_α :

$$\tau_\alpha = \frac{M_{S\alpha} \, M_L}{M_{L\alpha} \, M_S}$$

6) using directly the unique value of τ_α as an estimator of P_α.

7) plotting P_α against d_α.

This method, if implemented only once, gives a general idea of the sampler correctness. To reach objective, statistically valid conclusions, the method should be repeated a certain number of times.

23.4. EXAMPLES

We give in the present section a few examples of indirect estimation of the sampling bias, based on the estimation of the non-uniform sampling probability and on the use of the formulas resulting from the development of the discrete model (chapter 20). Under certain conditions, the methods described in the preceding sections allow a direct estimation of the sampling bias and this point will be developed in chapter 32. Experience, as well as a piece of reasoning, shows that the indirect method is more sensitive than the direct one.

23.4.1. Example No. 1 :

Implementation of the sampler characterized by fig. 23.3. on the iron ore characterized by the size analysis figuring in table 22.3.(last column). The relevant data have been gathered in table 23.1. (page 287).

From these data we easily compute :

$m(a_S)_1 = a_o = 55.43$ % Fe

The absolute bias introduced by the sole extraction error is :

$a_o - a_L = 55.43 - 55.13 = 0.30$ % Fe

In the iron mining industry such a bias is usually regarded as unacceptable. The relative extraction bias is :

$m(EE) = \dfrac{+ 0.30}{55.13} = + 5.4 \times 10^{-3}$

The relative variance is :

$\sigma^2(EE) = 38 \times 10^{-8}$

The relative standard deviation :

$\sigma(EE) = 6 \times 10^{-4}$

The $\pm 2\sigma(a_S)$ confidence interval :

$\pm 2\sigma(a_S) = \pm 0.07 \%$ Fe

Table 23.1. Sampling an iron ore by means of an incorrect device - Example No. 1.

Size fraction L_α mm	Weight proportion $\dfrac{M_{L\alpha}}{M_L}$	Average particle weight $M_{F\alpha}$ grams	Critical content $a_{L\alpha}$ Fe %	Extraction probability P_α
L_1 = 20-10	0.6130	9.42	54.40	0.56
L_2 = 10- 4	0.1446	1.12	54.95	0.63
L_3 = 4 - 2	0.0874	77.3×10^{-3}	56.85	0.92
L_4 = 2 - 1	0.0582	9.8×10^{-3}	58.22	1.00
L_5 = 1 - 0.5	0.0416	1.2×10^{-3}	56.14	0.88
L_6 = 0.5 - 0.25	0.0302	156×10^{-6}	59.64	0.55
L_7 = 0.25 - 0.12	0.0250	19×10^{-6}	55.26	0.35
L = 20 - 0.12	1.0000		55.13	0.635

23.4.2. Example No.2 :

Sampling a 50 mm barite ore, the size analysis of which, as well as the extraction probabilities are given in table 23.2. (column No.4) by means of a cross-stream sampler equipped with too narrow (W = 80 mm) and too fast (V_C = 2.0 m/s) a cutter.

Table 23.2. Sampling a barite ore by means of two different incorrect samplers

Size fraction L_α mm	Weight proportion $\dfrac{M_{L\alpha}}{M_L}$	Critical content $a_{L\alpha}$ $BaSO_4$ %	Extraction probabilities P_α	
			Example No. 2	Example No. 3
L_1 = 50-25	0.100	58.3	0.30	1.00
L_2 = 25-10	0.299	57.5	0.60	1.00
L_3 = 10- 5	0.120	58.3	0.90	1.00
L_4 = 5 - 1	0.167	52.8	1.00	1.00
L_5 = 1 - 0.1	0.193	44.2	1.00	0.80
L_6 = - 0.1	0.121	31.2	1.00	0.20
L	1.000	a_L = 51.15	a_o = 49.46 % $BaSO_4$	a_o = 53.72 % $BaSO_4$

From these data we can calculate :

$m(a_S)_1 = a_o = 49.46$ % $BaSO_4$ hence the absolute bias : $a_o - a_L = -1.69$ % $BaSO_4$
and the relative extraction bias :

$m(EE) = -3.3 \times 10^{-2}$

Such a bias, larger than 3 % in relative value, is obviously intolerable.

23.4.3. Example No.3 :

Same barite ore as in example 2. An ascending air draught alterates the extraction probability of the fine particles according to the figures given in table 23.2 (column 5). We can easily calculate :

$m(a_S)_1 = a_o = 53.72$ % $BaSO_4$ hence the absolute bias : $a_o - a_L = +2.57$ % $BaSO_4$
and the relative extraction bias :

$m(EE) = +5.0 \times 10^{-2}$

The dedusting effect introduces a 5 % relative bias.

23.5. CONCLUSIONS

The examples presented in this chapter show how dangerous can be the sampling devices when they are incorrectly designed and how important can be the corresponding biases.

As a result of more than 20 years of experience as a Consultant, trouble shooter and arbitrator in disputes involving sampling techniques, we can vouch that incorrect sampling devices are operating all over the world, including in the biggest Companies of the most developed Countries. If relative biases as large as 20 % tend to become rare though non-inexistent, 5 % biases are far from uncommon and 1 to 2 % biases are frequently observed. Such biases usually pass undetected except by an expert's eye. This is due to the following circumstances :

1) A number of equipment manufacturers still design, build, advertize and eventually sell on a large scale, samplers that do not respect, even approximately the rules of extraction and delimitation correctness developed in chapters 17 and 18.

2) The users of sampling equipment are very seldom aware of the importance of sampling correctness and of the conditions that should be fulfilled for a sampler to extract correct increments.

3) Points 1) and 2) are consequences of the fact that, but for a few notable exceptions, the teaching of the theory of sampling has been widely neglected, especially in English speaking Countries (except England).

We hope that the publication of this book will help manufacturers and their customers to understand the situation better.

FIFTH PART

SPLITTING PROCESS

In chapter 4 we described two basic sampling processes :

- The <u>increment sampling process</u> which was extensively studied in the second, third and fourth parts of this book. This process is adapted to the sampling of batches of particulate materials too heavy to be split. Increment sampling implements sampling ratios ranging from 1/1000 to 1/10.

- The <u>splitting process</u> which is going to be reviewed in the fifth part of this book. This process is adapted to the sampling of batches of particulate materials light or valuable enough to be handled in totality, at least once, for the sole purpose of their sampling. Splitting implements sampling ratios ranging from 1/20 to 1/2.

Devices sampling flowing streams by extracting part of the stream all of the time could be regarded as continuous splitters. Unfortunately, experience has proved for quite a long time (see for instance Warwick, 1903) that such devices were practically unable to carry out a correct extraction and always introduced uncontrollable biases. For this reason they have been rightfully discarded and we felt it unadvisable to give them unjustified credit by developing their theory. They have therefore been omitted.

As previously mentioned, we shall restrict our study to the splitting of "movable" lots, defined as follows :

A lot of particulate materials is said to be "movable" when it is small or valuable enough to be economically handled in totality for the sole purpose of its sampling.

The emphasis should be put here on the cost of splitting and on the value of the material to be sampled. Obviously enough, hand methods are restricted to the splitting of batches whose weight ranges from a few tons to a few grams but mechanical methods implementing for instance certain types of rotary sectorial splitters can be used with heavier batches (10 to 20 tons). Now, mechanical shovels can efficiently and economically split batches weighing thousands of tons by mere fractional shovelling. Just to give an idea of the possibilities of such a method, a single operator using first a 5-ton shovel and then a 1-ton shovel (or even smaller) can split several thousand tons down to a few tons in less than one day, i.e.

at a perfectly acceptable cost if the material to be sampled is valuable enough.

The development of fractional shovelling by means of mechanical shovels, as well as that of increment sampling methods, could and should revolutionize the sometimes medieval habits of commercial sampling in not too distant a future.

This fifth part is made of three chapters :

Chapter 24 : Splitting methods and devices
Chapter 25 : Models of the splitting processes
Chapter 26 : Practical implementation of the splitting processes -
 Reduction of drill core samples.

CHAPTER 24

SPLITTING METHODS AND DEVICES

24.1. INTRODUCTION

We shall first recall the definition of the splitting process as given in section 4.7. :

Any splitting process can be logically broken up into a sequence of four elementary and independent steps :

1) <u>Fraction delimitation</u> : in its relative movement through the domain (D_L) occupied by the lot L, the splitting tool or device delimits the geometrical boundaries of the domains occupied by the "<u>geometrical fractions</u>" of the lot. This relative movement can be realized in three different ways :

  - Stationary lot, moving tool : coning and quartering, alternate shovelling, etc... (the geometrical boundaries of the fractions are approximately defined),

  - Moving lot, stationary device : riffle divider, revolving feeder sectorial splitters, etc.. (geometrical boundaries defined with precision),

  - Moving lot, moving device : stationary feeder sectorial splitters, etc.. (geometrical boundaries defined with precision).

This purely geometrical operation does not take the fragment integrity into consideration. It corresponds to the delimitation step of the increment process.

2) <u>Fraction separation</u> : the particulate structure of the material to be sampled is taken into consideration in this second step. The "<u>material fractions</u>" actually separated coincide more or less with the sets of particles whose centre of gravity falls within the boundaries of the geometrical fractions. It corresponds to the extraction step of the increment process.

3) <u>Fraction dealing out</u> : the material fractions are then dealt out according to a usually systematic scheme between a certain number of "<u>potential samples</u>" (two with coning and quartering, alternate shovelling, riffling, more than two with the other splitting methods or devices) obtained by reunion of a certain number of fractions (two with coning and quartering, more and often much more than two with the other splitting methods or devices). It corresponds to the reunion step of the increment sampling process.

4) <u>Sample selection</u> : all potential samples are submitted to a selection process retaining one "<u>sample</u>" or several "<u>twin-samples</u>". This selection step may be regarded as probabilistic if and only if it is made at random.

These four elementary steps are, potentially at least, error generating.

Whenever a low sampling ratio is to be achieved, splitting is repeated as many times as necessary to obtain the required sample weight.

Splitting methods and devices belong to a variety of types but most of these fall within one of the four following categories :

- Coning and quartering,
- riffling,
- fractional and alternate shovelling,
- sectorial dividers.

These four categories will be reviewed in sections 24.3. to 24.6.

24.2. TRUE AND DEGENERATE SPLITTING PROCESSES

From a theoretical standpoint, it is appropriate to distinguish :

1) <u>True splitting processes</u> : they divide the lot to be sampled into a certain number of twin potential samples <u>of equal bulk</u>. True splitting processes fall within the province of the present part of this book.

2) <u>Degenerate splitting processes</u> : they divide the lot to be sampled into a certain number of predetermined potential samples and a certain number of predetermined sampling rejects, samples and rejects being <u>of unequal bulk</u>. Degenerate splitting falls within the province of the increment process studied in earlier parts of this work.

Any splitting process (true or degenerate) can be regarded as a sequence of :

1) A technical partitioning process which can be biased,

2) A selecting process that can be either deterministic or probabilistic. We shall assume it to be probabilistic and correct. Then :

- with true splitting, it cancels the effects of an eventual "technical" bias,
- with degenerate splitting, it cannot cancel the effects of such a bias due to the fact that all fractions cannot be considered as potential samples.

This distinction does not mean to condemn degenerate splitting methods or devices as such : they can and often do achieve a perfectly correct and unbiased sampling. Its only purpose is to underline the fact, which will be developed in chapter 30 dedicated to the notion of "<u>equity</u>" of commercial sampling, that :

- true splitting processes can be made equitable even if they are technically biased,

- degenerate splitting processes, like increment sampling processes to which they actually belong, may be regarded as equitable if and only if they are technically unbiased.

The distinction between true and degenerate splitting is very important in commercial sampling, the financial interests at stake demanding an equitable procedure.

24.3. CONING AND QUARTERING

This is probably the oldest of all probabilistic sampling methods. It is also referred to as "Cornish quartering" which seems to fix its origin in the tin mines of Cornwall, probably at the beginning of the 19 th century. It used to be implemented with batches weighing up to 50 tons but is now restricted to the sampling of lots of - 50 mm materials weighing less than one ton. There are a number of variants to this method. We shall just describe the most typical one, illustrated by fig. 24.1. to 24.3.

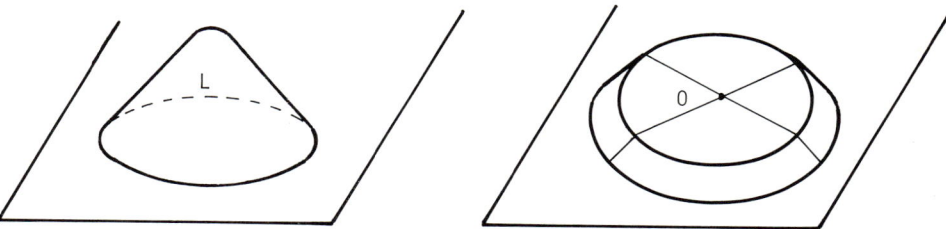

Fig. 24.1. Coning and quartering - First step : coning and mixing

Fig. 24.2. Coning and quartering - Second step : flattening

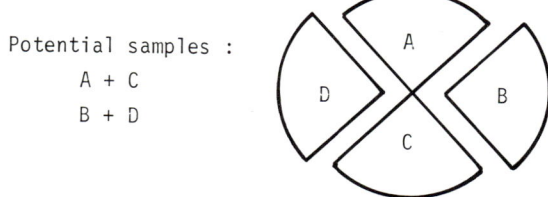

Fig. 24.3. Coning and quartering - Third step : quartering.

First step, the material is spread on a smooth steel or concrete surface that can be easily cleaned. It is piled into a conical heap, each shovelful being directly dropped onto the apex. This operation is repeated two or three times, its purpose being to give the particle distribution a revolution axis (revolution homogeneity or symmetry towards the vertical axis described in section 19.3.4.5. This first step is schematized in fig. 24.1.

Second step : the material is dragged down by means of a shovel so as to form first a truncated cone, then a flat circular cake, respecting the symmetry achieved in the first step as much as possible. This second step is illustrated in fig. 24.2.

Third step : the circular cake is divided into four quarters along to two diameters at right angles. Two opposite quarters (e.g. A and C or B and D) are retained to form the sample while the remaining quarters are rejected. This third step is illustrated in fig. 24.3. The choice between the two potential samples (A + C) and (B + D) can be left to the operator's initiative, which is acceptable with technical sampling, or made at random by tossing up a coin, which is always advisable with commercial sampling (see chapter 30).

The interest of this method is purely historical, even though it remains a favorite with a number of professional or sworn samplers. It is uselessly time-consuming and costly. Experience shows that it is neither more accurate, nor more precise, nor cheaper than alternate shovelling which achieves the same splitting ratio (1/2) by means of the same tool : shovel or scoop.

24.4. RIFFLING

24.4.1. Description :

The "riffle splitter", also known as "Jones riffle" or "jones splitter" consists of an assembly comprising an even number of identical adjacent chutes, numbering usually between 12 and 20, seldom more. The device with 7 chutes we have met somewhere on the American Continent must be regarded as an accidental exception not to be imitated. These chutes form an angle of about 45° or more with the horizontal plane, they lead the material alternately towards the left bucket and the right bucket of fig. 24.4. The material is fed from a rectangular scoop after an even distribution of the material on the surface of the scoop.

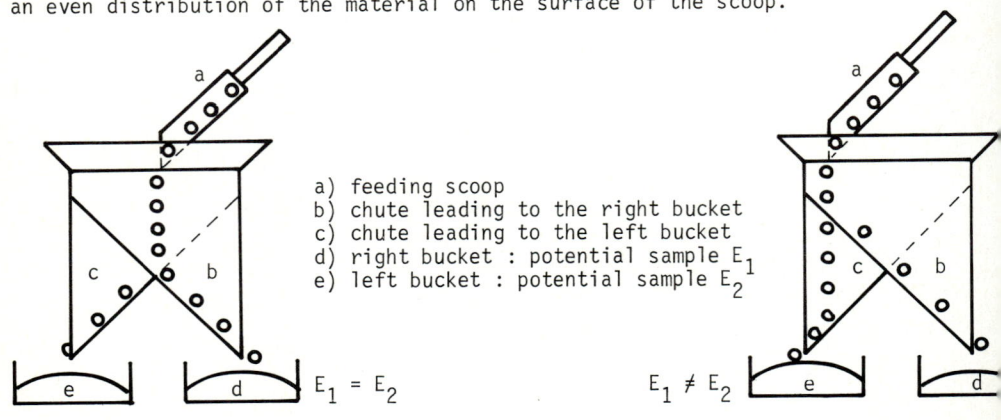

a) feeding scoop
b) chute leading to the right bucket
c) chute leading to the left bucket
d) right bucket : potential sample E_1
e) left bucket : potential sample E_2

Fig. 24.4. Riffle splitter - Correct use

Fig. 24.5. Riffle split Incorrect use

Each of the two buckets receives a potential sample. It is always advisable either to toss up a coin (commercial sampling) or to alternate the bucket from which the sample is taken (technical sampling), rather than to leave its choice to the operator's initiative.

24.4.2. Possibility of an operating bias :

If we except a few home made devices, riffle splitters to be found on the market are usually correctly designed, but they can be incorrectly utilized. As early as 1953, in an experiment related also in Gy (1971, section 26.1.4.2.), we showed that a bias was likely to take place when using a riffle splitter in a dissymmetrical way illustrated in fig. 24.5. (preceding page). When the scoop is discharged too near one of the sides and too fast, one of the two sets of chutes is likely to overflow in the other : then, one of the potential samples is systematically heavier than the other one. Experience shows that this overflowing is selective, the coarse particles overflowing more readily than the fines.

This bias is cumulative, which made it possible to detect it and to ascertain its causes. In an experiment primarily devised in order to estimate the sampling variance and carried out by an experienced operator, a batch of 8 kg of a 8.3 % Pb galena ore crushed to 2.5 mm was partitioned into 16 twin samples by four consecutive halving stages. Each sample was weighed and assayed and its filiation carefully recorded. To the general surprise we observed, when analysing the results, the deviations presented in tables 24.1. and 24.2.

Table 24.1. Results of a four-stage riffling scheme - Weight : deviation % from the mean

Genealogy of the sample				
4 x left 0 x right	3 x left 1 x right	2 x left 2 x right	1 x left 3 x right	0 x left 4 x right
+ 11.35 %	+ 7.84 % + 8.16 % + 3.90 % + 6.56 %	+ 6.98 % + 4.41 % - 0.39 % + 2.75 % - 6.01 % - 6.12 %	- 3.46 % - 9.99 % - 6.99 % - 3.05 %	- 15.96 %

On each riffling stage, the left bucket contained a systematically heavier and lower grade sample :

Average relative difference (left - right) : weight : + 6.36 %
grade : - 2.01 %

Further experiments showed the cause of these systematic differences to lie in a dissymmetrical use of the splitter (fig. 24.5.). As soon as the scoop is slowly discharged towards the middle of the openings (fig. 24.4), these differences in weight and grade vanished. This experiment shows how important it is to use the riffle splitter in a symmetrical way.

Table 24.2. Results of a four-stage riffling scheme - Pb % - Deviation % from the mean.

Genealogy of the sample				
4 x left 0 x right	3 x left 1 x right	2 x left 2 x right	1 x left 3 x right	0 x left 4 x right
- 2.32 %	- 0.52 % - 2.56 % - 4.00 % - 3.28 %	+ 3.33 % + 2.19 % + 1.11 % - 2.92 % - 1.11 % + 2.37 %	+ 2.13 % + 1.77 % + 1.23 % + 1.41 %	+ 2.67 %

24.4.3. Practical conclusions :

The normal range of use of riffle splitters is :

- maximum particle size : about 15 mm
- lot weight : a few hundred kilograms to hundred grams
- sample weight : down to a few grams
- nature of the material : dry solids.

Riffle splitters always achieve a true splitting.

When correctly used, the riffle splitter is a very convenient device. It is cheap and can be used everywhere. It is built in various sizes, usually referred to by the width of the chutes, say from 30 mm to 6 mm. It is not advisable to use a riffle splitter for the sampling of materials with a maximum particle diameter larger than the half of the chute opening as bridging would be likely to take place.

Riffles should be carefully cleaned before use (dust rejected) and after use (dust recovered with the samples). Electrostatic sticking of dry fine particles requires the use of a small brush, which makes it unadvisable to miniaturize riffle splitters as one might be tempted to do for the preparation of one-gram assay portions.

Riffles have been used in some instances for the continuous sampling of flowing streams or arranged in a cascade of three or four, in order to obtain 1/8 or 1/16 in a single operation. Such uses are definitely unadvisable as they have been shown to be likely to introduce a bias.

When using riffle splitters for technical purposes, the "alternation rule" should be observed in order to suppress any eventual bias. It consists in retaining one of the buckets (the right one for instance) at the first and all odd riffling stages, and the other (the left one) at the second and at all even riffling stages.

When using riffle splitters for commercial sampling, the sample must be selected at random at each splitting stage, by tossing up a coin for instance.

24.5. FRACTIONAL SHOVELLING

24.5.1. Procedure :

Fractional shovelling is certainly the simplest and cheapest of all bulk sampling methods. It consists in moving the whole of the batch by means of a hand or mechanical shovel, retaining a sample made of one shovelful out of N , thus achieving a splitting ratio $\tau = 1/N$.

Fractional shovelling can be performed in two ways illustrated by fig. 24.6 and 24.7 respectively.

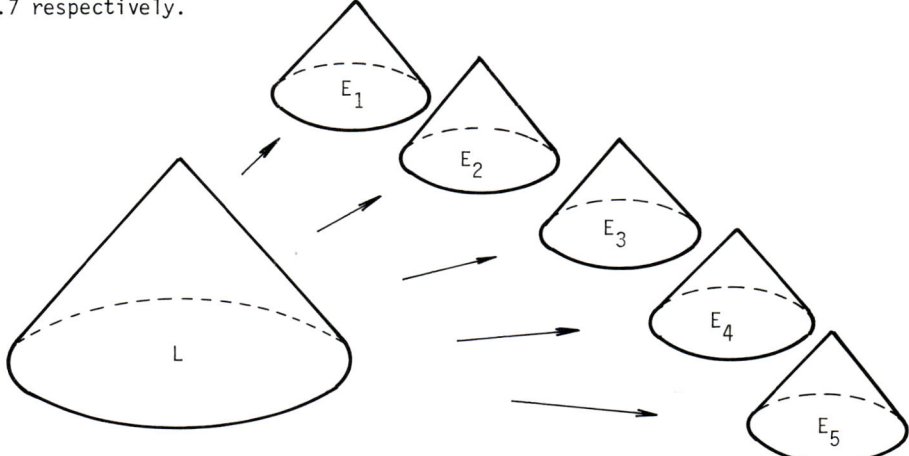

Fig. 24.6. True fractional shovelling with N = 5.

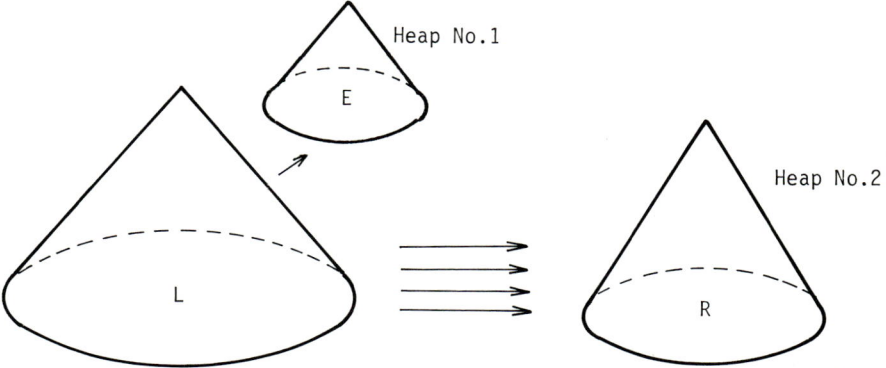

Fig. 24.7. Degenerate fractional shovelling with N = 5.

1) <u>True fractional shovelling</u> (fig. 24.6.) : the shovelfuls extracted from the lot are deposited on top of N distinct heaps, which after completing the dealing out of the whole lot become N twin potential samples of equal bulk. One or several actual samples are selected at random from this set of N units. True fractional shovelling falls within the province of the splitting process presently studied.

2) <u>Degenerate fractional shovelling</u> (fig. 24.7.) : the shovelfuls are dealt out as follows : every N th shovelful is deposited on heap No.1 and the remaining (N - 1) shovelfuls of the cycle are deposited on heap No.2. According to this procedure, heap No.1 is a predetermined sample and heap No.2 a predetermined reject. Degenerate fractional shovelling falls within the province of increment sampling when used for technical sampling but, and this is very important, may be regarded as a non-probabilistic process when used in commercial sampling where voluntary biases are liable to take place.

3) <u>Alternate shovelling</u> (fig.24.8) is a fractional shovelling characterized by $N = 2$ and a sampling ratio $\tau = 1/2$. It always achieves a true splitting but can be regarded as probabilistic, like all true splitting processes, only inasmuch as the sample is selected at random.

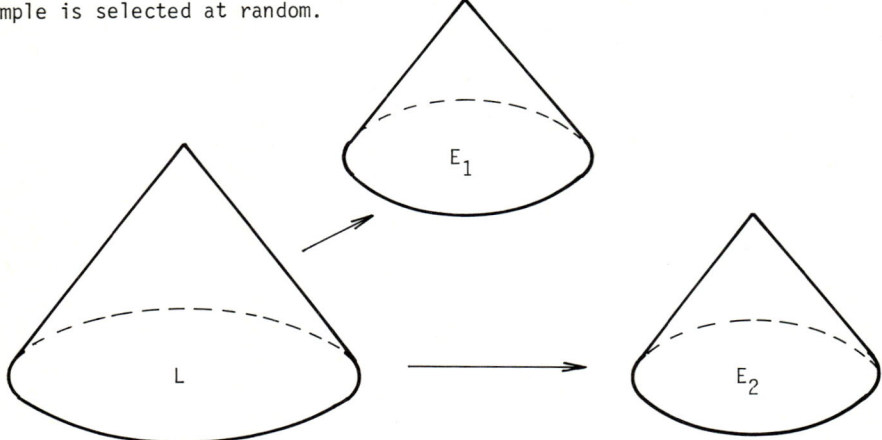

Fig. 24.8. Alternate shovelling - N = 2.

24.5.2. Possibility of a bias :

The process is technically correct and therefore unbiased whenever the operator works blindly. When sampling coarse materials however, an astute operator can always manage to put a larger (or smaller) proportion of coarse fragments into one of the fractions. When the sample is predetermined (degenerate shovelling), this affects the correctness, therefore the accuracy and, in commercial sampling, the equity of the process. But when all potential samples have equal chances of being selected (true shovelling), the technical bias introduced by the clever operator may turn as well to the detriment of the cheat.

This illustrates the main difference between true and degenerate splitting methods and justifies the compulsory implementation of true methods when splitting by hand for commercial purposes.

The very important notion of equity of a commercial sampling will be developed in chapter 30.

24.5.3. Practical conclusions :

The range of utilization of fractional shovelling is different according as we deal with hand or mechanical shovels :

1) <u>Hand shovelling</u> :

- nature of the material : dry, wet or even sticky solids,
- maximum particle size : seldom used for fragments coarser than 100 mm (4"). This is due to the usual size of hand shovels and to the average capacity of men for handling heavy loads.
- lot weight : up to a few tons, may vary widely from one Country to the next.
- sample weight : alternate shovelling can be implemented by means of a chemist's spatula and provide samples as small as one gram.
- splitting ratio : from 1/2 with alternate shovelling to 1/10 and exceptionally 1/20 with fractional shovelling.
- shovel sizes : for theoretical reasons, a shovelful should contain less than $M_L/30N$. This rule ensures that each potential sample is made of at least 30 shovelfuls. Example : lot weight M_L = 800 kg. N = 5. 30 N = 150. $M_L/150$ = 5.3 kg. As the common shovel can carry more than 10 kg, it is advisable to use a smaller shovel. A sampling laboratory should always have several sets of shovels and scoops with capacities ranging from 10 kg to 100 g and even less.

Fractional shovelling is usually restricted to the preparation of samples that can be split by means of riffles but can be used all the way down to a few grams when no riffles are available.

According to our experience, fractional shovelling and especially alternate shovelling is always easier, quicker, cheaper and eventually more reliable than coning and quartering.

2) <u>Mechanical shovelling</u> :

- Nature of the material : dry, wet or even sticky solids,
- maximum particle size : up to 250 or 300 mm,
- lot weight : up to several thousand tons,
- sample weight : down to a few tons when a small capacity shovel is available,
- splitting ratio : from 1/2 to 1/10 (exceptionally 1/20).
- shovel capacity : here again, a shovelful should contain less than $M_L/30N$. Mechanical shovels of various capacities are available, knowing that it is easier

to find shovels of high capacity than shovels of low capacity. A 100 kg shovel would be very useful in sampling facilities.

Fractional shovelling by means of mechanical shovels is the only probabilistic sampling method applicable to lots of valuable materials extending in the three dimensions of space and weighing up to a few thousand tons. In commercial sampling, equity is (or rather should be) of the essence of the contract. Fractional shovelling should obviously be preferred to non-probabilistic methods such as the hammer and shovel method but it will probably take another quarter century before this evidence is recognized.

The same principle is applicable and has indeed been applied to the sampling of pulps, the bucket replacing the shovel (see for instance Gy, 1971, sections 23.1.2. and 23.2.2.).

24.6. SECTORIAL SPLITTERS

These belong to two categories :

1) revolving feeders,
2) stationary feeders.

Both can achieve true or degenerate splitting.

24.6.1. Revolving feeder sectorial splitter :

This device is schematized in fig. 24.9.

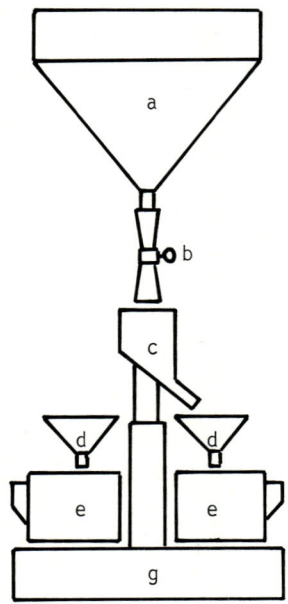
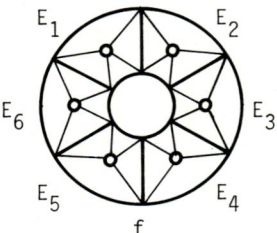

a) conical tank with a stirrer
b) pinch valve
c) revolving feeder
d) set of 6 adjacent sectors
e) set of 6 containers receiving 6 potential samples E_1 to E_6
f) horizontal view of the sectorial divider
g) motor and gear

Fig. 24.9. Revolving feeder sectorial divider

This splitter was designed by Minemet flotation engineers in order to obtain a certain number (6 to 12) of twin pulp samples for comparative flotation tests. It is made of :

a) a cylindro-conical tank equipped with a stirrer,

b) a pinch-valve regulating the rate of flow,

c) a revolving feeder (about 10 rpm),

d) a set of N adjacent sectors,

e) a set of N containers.

The rate of flow is so regulated as to empty the tank in more than three minutes, each container receiving then more than 30 increments. The tank walls are progressively washed by means of a wash-bottle, as well as the stirrer at the end of the flow.

The same device can be used with dry, free flowing sands.

The only source of bias would be a non-uniformity of the revolving motion of the feeder. This can easily be checked by weighing the fractions obtained. As long as the weights vary little and at random, the splitter is safe and reliable. If the weights vary cyclically or if one of the weights is systematically higher or lower than the others, the mechanical part of the splitter must be checked and fixed.

- nature of the material : pulps of finely ground materials, fine dry sands,
- maximum particle size : depends on the diameter of the feeder outlet, can be as large as a few mm.
- lot weight : as the receiving containers are stationary, there is no limit to the lot weight. In fact, such a device has been used in a continuous way for the splitting of a continuous stream into N quasi-continuous twin streams.
- sample weight : by definition 1/N th of the lot weight.

24.6.2. Stationary feeder sectorial splitter :

In these splitters, the feeder is stationary and the splitting ring revolves at uniform speed beneath the feeder. A typical example of degenerate stationary feeder sectorial splitter is represented in fig. 24.10. (by courtesy of Minemet-Industrie, Trappes, France). In this splitter four potential samples are directly received in screwed-up jars.

- nature of the material : dry solids,
- maximum particle size : up to a few mm,
- lot weight : it depends on the capacity of the jars and on the sampling ratio.

Fig. 24.10. *Stationary feeder sectorial splitter* - *Four two-liter jars* - *Minemet-Industrie, Trappes, France.*

The splitter represented in fig. 24.10 is equipped with four two-liter jars, each of these receiving 1/50 of the feed. The volume of the lot should not exceed 100 liters.

- speed : usually about 10 rpm.
- optional : a time switch that makes it possible to work intermittently as an increment sampler.

Such splitters, with or without time switch, actually fall within the province of the increment process. Primarily designed as true splitters, they evolved in such a way that they became increment samplers.

In automated and semi-automated sampling plants such as those to be found in processing plants, smelters, rail or sea loading or unloading facilities, in cement plants, etc.. splitters of this type tend to equip the last two sampling stages more and more. One or two two-litre jars receive for instance the samples representing a shift production. After an eventual comminution, this sample can be fed to another splitter of the same type equipped with six 200 ml-jars, each receiving 1/20 of the feed. These are the twin laboratory samples to be sealed, weighed and, in commercial sampling, attributed at random to the various parties of the contract : seller, buyer, shipping company, umpire or kept in reserve.

Fig. 24.11. represents such a splitter used for the ultimate sampling stage of uranium concentrates before shipment. The tight room can be locked-up as a safety against an eventual tampering with the samples. This device being used in equatorial Countries, the atmosphere is carefully controlled in the room in order to prevent any moisture pick-up. The jars containing the samples can be sealed in the tight room.

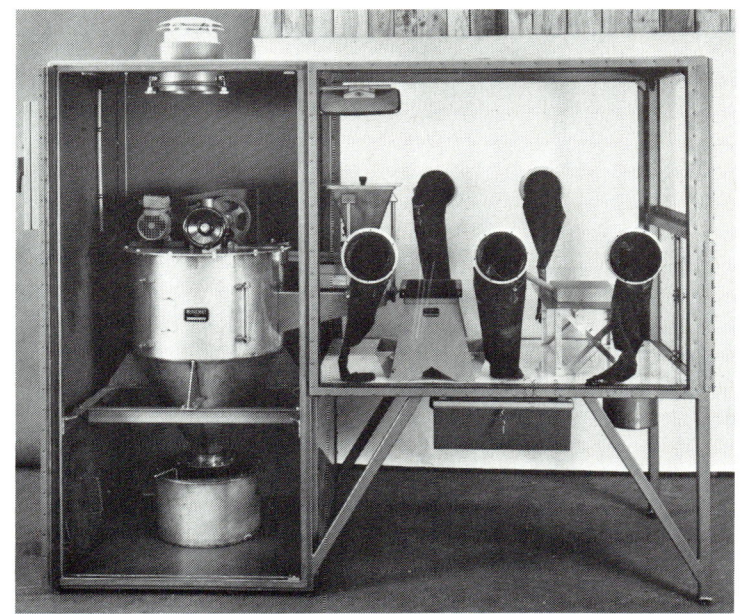

Fig. 24.11. Final sampling stage of uranium concentrates - Minemet-Industrie.

The fact that these splitters usually achieve a degenerate splitting is no inconvenience, even in commercial sampling, on condition however that the increment delimitation and extraction be correct (see chapters 17 and 18) which condition is usually respected.

CHAPTER 25

MODEL OF THE SPLITTING PROCESS - SPLITTING ERRORS

25.1. LINKING UP WITH THE EXISTING MODELS

According to the definition of true splitting processes, the lot L is first divided into a set of N_G groups of particles, the "fractions" defined in section 24.1., after a procedure that may vary from one device or one method to the next but which is irrelevant when looking for a model.

These fractions are then dealt out among a certain number N_p of potential samples (N of chapter 24), always according to a systematic pattern, exactly in the same way a pack of cards are dealt out to a certain number of players, each potential sample receiving Q fractions. We must distinguish two cases :

1) <u>Q is definite</u> and the quotient $N_G/N_p = Q$ is always an integer. This case covers coning and quartering (N_G = 4. N_p = 2. Q = 2) and riffling (N_G = 2k. N_p = 2. Q = k - usually between 6 and 10).

2) <u>Q is indefinite</u> and the quotient N_G/N_p is not necessarily an integer. This case covers fractional shovelling and splitting by means of sectorial splitters. If Q' is the Euclidean quotient of N_G by N_p (Q' is an integer, see definition of the Euclidean division in section 7.2.2.), the number Q of fractions is Q = Q' for a certain number of potential samples and Q = Q' + 1 for the other ones. This is one of the reasons for which Q (or rather Q') must be large enough (say 30 to 50). Whenever it is, the difference between Q' and Q' + 1 can be regarded as negligible.

<u>From a theoretical standpoint</u>, the continuous model of systematic selection developed in the second part of this book is applicable. We know that in such a case, the continuous selection error CE is the sum of four components :

$$CE = QE_1 + QE_2 + QE_3 + WE$$

studied in detail in chapters 8 to 13. The estimation of the moments of these errors is based on the determination of the variograms characterizing the material. Such a determination is possible, by application of the results of chapter 14, but is usually not necessary.

<u>From a practical standpoint</u>, experience shows that with batches as small as

those to which splitting is applicable, the long-range quality fluctuation error QE_2, the periodic quality fluctuation error QE_3 and the weighting error WE are likely to be negligible. Furthermore, a variographic experiment would be too costly.

We are therefore led to the conclusion that, with the possible exception of fractional shovelling of very heavy batches by means of mechanical shovels, the continuous selection error CE reduces itself to the short-range quality fluctuation error QE_1. Now, according to the results of chapter 21, we know that this error can be broken up into a sum of two components :

$CE = QE_1 = FE + GE$ with :

- FE : fundamental error,
- GE : grouping and segregation error.

We are therefore in possession of all the necessary theoretical results.

25.2. MOMENTS OF THE CONTINUOUS SELECTION ERROR CE

According to the results of section 20.5. and replacing the uniform selecting probability P by its a posteriori equivalent, the splitting ratio $\tau = 1/N_P$ (assuming that only one potential sample is retained), we obtain :

$m(CE)_1 = 0$ (first approximation)

$m(CE)_2 = - \dfrac{N_P - 1}{a_L M_L^2} \sum_n (a_n - a_L) M_n^2$ (second approximation) with $n = 1, 2, .. N_G$

$\sigma^2(CE)_1 = \dfrac{N_P - 1}{a_L^2 M_L^2} \sum_n (a_n - a_L)^2 M_n^2$ (first approximation)

with :

N_P : number of potential samples,
M_n : weight of the group of particles, the fraction G_n, with $n = 1, 2, .. N_G$
a_n : critical content of the fraction G_n,
M_L : weight of the lot L to be sampled,
a_L : critical content of the lot L.

Now, by definition of the true splitting processes, the weights M_n never deviate very much from their mean M_G defined as :

$M_G = \dfrac{M_L}{N_G}$

It is therefore legitimate to write (which the result will justify) :

$M_n^2 = M_n M_G = \dfrac{M_n M_L}{N_G}$

which involves :

$$m(CE)_2 = -\frac{N_P - 1}{N_G a_L M_L} \sum_n (a_n - a_L) M_n = -\frac{N_P - 1}{N_G a_L M_L}\left[\sum_n a_n M_n - a_L M_L\right] = 0$$

When implementing a true splitting process, we may always safely assume that the selection is unbiased, even in second approximation. If in the expression of the variance we write :

$$M_n^2 = M_G^2 = \frac{M_L^2}{M_G^2}$$

we obtain :

$$\sigma^2(CE)_1 = \frac{N_P - 1}{N_G^2 \, a_L^2} \sum_n (a_n - a_L)^2 = \frac{N_P - 1}{N_G \, a_L^2} \sigma^2(a_n) = \frac{N_P - 1}{N_G} u^2(a_S)$$

This variance can be estimated when $\sigma^2(a_n)$ is known. It can also be linked up with the distribution and constitution heterogeneities :

$$\sigma^2(CE)_1 = \frac{N_P - 1}{N_G} DH_L = \frac{N_P - 1}{N_F - 1} \times \frac{N_G - 1}{N_G} (1 + \gamma\xi) \, CH_L$$

or with the fundamental error FE and the grouping and segregation error GE :

$$\sigma^2(CE)_1 = \sigma^2(FE) + \sigma^2(GE) = (1 + \gamma\xi) \, \sigma^2(FE) = \left[\frac{1}{M_S} - \frac{1}{M_L}\right](1 + \gamma\xi) \, c \, \ell \, f \, g \, d^3$$

with :

N_F : number of fragments in the lot L,
CH_L : constitution heterogeneity of L,
DH_L : distribution heterogeneity of L,
M_S : sample weight,
γ : grouping factor defined as $\gamma = \frac{N_F - 1}{N_G - 1}$
ξ : segregation factor,
c : mineralogical composition factor of the material to be sampled,
ℓ : liberation factor of the material,
f : particle shape factor,
g : size range factor,
d : maximum particle diameter.

For more detail concerning these factors, the reader will usefully refer to chapter 22.

25.3. MINIMIZING OF $\sigma^2(CE)$

In the latter expression of $\sigma^2(CE)$, the following factors must be regarded as intangible data of the problem : M_L, c, ℓ, f and g. We can therefore act, theoretically at least on : M_S, d, γ and ξ.

1) <u>Sample weight M_S</u> : the larger the sample weight M_S and the smaller the variance. When $N_P = 2$ (coning and quartering, riffling, alternate shovelling),

$$M_S = \frac{M_L}{2} \qquad \text{and} \qquad \frac{1}{M_S} - \frac{1}{M_L} = \frac{1}{M_L}$$

When $N_p > 2$ (fractional shovelling, sectorial splitting), several (say N) potential samples can be gathered to obtain a single sample. Then :

$$M_S = N\frac{M_L}{N_G} \qquad \text{and} \qquad \frac{1}{M_S} - \frac{1}{M_L} = \frac{N_G}{N\,M_L} - \frac{1}{M_L} = \frac{1}{M_L}\left[\frac{N_G}{N} - 1\right]$$

Then, obviously, the larger N, the smaller the variance.

2) <u>Maximum particle diameter d</u> : the smaller the maximum particle diameter, the smaller the variance, hence the interest of comminution prior to sampling. A relatively modest comminution may reduce the variance considerably since it is proportional to the cube of d. But it is not always feasible and in some instances, the size analysis must be regarded as intangible.

3) <u>Grouping factor γ</u> : the variance is proportional to γ which is itself proportional to the average bulk of the groups G_n. Then, the larger the number N_G of groups and the smaller the variance. If the total sample weight is predetermined, it is always more precise to take a large number of small groups rather than a small number of large groups. It is for this reason that alternate shovelling with a sample made of say 50 groups (the shovelfuls) involves a value of γ which is practically 25 times smaller than with coning and quartering with a sample made of only two groups (the quarters).

4) <u>Segregation factor ξ</u> : mixing reduces ξ, hence the widely recognized interest of homogenizing prior to sampling. This property is supposed to work in favour of coning and quartering since the first step of this method consists in a sort of blending. According to various experiments, the value of ξ can be reduced by 2 or even 4, but if we consider the product $\gamma\xi$ involved in the expression of the variance, it is at least 6 to 8 times smaller with alternate shovelling than with coning and quartering. This observation justifies the statement made in section 24. that coning and quartering is never to be preferred to alternate shovelling.

25.4. DELIMITATION ERROR DE

We showed that in addition to the continuous selection error CE covered by the selection model two materialization errors might arise :

- DE : increment delimitation error,
- EE : increment extraction error.

We find exactly the same errors when studying the splitting processes. As far the delimitation error is concerned (the extraction error will be dealt with in the next section), we must distinguish :

1) <u>True splitting devices</u> (all riffle splitters, most sectorial dividers) : these are usually designed and manufactured in such a way that delimitation errors are non-existent, at least when they are new or well maintained. In any case, the delimitation error is negligible.

2) <u>Degenerate splitting devices</u> (some sectorial splitters) : they are liable to introduce a delimitation error when the sectors are not delimited by <u>radial edges</u>. Some manufacturers (and also some users) design and build sectorial splitters with rectangular openings which achieve an incorrect delimitation and are liable to generate a bias (see chapter 17). This defect has been repeatedly observed on home-made devices.

3) <u>True splitting methods</u> (coning and quartering, true fractional shovelling, alternate shovelling) : these may generate small delimitation errors, due to the fact that all splitting methods involving shovels and scoops do not achieve a precise delimitation of the fractions (scoopfuls or shovelfuls), but they are unlikely to generate a systematic difference between potential samples. If the actual sample is selected at random, the possibility of a delimitation bias vanishes completely. Hence the necessity, at least in commercial sampling, of always selecting the sample at random.

4) <u>Degenerate splitting methods</u> (degenerate fractional shovelling) : these offer the operator a possibility of voluntarily adulterating the composition of a sample known to be predetermined. This can be achieved for instance by taking heavier shovelfuls of fine (or coarse) material when they go into the sample. Such an opportunity arises only in commercial sampling. For this reason, degenerate splitting methods are unadvisable when carrying out a commercial sampling.

25.5. EXTRACTION ERROR EE

Here again, we must distinguish :

1) <u>true splitting devices</u> : when the chutes or the sectors of the splitter are correctly delimited, the symmetry of the system works in such a way that no extraction error is liable to take place. The bias observed with the riffle splitter and described in section 24.4.2. may however be regarded as an exceptional extraction bias resulting from an incorrect use of a correct device.

2) <u>Degenerate splitting devices</u> : degenerate sectorial splitters do not usually work in a symmetrical way and extraction errors, more often than not extraction biases, are susceptible to arise. More specifically, the manufacturers (whether or no they are professional ones) should be reminded that the rules concerning the cutter width and speed set forth in section 18.4.5. must be respected with degenerate splitters as well as with cross-stream samplers.

3) <u>True splitting methods</u> : they may generate extraction errors but are unlikely to generate systematic differences between potential samples. If the actual sample is selected at random, the possibility of a bias vanishes.

4) <u>Degenerate splitting methods</u> : here again, these offer the operator a possibility of voluntarily adulterating the composition of a sample known to be predetermined, for instance by putting a larger (or a smaller) proportion of coarse fragments in the shovelfuls going to the sample. To provide against such a possibility of fraud in commercial sampling, it is always advisable to implement only true splitting methods.

CHAPTER 26

PRACTICAL IMPLEMENTATION OF SPLITTING PROCESSES - EXAMPLE -
REDUCTION OF DRILL CORE SAMPLES

26.1. INTRODUCTION

The purpose of this chapter is to show on a particular but very important example how splitting processes can be implemented.

The object of mineral exploration is to estimate the grade and tonnage of the various qualities of ore contained in a given mineral deposit. This estimation cannot be directly carried out on the whole of the orebody. It requires :

1) a core drilling campaign : the orebody is represented by a primary sample made of a series of cores, further divided into core sections of equal or unequal length.

2) core halving : the core sections are then longitudinally "split" (the word splitting is then used in a different sense) into two halves by means of a chisel or of a diamond saw. The first half is kept for various examinations, physical measurements and stored for further reference. The second half, referred to as the "core sample" is used for chemical assaying.

3) core sample reduction : the core sample bulk and particle size must then be reduced to the size of a laboratory sample and ultimately to that of a few one-gram assay portions.

The present chapter deals with the second and third steps of this sequence.

26.2. CORE SAMPLE REDUCTION METHODOLOGY

Core sections are usually delivered to a specialized laboratory for longitudinal halving (usually referred to as "core splitting" but in order to avoid any confusion with the "splitting processes" studied in this fifth part of our work we shall call "halving" this operation), reduction and analysis.

26.2.1. Longitudinal core halving :

This operation is seldom carried out by hand but rather by means of a "core splitter", a mechanical tool using either a chisel or better a diamond saw to cut the core into two longitudinal halves. This step is liable to introduce a random error and a bias. Random errors are minimized when both halves are as similar as possible which occurs when the cutting plane is perpandicular to the mineralization strata (if any). Bias is suppressed when the half retained for assaying is selected at random (by tossing up a coin). In no case should the mineralogist or the chemist be allowed to choose "his" half. The device should be carefully cleaned before use (dust rejected) and after use (dust and other particles recovered and equitably divided between the two halves).

26.2.2. Drying :

Core samples are often dry enough to be directly fed to the crushing and pulverizing equipment but when the moisture content exceeds a few percents, it is necessary to dry the samples prior to pulverizing. Stove or hot-plate drying are to be avoided since overheating is liable and often likely to decompose some of the mineralogical constituents, thus introducing an estimation bias. This point will be developed in chapter 27. It is always advisable to use a drying oven set to 105 to 110°C (a few exceptions such as native sulphur).

26.2.3. Size reduction :

Size reduction is achieved by crushing, grinding and ultimately pulverizing. The core sample, made of roughly broken core halves can be comminuted by means of conventional laboratory equipment such as jaw, cone or rolls crushers, disk pulverizers and batch vibrating mills. When available, the best combination consists of a small adjustable jaw crusher reducing the grain size to about 10 or 5 mm, followed by an adjustable rolls crusher reducing the grain size to about 2 to 3 mm (in closed circuit). Then a disk pulverizer, McCool type, grinds the material to minus 150 or 100 microns (100 to 150 mesh) in one or better two stages. It may be convenient to use two pulverizers : a first one set to about 0.8 mm (plate gap setting) and a second one set to the size required by assaying (usually between 100 and 150 microns). Batch vibration mills can also be used between 2 or 3 mm and 100 microns or even below if a finer grind is necessary.

Size reduction proper is a non-random operation as long as the sample integrity is respected. This point will be dealt with in chapter 27.

26.2.4. Sampling :

Weight reduction is usually carried out by splitting. Riffle splitters are very convenient, cheap and most generally utilized. Every sampling laboratory should have a set of riffles of various sizes. Their use is practically foolproof if the recommendations made in chapter 24 are observed.

26.2.5. Mixing :

Mixing is supposed to suppress or at least reduce the grouping and segregation error presented in chapter 21. Twin-shell or Vee-type mixers are well known and very satisfactory. Blending time is usually 10 to 60 mn. One may however wonder whether the suppressed random grouping and segregation error (which is usually likely to be small) compensates the risks of a bias resulting from losses and/or contamination attached to any preparation. For this reason, we do not usually recommend mixing. This point will be developed in chapter 27.

26.3. SELECTION OF A CORE SAMPLE REDUCTION SCHEME

26.3.1. Sample reduction diagram :

It is convenient to represent a sample reduction scheme on a sheet of log-log-paper and to plot log-weight vs. log-particle-size (log M_S vs. log d). This is represented in fig. 26.1.

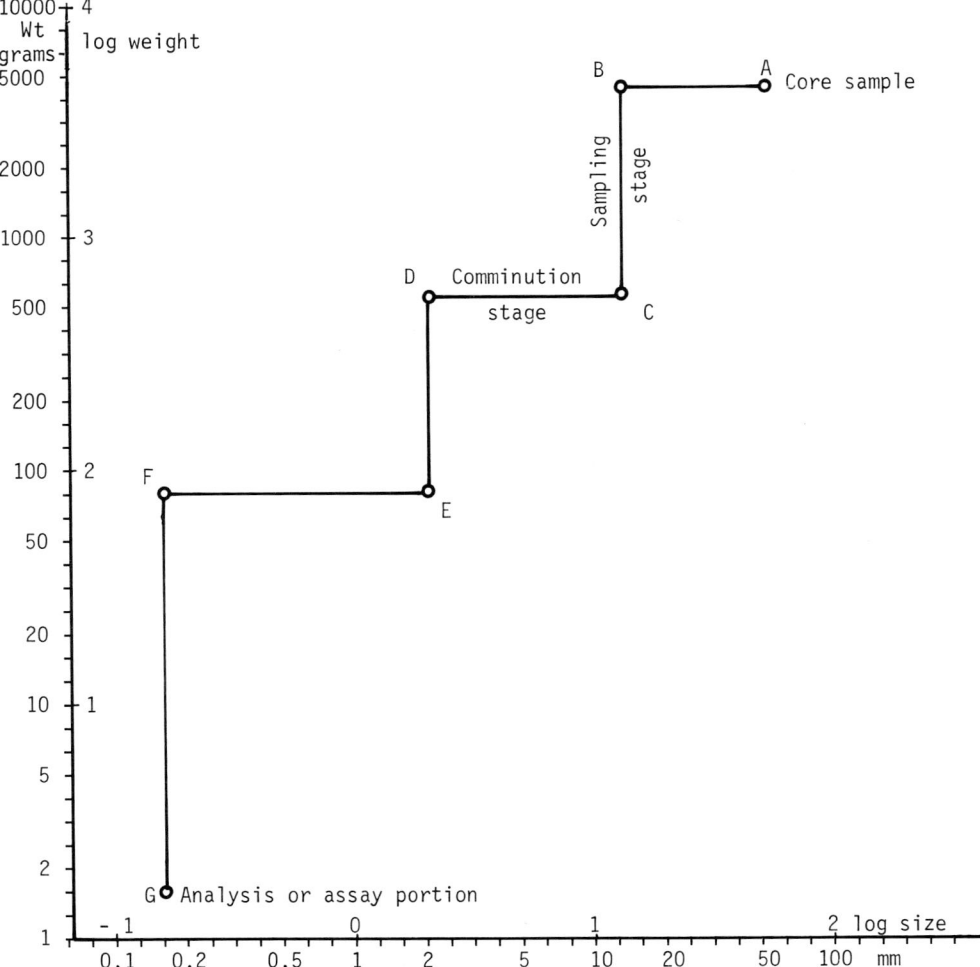

Fig.26.1. Sample reduction diagram - Example : core sample reduction

On this diagram, any batch of particulate material is represented by a point. Any comminution stage (size reduction at constant weight) is represented by a horizontal segment. Any sampling stage (weight reduction at constant particle size) is represented by a vertical segment. Any sample reduction scheme is represented by a step-like broken line joining the point A representing the original batch (the core sample) to the point G representing the ultimate sample (assay portion).

26.3.2. Safety rule :

This safety rule is represented in fig. 26.2. by the "safety line".

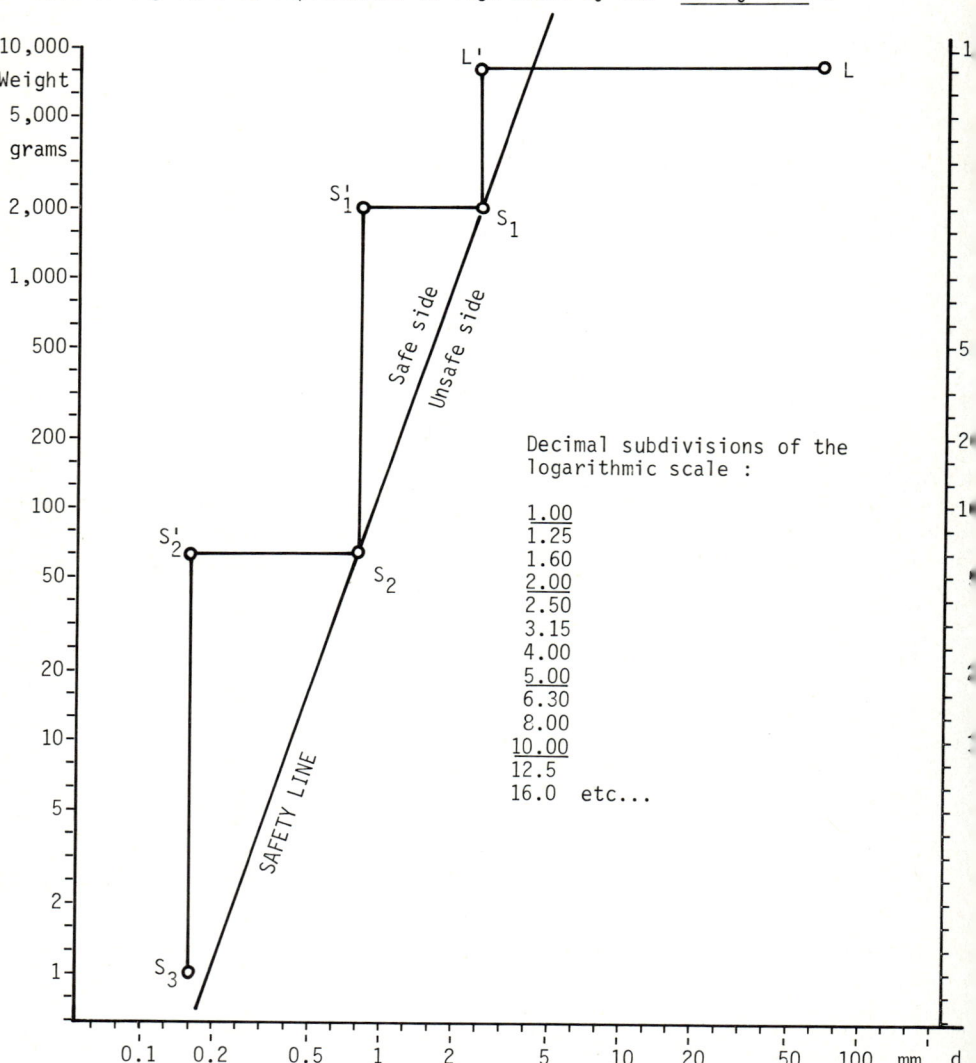

Fig. 26.2. Safety rule - Safety line - Example of safe core sample reduction.

This rule is derived from the formula expressing the fundamental variance $\sigma^2(FE)$ in terms of the sampling constant C, maximum particle diameter d, and sample weight M_S : it states that the sample weight should always be larger than a certain value M_{So} with (M_S imperatively expressed in grams and d in cm) :

$$M_S \geq M_{So} = 125{,}000 \; d^3$$

It has been devised in order to help geologists who do not wish to use complicated formulas. It is valid for all geological materials such as core samples, with the exception of gold ores, irrespective of the critical content, with the exception of very low grade ores. This rule is obviously the result of a compromise between cost and reproducibility. When low cost is regarded as more important than reproducibility, then the factor 125,000 may be reduced to 60,000. When on the contrary precision is regarded as more important than cost, which should be nearly always the case, the factor can be doubled to 250,000. The reader should understand that what is important is the order of magnitude of the sample weight M_S.

The safety rule is represented on the diagram by the safety line. It is the straight line joining on a log-log-graph paper the two points :

d_1 = 0.02 cm (or 0.2 mm)   - M_{S1} = 1 gram
d_2 = 0.2   cm (or 2 mm)   - M_{S2} = 1000 grams (or 1 kg).

The safety rule can be expressed as follows : any sampling stage represented by a segment belonging entirely to the upper (or left) part of the graph can be regarded as safe. Any sampling stage represented by a segment belonging partly or entirely to the lower (or right) part of the graph must be regarded as unsafe, i.e. likely to involve unacceptable errors. Examples of a safe and of an unsafe sample reduction schemes are given in section 26.4.

26.3.3. Core sample L :

Point L representing a core sample always falls on the wrong side of the safety line. For instance, the core sample made of core halves covering a one-meter section of a 3" core weighs about 8 kg. After breaking roughly the elongated core fragments, the core sample can be regarded as a 7.5 cm (or 75 mm) material. Such a sample is represented by point L of fig. 26.2.

In order to avoid any misunderstanding, we must emphasize the fact that a core is a compact sample extracted from a compact object, the orebody, even though it is eventually broken in the process of its recovery. Our safety rule, established for the sampling of particulate materials becomes applicable only when reducing the core sample weight. The fact that point L lies on the wrong side of the line merely means that the core sample may not be sampled as it is and requires comminution prior to any weight reduction.

26.3.4. Primary comminution stage L L' :

Point L' of fig 26.2. representing the crushed core sample should be on the left side of the safety line, far enough to allow at least for a two-step splitting (sampling ratio 1/4). In our example, that can be regarded as typical, the core sample has to be reduced to a 0.25 cm (2.5 mm) grain size prior to any sampling. This can be achieved for instance by means of a jaw crusher followed by a rolls crusher.

26.3.5. Primary sampling stage L' S_1 :

Point S_1 representing the primary sample should be on or above the safety line In our example, a two-step splitting brings point S_1 on the line. This splitting can be achieved by means of a riffle splitter. After the first splitting step, we shall for instance retain the half collected in the right bucket of the splitter. This half is fed again to the splitter (after rejecting the half contained in the left bucket). After this second step, we shall retain the half contained in the left bucket, in order to respect the alternation rule stated in section 24.4.3. Our primary sample weighs about 2 kg.

26.3.6. Secondary comminution stage $S_1 S_1'$:

The fact that point S_1 lies on the safety line means that the primary sample must be comminuted again prior to any new sampling stage. The grain size will depend on the available equipment. Assuming that we use a disk pulverizer set to 0.08 cm (0.8 mm or 800 microns), we obtain point S_1' of fig. 26.2. well above the safety line, which allows a multi-step splitting.

26.3.7. Secondary sampling stage $S_1' S_2$:

A five-step splitting by means of a riffle splitter, respecting the alternation rule, will produce a 60 to 65 gram sample represented by point S_2, just on the safety line.

26.3.8. Tertiary comminution stage $S_2 S_2'$:

The next comminution stage must bring the material to the grain size required by assaying, usually between 100 and 150 microns. This can be achieved by means of a second disk pulverizer set to 150 microns for instance. We thus obtain the laboratory sample S_2' which is delivered to the analytical laboratory. One might question the advisability of separating the secondary and tertiary comminution stages by a sampling stage and wonder whether it would not be more convenient to grind in a single stage from 2.5 mm to 150 microns. This mainly depends on equipment availability. The use of two distinct disk pulverizers with an intermediary sampling stage should reduce the sample reduction time to a minimum as well as the sample reduction cost.

26.3.9. Tertiary sampling stage $S_2' - S_3$:

Though this fact is often overlooked, the taking of a one-gram assay portion S_3 out of the laboratory sample S_2', usually after drying, is a sampling stage liable to introduce random errors and biases. The smaller the grind, the smaller the risk of sampling bias due to a possible segregation of certain minerals, either because of their density or because of their shape (e.g. micas). It is always advisable at this stage of the procedure to mix the laboratory samples by putting them in cylindrical jars that can roll on two revolving rollers. Segregation bias is then unlikely to take place. It should be pointed out that when analytical reproducibility is experimentally determined by repeating a given assay on a series of assay portions extracted from the same laboratory sample, the variance which is calculated includes the tertiary sampling variance which very often represents a large part of the total variance.

It is satisfactory to observe that according to the analytical habits point S_3 is usually (but not always) on the safe side of the line.

26.4. EXAMPLES

In order to illustrate the preceding sections we shall estimate the total sample reduction variance and the corresponding 95 % probability confidence interval when implementing safe and unsafe reduction schemes with four different materials :

- 65 % Fe magnetite ore,
- 5 % Zn sphalerite ore,
- 0.5 % Cu chalcopyrite ore,
- 0.1 % U pitchblende ore

covering the natural range from very rich to very low-grade materials. We shall apply to each sampling stage the expression :

$$\sigma^2(SE) = \sigma^2(QE_1) = \sigma^2(FE) + \sigma^2(GE)$$

The grouping and segregation variance $\sigma^2(GE)$ is usually small enough to be neglected. It is always safe to admit that it is smaller than $\sigma^2(FE)$ and to base our calculations on :

$$\sigma^2(SE) = 2\,\sigma^2(FE) = 2\left[\frac{1}{M_S} - \frac{1}{M_L}\right] C\,d^3 \qquad \text{with :}$$

M_S : sample weight (in grams),
M_L : lot weight (in grams),
d : maximum particle size (in cm),
C : sampling constant (in g/cm^3). Assuming that the mineralogical components are completely liberated at a size smaller than 0.25 cm (a safe assumption), C can be regarded as a true constant for the three sampling stages (liberation factor $\ell = 1$).

According to the results of chapter 22, we easily calculate :

$C = 0.04$ g/cm^3 : for the 65 % Fe magnetite ore (a_L = 0.9 g Fe$_3$O$_4$ / g solids)
$C = 7.5$ g/cm^3 : for the 5 % Zn sphalerite ore (a_L = 0.075 g ZnS/ g solids)
$C = 45$ g/cm^3 : for the 0.5 % Cu chalcopyrite ore (a_L = 0.015 g CuFeS$_2$/g so
$C = 600$ g/cm^3 : for the 0.1 % U pitchblende ore (a_L = 0.0011 g UO$_2$/g solids

26.4.1. Safe reduction scheme :

We shall calculate the total sampling variance involved when implementing the scheme presented in the preceding section and represented in fig. 26.2. Such a scheme is typical of what is done in serious laboratories. We can calculate the value of the factor F :

$$F = \left[\frac{1}{M_S} - \frac{1}{M_L}\right] d^3$$

which does not depend on the material but only on the sampling stage :

$F_1 = 5.8 \times 10^{-6}$ (primary sampling stage)
$F_2 = 8.3 \times 10^{-6}$ (secondary sampling stage)
$F_3 = 4.0 \times 10^{-6}$ (tertiary sampling stage).

Since C is a constant, we can write the total sampling variance σ^2(TE) :

$\sigma^2(TE) = \sigma^2(SE_1) + \sigma^2(SE_2) + \sigma^2(SE_3) = 2 C (F_1 + F_2 + F_3) = 2 C F_0$ with :

$F_0 = F_1 + F_2 + F_3 = 18 \times 10^{-6}$

$\sigma^2(TE) = (36 \times 10^{-6}) C$

Table 26.1 gives the value of σ^2(TE), that of the 95 % probability relative confidence interval and that of the 95 % probability absolute confidence interval (expressed in % metal). This table shows that the richer the sampled material the safest the rule. For this reason, the factor 125,000 of the safety rule may, if necessary be divided by 2 with high grade and very high grade materials or multiplied by 2 with low grade and very low grade materials.

Table 26.1. Safe reduction scheme - Total sampling variance - Confidence interv

Sampled material	Total sampling variance σ^2(TE)	Relative confidence interval $\pm 2\sigma$(TE)	Absolute confidence interval $\pm 2\sigma(a_S)$
65 % Fe	1.4×10^{-6}	$\pm 0.24 \times 10^{-2}$	± 0.16 % Fe
5 % Zn	2.7×10^{-4}	$\pm 3.3 \times 10^{-2}$	± 0.16 % Zn
0.5 % Cu	16.2×10^{-4}	$\pm 8.0 \times 10^{-2}$	± 0.04 % Cu
0.1 % U	2.1×10^{-2}	$\pm 29.0 \times 10^{-2}$	± 0.03 % U

26.4.2. Unsafe reduction scheme :

We have represented in fig. 26.3. the diagram of a reduction scheme actually implemented in the sampling laboratory of a well known mining Company.

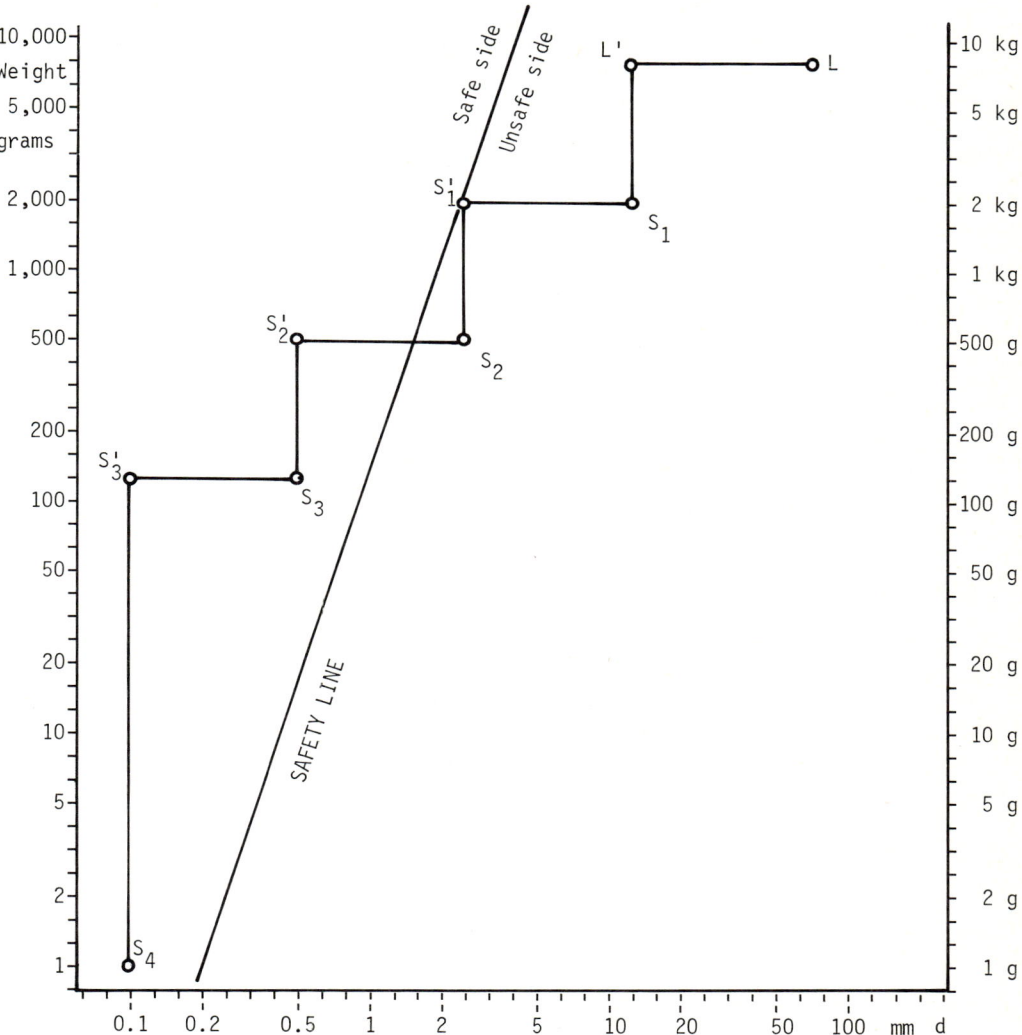

Fig. 26.3. *Safety line - Example of unsafe core sample reduction*.

The 8 kg-core sample represented by point L (the lot to be sampled in our example) was first crushed to 1/2" (1.25 cm) corresponding to point L', then reduced by splitting to 2,000 grams (point S_1), crushed again to about 0.25 cm (point S_1'), reduced by splitting to 500 grams (point S_2), pulverized to 0.05 cm (point S_2'), reduced by splitting to 125 grams (point S_3) and pulverized to 100 microns (point S_3').

From this laboratory sample was extracted a one-gram assay portion represented by point S_4. Although this four step reduction scheme seems very well balanced, at least on the graph, we shall see that it usually generates unacceptable errors : this is due to the fact that two of the four sampling stages fall completely on the wrong side of the safety line. The factors F can be calculated :

$F_1 = 732 \times 10^{-6}$ (primary sampling stage L' S_1)
$F_2 = 23 \times 10^{-6}$ (secondary sampling stage S_1' S_2)
$F_3 = 1 \times 10^{-6}$ (tertiary sampling stage S_2' S_3)
$F_4 = 1 \times 10^{-6}$ (ultimate sampling stage S_3' S_4).

Now, due to the fact that the primary sampling stage takes place at a rather coarse size (12.5 mm) it was not possible to assume a total liberation of the critical mineral, so that we chose to take a liberation factor of 0.2 for the primary sampling stage and $\ell = 1$ for the three other stages. The corresponding values of C are given in table 26.2. for the four sampling stages and for the four materials :

Table 26.2. *Unsafe sample reduction scheme - Sampling constant C* (g/cm^3).

Material	C_1	C_2	C_3	C_4
65 % Fe	0.008	0.04	0.04	0.04
5 % Zn	1.5	7.5	7.5	7.5
0.5 % Cu	9.0	45.0	45.0	45.0
0.1 % U	120.0	600.0	600.0	600.0

The total sampling variance is the sum :

$\sigma^2(TE) = 2 \, (F_1 C_1 + F_2 C_2 + F_3 C_3 + F_4 C_4)$

Table 26.3 gives the value of $\sigma^2(TE)$, that of the relative confidence interval $\pm 2\sigma(TE)$ and that of the absolute confidence interval $\pm 2\sigma(a_S)$ expressed in metal %

Table 26.3. *Unsafe reduction scheme - Variance - confidence intervals.*

Sampled material	Total sampling variance $\sigma^2(TE)$	Relative confidence interval $\pm 2\sigma(TE)$	Absolute confidence interval $\pm 2\sigma(a_S)$
65 % Fe	14×10^{-6}	$\pm \, 0.75 \times 10^{-2}$	$\pm \, 0.50$ % Fe
5 % Zn	27×10^{-4}	$\pm \, 10 \times 10^{-2}$	$\pm \, 0.51$ % Zn
0.5 % Cu	1.6×10^{-2}	$\pm \, 25 \times 10^{-2}$	$\pm \, 0.12$ % Cu
0.1 % U	21×10^{-2}	$\pm \, 91 \times 10^{-2}$	$\pm \, 0.09$ % U

This table shows that the variance is about ten times larger than with the safe scheme. Accordingly, the confidence intervals are more than three times larger and become unacceptable, especially with low grade materials.

We might have presented a still unsafer reduction scheme starting with coning and quartering of the roughly broken core sample. The confidence interval becomes so wide that the whole operation becomes meaningless : 65 % ± 8 % Fe for instance, to say nothing of the lower grade materials for which the relative confidence intervals exceed ± 100 % (which means more particuliarly that the normal approximation is no more acceptable). These examples should illustrate well enough the necessity of respecting the safety rule and of following the recommendations summarized in the next section.

26.5. RECOMMENDATIONS

On the theoretical side, the most important part of this chapter is that no weight reduction should take place on the core sample prior to crushing to a grain size of about 2 to 3 mm.

At each sampling stage, the weight M_S of the sample, expressed in grams should never be smaller than $125,000 \, d^3$, where d, expressed in centimeters (cm), is the maximum particle diameter.

This safety rule is valid for all materials and constitutes a usually satisfactory compromise between reproducibility and cost. When the emphasis is put on reproducibility or when sampling low grade materials, the factor 125,000 can be doubled. When on the contrary the emphasis is put on low cost or when sampling high grade materials, the factor 125,000 can be divided by two.

On the practical side, all operations carried out on samples : handling, comminution, drying, mixing, splitting, etc .. should be performed with analytical care and accuracy. This point will be developed in the next chapter.

The respect of these rules will always ensure that the core sample reduction is unbiased and conveniently reproducible.

SIXTH PART

LOT AND SAMPLE PREPARATION

All non-selective operations carried out on the lot and on the successive samples are referred to as "preparation stages". Their object is to bring the object to be sampled to the place and under the form required for the next sampling stage and eventually for assaying or analysis. These operations belong to the following categories :

- <u>transfer</u> : continuous or discontinuous, horizontal, oblique or vertical, ascending or descending, etc...

- <u>comminution</u> : crushing, grinding, pulverizing, etc...

- <u>screening</u>, often in connexion with comminution : wet or dry,

- <u>mixing</u> or blending, stirring of pulps,

- <u>drying</u> of solids,

- <u>filtration</u> of pulps, etc ...

Though non-selective, all these operations are, potentially at least, error-generating inasmuch as they can adulterate the critical content of the object submitted to preparation.

Strictly speaking, these are not sampling errors but they fall nevertheless within the province of this book, if only for the following reasons :

1) They alter the critical content to be eventually estimated by analysis or assaying,

2) They usually result from ignorance, negligence, carelessness, awkwardness unintentional mistakes or deliberate frauding, or even sabotage, these being the fact of sampling operators.

3) They are accounted for as sampling errors in the wider sense of the word sampling.

The sixth part of this book is dedicated to the study of such errors and is made of a single chapter :

Chapter 27 : Preparation errors PE.

CHAPTER 27

PREPARATION ERRORS PE

27.1. INTRODUCTION

Preparation errors fall in the following categories :
1) Errors by contamination PE_C (section 27.2.),
2) Errors by loss PE_L (section 27.3.),
3) Errors by alteration of the chemical composition PE_A (section 27.4.),
4) Errors by alteration of the physical composition PE_P (section 27.5.),
5) Errors resulting from unintentional mistakes PE_M (section 27.6.),
6) Errors resulting from frauding or sabotage PE_F (section 27.7.).

Contrary to the errors reviewed in the preceding chapters, preparation errors can hardly be regarded as random errors with stationary properties such as mean or variance. Most of these may however, in routine operations, result in a bias.

With such errors, the strategy does not consist in trying to estimate their moments and the purpose of this chapter is merely to show how they can occur and what should be done in order to prevent them.

27.2. ERRORS RESULTING FROM CONTAMINATION PE_C :

Such errors take place as soon as foreign materials contaminate the lot or one of its samples. This may happen for instance in the following circumstances :

27.2.1. Contamination by dust :

When handling materials containing dry fine particles, it is practically impossible to prevent the formation of dust clouds that tend to fly everywhere. This dust is likely to contaminate any sample which is not adequately protected. The solution is triple :
- prevent the formation of dust by reducing free falls as much as possible,
- enclose the inevitable sources of dust in tight boxes, using a dust collecting system to create a <u>slight</u> depression,
- protect the sampling circuit and each sampling device.

This error is of the same kind as the error mentioned in section 17.4.5.

27.2.2. Contamination by material present in the sampling circuit and equipment :

Any sampling circuit or device working intermittently, whether in a plant or in a laboratory, should be carefully cleaned by means of nozzles or vacuum cleaners according to the nature of the material to be sampled, prior to a new sampling operation. A sampling system handling the same material in a routine way does not need a daily cleaning. A weekly inspection is usually enough (e.g. samplers operating in mineral processing facilities). Things are obviously different in a laboratory receiving samples of very different grades such as feed, concentrates and tailings of a processing plant. In such a case, a distinct equipment (splitters, crushers, pulverizers, etc..) and whenever possible distinct rooms should be utilized for each grade of products. If the pulverizing of a concentrate after that of a reject does not result necessarily in an appreciable error, the pulverizing of a tailings sample after that of a concentrate will always result in a positive bias due to contamination. When the same pulverizer must be used for high grade as well as for very low grade materials, in addition to wet or dry cleaning, it is advisable to pulverize a neutral sample, for instance the sampling reject of a previous operation involving the same kind of product (feed, concentrate or tailings) as the product to be pulverized next. All materials collected in the course of preventive cleaning must be rejected.

27.2.3. Contamination by abrasion :

Crushing, grinding, pulverizing and to a lesser extent all handling operations carried out on abrasive materials are likely to introduce into the lot or samples minute particles of material abraded from the equipment. This problem becomes very serious when iron and allied metals are penalized impurities (e.g. handling silica samples in the glass making industries). The solution may consist either in selecting convenient anti-abrasion or non-critical materials (to be determined case by case) or in using conventional steel machinery after assaying blank samples in order to evaluate the contamination which is likely to be comparable from one sample of a given material to the next, and correcting the assays accordingly.

27.2.4. Contamination by corrosion :

Corrosion of the sampling and sample reduction equipment is likely to take place when handling the following agressive materials :

- wet materials developing acid reactions such as certain ores containing sulphides (especially pyrrhotite, pyrite, etc..),
  - acid flotation pulps,
  - flotation pulps in sea water,
  - hydrometallurgical pulps or solutions,
  - very corrosive minerals such as potash, etc..

In each particular case, the solution must be carefully studied with the help of a specialist of anti-corrosion metals, alloys and materials. When handling "normal" materials, stainless steel is recommended for all parts of the machinery in contact with the material to be sampled.

27.2.5. Contamination by salting :

This point will be dealt with in section 27.7.

27.3. ERRORS RESULTING FROM LOSSES PE_L

Such errors take place as soon as particles are withdrawn from the lot to be sampled or from one of its samples. This may happen for instance in the following ciecumstances :

27.3.1. Loss of fines as dust :

As already mentioned, when handling dry, fine materials, any free fall is likely to generate dust. If this dust belongs to a sample its loss introduces an error of the PE_L type.

Now, the very popular solution which consists in enclosing the sampling equipment in a clean tight box connected to a very efficient dust collecting system is undoubtedly satisfactory from the environmental standpoint, but much less so when the sampling and preparation errors are taken into consideration.

Generally speaking, in sampling plants and laboratories, more specifically in the vicinity of sampling or splitting devices, dust collecting systems must be utilized with great prudence and moderation as they generate draughts which may achieve an artificial dedusting of the samples and result in a bias as soon as the depression is too high.

27.3.2. Loss of material remaining in the preparation or sampling circuit :

After any sampling operation, the sampling and preparation equipment must always be carefully cleaned and the recovered material added to the sample whenever it belongs to it. Material recovered on or in the outer parts of a sampler obviously belong to the sampling reject and must be added to it.

27.3.3. Loss of certain fractions of the sample :

When preparing samples for assaying, the laboratory sample is usually pulverized in closed circuit with a 100 to 150 microns screen. As this process must be repeated several times, an impatient or careless operator may be tempted to throw away the second or third oversize, on the pretext that it is not worth mentioning or that time is money. This amounts to forgetting that this refractory oversize has every chance of being a concentrate of one of the mineralogical components : the technique of selective comminution is founded on this property.

The solution to this important but often neglected problem must be specific. When dealing with native gold ores, for example, such a practice is disastrous as gold flakes tend to flatten down upon pulverizing. In one instance, we found it to be responsible for a negative bias ranging about 50 %, one of the highest biases ever met with. Native gold ores should not be pulverized by means of such devices as disk pulverizers, pestles and mortars or the like unless your idea be to plate your equipment with 24-carat gold. This is one of the numerous difficulties of gold mining and metallurgy. Here again, there is no ready-made solution. The only good one is expensive : it consists in handling very large samples, in concentrating these by means of jigs or shaking tables, super-panners or simple pans, weighing concentrates and tailings separately, melting the whole of the concentrate and as large a sample of the tailings as possible and eventually calculating the gold content of the processed sample.

27.3.4. Deliberate loss of fractions of the sample :

Losses such as those mentioned in the preceding paragraph may be unintentional, sometimes they are deliberate. This point will be dealt with in section 27.7.

27.4. ERRORS RESULTING FROM ALTERATION OF THE CHEMICAL COMPOSITION PE_A

In sections 27.2 and 27.3. we reviewed the possibilities of contamination or loss by addition or substraction of particles. We must now study the possibilities of contamination or loss by addition or substraction of atoms or molecules, or in other words the possibilities of adulteration of the chemical composition of one of the mineralogical components, critical or not.

As the definition of the critical content is (section 1.1.2.) :

$$\text{Critical content} = \frac{\text{Weight of critical component}}{\text{Weight of active components}}$$

this critical content is altered whenever one of the terms of the quotient is adulterated.

27.4.1. Errors by addition or fixation :

Typical examples of such errors are :

1) <u>Oxidation of sulphides</u> : pyrrhotite, marcasite, pyrite, chalcopyrite, etc.. may be very reactive, especially when wet, finely divided (flotation concentrates for instance) and gathered in large bulks. Oxidation of sulphides is an exothermic reaction and accelerates as the temperature increases. Such a vicious circle may end in blazing up of the whole mass when enough oxygen is present. But before that point is reached, sulphides are more or less slowly transformed into sulphates by fixation of oxygen. When the sulphide is not a critical component, the numerator of the quotient remains unchanged whereas the denominator increases : the critical content decreases. This oxidation results in a negative bias.

2) <u>Fixation of water or carbon dioxide by oxides or calcined minerals</u> : as atmosphere always contains H_2O and CO_2 molecules, some materials such as for instance quicklime are likely to pick up water or carbon dioxide very quickly.

Some minerals are very stable : they can stand overdrying without any oxidation or exposure to the weather without hydration or carbonation. Others are not and it is always advisable to take a minimum of precautions. The most critical points are drying and handling in a damp atmosphere.

To prevent errors by fixation as well as by elimination (next paragraph), drying should always be carried out in a drying oven set to about 105 - 110°C (221 - 230°F) and well regulated. In tropical or even in temperate but wet countries, samples of certain minerals tend to pick up moisture as soon as they are out of the drier. These should always be placed in a dessicator until the next operation. Such errors are efficiently provided against by equipping the sampling and assaying laboratories with air conditioning systems. Glove boxes such as the one represented in fig. 24.11. are also a good solution in tropical Countries.

27.4.2. Errors by substraction or elimination :

Typical examples of such errors are :

1) <u>Elimination of combined water by overdrying</u> : a large number of minerals (especially gangue minerals) contain water molecules in their crystal lattice. Materials containing such minerals should be dried up with peculiar care as they are liable, some of these likely, to loose part of this water at low temperature. Gypsum for instance, whose molecule writes $CaSO_4$, $2\ H_2O$, looses 3/4 of its water (one and a half molecule) between 110 and 130°C and all of it at 143°C. As the weight of this water represents 21 % of the gypsum molecular weight, overdrying of a gypsum-bearing material may alter the critical content to be estimated. As the denominator of the fraction is systematically reduced, the critical content is systematically increased by a factor that may reach 1.26.

If moisture is the critical component, overdrying of a material containing 5 % free moisture and a gypsum gangue may result in a loss on drying (accounted as moisture) reaching 30 % (or 500 % positive bias).

2) <u>Elimination of carbon dioxide by overdrying</u> : carbonates are usually regarded as stable minerals but drying conditions tend sometimes towards calcination in such a way that a non-negligible amount of CO_2 may be eliminated.

In sampling and assaying laboratories, whether the purpose of drying is to estimate the moisture content or to eliminate this moisture prior to pulverizing or any other operation, the drying conditions are always critical. Temperatures as high as 250 to 300°C (480 to 570°F) can be observed in materials undergoing infrared or hot-plate drying, which definitely condemns all kinds of driers except the well regulated and ventilated drying oven set to 105°C (220°F).

27.5. ERRORS RESULTING FROM ALTERATION OF THE PHYSICAL COMPOSITION PE_p

Such errors arise when the critical component is moisture, a size fraction or sulphur in native sulphur ores.

27.5.1. Addition or creation of critical component :

If we except salting, which will be dealt with in section 27.7., addition of critical component is restricted to sampling for moisture or size analysis.

Believe it or not, we have seen moisture samples exposed to rain. Moisture samples should always be well protected against accidental addition of water by exposure to rain, spindrift or simply fog as certain minerals are liable to pick up moisture from a damp atmosphere very quickly.

When sampling for size analysis, a non-critical component, coarse particles for instance, may very readily be transformed into critical component when the latter is the undersize to a certain mesh, by breakage. As the percentage passing a certain screen is often limited by sales agreements and penalized when too high, it is therefore important to prevent the breakage of particles and the production of penalized fines. The reader should not forget that autogeneous grinding is a very efficient technique based on the grinding effect of free fall of the ore. On one occasion we had to act the part of an arbitrator between a mining Company and a steel plant about a dispute concerning the percentage of ore passing a 1/2" (12.5 mm) screen. On its way from the mine to the blast furnace, the ore underwent the following operations :

- loading in trucks,
- transfer into a surge bin,
- transfer into railway cars,
- unloading onto a quaiside pile,
- reclaiming by means of a bucket-wheel,
- transfer on various conveyor belts,
- loading into the holds of a sea-going ship,
- reclaiming from these holds,
- transfer into a battery of surge bins,
- extraction and transfer on various conveyor belts,
- loading into river barges,
- reclaiming from these barges,
- transfer into a battery of surge bins,
- extraction and transfer on a conveyor belt,
- sampling for size analysis.

Practically all these operations involved free falls ranging from a few feet (say 1 meter) to 20 meters. Samples taken upon loading of sea-going vessels showed the proportion of - 12.5 mm to range between 4 and 8 %. On the same material, at the other end of the journey, this proportion had raised to 12-16 %. As the undersize material over 10 % was heavily penalized, this particular error was economically disastrous for the seller, which was unequitable since an important part of the degradation took place in the buyer's facilities, under the buyer's supervision and responsibility.

Such a problem arises with commodities such as coal, coke, iron ore, manganese ore, bauxite, etc... the transport of which involves very high rates of flow, deep-draught vessels, etc.. and where cost and time are always of the essence of the contract. Free falls can be reduced to some extent but never suppressed. We do not see any economical solution to this very arduous problem.

27.5.2. Substraction or destruction of critical component :

This case may arise when sampling for moisture or size analysis or in the very peculiar case of native sulphur ores.

1) Moisture samples should not be kept in the sun or near a heat source prior to their weighing before drying,

2) When the critical component is the oversize to a given mesh, breakage is a destruction of critical component (see overleaf),

3) Sulphur begins subliming at temperatures as low as 80°C (176°F). For this reason, sulphur ores and concentrates should not be dried up even in a well regulated drying oven but in the open air at room temperature.

27.6. ERRORS RESULTING FROM UNINTENTIONAL MISTAKES PE_M

As things are, sampling operators are liable to commit a great variety of unintentional mistakes due to ignorance, carelessness, awkwardness, lack of experience, etc.., the emphasis being put on the word "unintentional".

We shall just mention a few of these errors : dropping of samples, loss of fractions, mixing of sub-samples belonging to different samples, labelling mistakes, etc.. this list is far from being limitative.

With experienced labour such mistakes are but accidental : they may account for rogue results, they can be reduced but certainly not suppressed. Economically these errors are unimportant.

When sampling and preparation are carried out by unspecialized labour as it is too often the case, these errors can become economically dangerous.

27.7. ERRORS RESULTING FROM FRAUDING OR SABOTAGE PE_F

However sad such a subject may be, it cannot be ignored in a work such as this. Deliberate frauding such as salting does exist, deliberate sabotage too but fortunately enough they remain the exception.

They nearly always take place in commercial sampling. Frauding and sabotage can be practically eliminated by systematic suppression of manual operations. Modern commercial sampling facilities work in a continuous and automatic way and the only human intervention consists in screwing up the caps of sample jars, which in commercial sampling can be carried out in turn by each party or its representative in the presence and under the control of the other one (see section 30.4.).

This control should always be carried out by clever people with a critical mind apt to detect any deviation from the normal procedure. Automatic sampling facilities can be designed in such a way that they can be put under a double system of lock and key and operated from an external control room.

The solution to the problem of frauding does exist. It is definitely cheap in comparison with the cost of frauding, especially as automatic facilities which require investments reduce the cost of labour considerably. Unfortunately, those whose interests are at stake choose too often to ignore this solution.

27.8. CONCLUSIONS

In order to prevent all preparation errors, sampling and preparation should always be carried out by a specialized staff placed under the supervision and responsibility of the quality control department. NOT of the production department.

The qualities required of production people are completely different from those required of control operators, and sampling definitely belongs to control, not to production.

It is perfectly incongruous to see in the same works, as one too often does, meticulous people performing complex analyses and assays with amazing dexterity and remarkable accuracy on samples carelessly obtained, handled and prepared by unqualified labour unconscious of the numerous errors that may take place and unaware of how to reduce or suppress these. This amounts to estimating at great expense the third decimal of a result when the first one is already uncertain.

Sampling and sample preparation are, in the chain of operations leading to quality estimation, exactly on the same importance level as chemical analysis and require exactly the same care.

SEVENTH PART

RESOLUTION OF SAMPLING PROBLEMS

All elements of resolution of sampling problems have been patiently gathered along the first 27 chapters of this book but are scattered in such a way that at this point of our work it seems appropriate to make a practical synthesis.

This synthesis is made of two chapters :

Chapter 28 : Recapitulation of the sampling errors
Chapter 29 : Solvable and unsolvable sampling problems

It will be completed by the eighth and ninth parts which will deal with :

Eighth part : Commercial sampling
Ninth part : Automatic sampling plants

We chose to include chapters 31 and 32 :

Chapter 31 : Testing the agreement between two series of independent estimates of a same characteristic - Discrepancies between Seller and Buyer
Chapter 32 : Testing the agreement between an estimate and the true value of a given characteristic - Check of sampling bias

in the eighth part because their subjects are of extreme importance in commercial sampling but the statistical approach (chapter 31) and the practical approach (chapter 32) to the problems of systematic differences and biases are also relevant in technical sampling and for instance when establishing a balance between :

- the material (metal or mineral) fed to a processing plant or to a smelter,
- the material (metal or mineral) recovered in concentrates or ingots and lost in tailings or slags.

CHAPTER 28

RECAPITULATION OF THE SAMPLING ERRORS

28.1. ANALYSIS OF THE OVERALL ESTIMATION ERROR

The resolution of a sampling problem consists in the first place in making a complete census of all the sampling errors liable to take place and design a sampling scheme ensuring :

- complete elimination of a certain number of errors,
- practical elimination of a certain number of others,
- minimizing and estimation of the remainder.

As the properties of the seven categories of sampling errors, in the proper sense and those of the six classes of preparation errors are scattered according to the logical order of the theoretical developments, it seems appropriate to make a quick review of all these errors and of their properties (see table 28.1.).

28.1.1. Breaking up of the overall estimation error OE :

The estimation of the proportion (critical content) of a given component (critical component) in a certain batch of particulate material (lot L) involves two broad categories of error-generating operations :

1) <u>Sampling operations</u>, in the wider sense of the word, generating the "<u>total sampling error</u>" TE (see section 28.1.2.),

2) <u>Analytical operations</u> such as assaying, moisture analysis, size analysis, estimation of the concentration of a pulp, etc... generating the "<u>analysis error</u>" AE, which is not studied in this book.

Then the "<u>overall estimation error</u>" OE is :

$$OE \equiv TE + AE$$

28.1.2. Breaking up of the total sampling error TE :

Sampling, in the wider sense, is usually carried out as a sequence of particle size and bulk reduction stages and the "<u>total sampling error</u>" TE is the resultant of the errors TE_n arising at stage No.1, ... No.n, ... No.u (u for ultimate). Then :

$$TE \equiv TE_1 + ... + TE_n + ... + TE_u$$

28.1.3. Breaking up of the total sampling error TE_n arising at stage No.n :

Any sampling stage, again in the wider sense, is a sequence of two series of operations :

1) <u>Preparation stages</u> : generating the "<u>preparation errors</u>" PE, studied in chapter 27 and broken up in section 28.1.9.

2) <u>Sampling stages</u>, in the proper sense of bulk reduction by selection : generating the "<u>sampling errors</u>" SE, studied in chapters 6 to 26 and broken up in section 28.1.4. From now on, each sampling stage in the wider sense will be regarded as an independent operation generating a total sampling error TE with :

$$TE \equiv PE + SE$$

28.1.4. Breaking up of the sampling error SE :

Any sampling in the proper sense consists of two categories of operations, taking place in inverse order according as we are considering increment sampling (direct order) or splitting (inverse order) :

1) <u>An immaterial selection process</u> : generating the "<u>selection error</u>" CE studied by means of two models in the second, fourth and fifth part of this book and broken up in section 28.1.5.

2) <u>A materialization process</u> : generating the "<u>increment materialization errors</u>" ME, studied in the third part of this book and broken up in section 28.1.8.

Then, the sampling error SE is :

$$SE \equiv CE + ME$$

28.1.5. Breaking up of the selection error CE :

The selection process involves two categories of properties of the material to be sampled (see chapter 9) :

1) <u>Qualitative properties</u> : represented by the critical content function $a(t)$ and generating the "<u>quality fluctuation error</u>" QE, broken up in section 28.1.6.

2) <u>Quantitative properties</u> : represented by the rate-of-flow or weighting function $\mu(t)$ and generating the "<u>weighting error</u>" WE, studied in chapter 13 and recapitulated in section 28.6.

Then, the selection error CE is :

$$CE \equiv QE + WE$$

28.1.6. Breaking up of the quality fluctuation error QE :

The critical function $a(t)$ has been shown to be the sum of four independent components :

1) <u>A constant term</u> a_0 characterizing the average properties of $a(t)$

2) <u>Short-range quality fluctuations</u> $a_1(t)$ responsible for the "<u>short-range quality fluctuation error</u>" QE_1, studied in chapters 10 and 21 and broken up in section 28.1.7.

3) <u>Long-range quality fluctuations</u> $a_2(t)$ responsible for the "<u>long-range quality fluctuation error</u>" QE_2, studied in chapter 11 (see section 28.4.)

4) <u>Periodic quality fluctuations</u> $a_3(t)$ responsible for the "<u>periodic quality fluctuation error</u>" QE_3, studied in chapter 12 (see section 28.5.)

Then, the quality fluctuation error QE is :

$$QE \equiv QE_1 + QE_2 + QE_3$$

28.1.7. Breaking up of the short-range quality fluctuation error QE_1 :

The component $a_1(t)$ of $a(t)$ takes into account all the local perturbations resulting from the particulate structure of the material submitted to sampling. It has been shown in chapters 19 and 21 that the resulting error QE_1 depends on :

1) <u>The constitution</u> of the particulate material submitted to sampling, generating the "<u>fundamental error FE</u>", studied in section 21.3.2. (see section 28.2.),

2) <u>The distribution</u> of the particles throughout the domain (D_L) occupied by the lot L to be sampled, generating the "<u>grouping and segregation error</u>" GE, defined and studied in section 21.3.3. (see section 28.3.).

Then, the short-range quality fluctuation error QE_1 is :

$$QE_1 \equiv FE + GE$$

28.1.8. Breaking up of the materialization error ME :

By analysing the differences between the continuous model (studied in the second part, chapters 6 to 15) and the discrete reality, we have been able to show in chapter 16 that during the materialization process, two independent operations took place :

1) <u>Geometrical delimitation</u> of the punctual increments, liable to generate the "<u>increment delimitation error</u>" DE studied in chapter 17 (see section 28.7.),

2) <u>Material extraction</u> of the particles contained in the extended increment resulting from this geometrical delimitation, liable to generate the "<u>increment extraction error</u>" EE studied in chapter 18 (see section 28.8.).

Then, the materialization error ME is :

$$ME \equiv DE + EE$$

28.1.9. Breaking up of the total preparation error PE :

We have shown in chapter 27 that the components of the total preparation error PE were :

1) Errors by contamination PE_C, studied in section 27.2.
2) Errors by loss PE_L, studied in section 27.3.
3) Errors by alteration of the chemical composition PE_A, studied in section 27.
4) Errors by alteration of the physical composition PE_P, studied in section 27.
5) Errors resulting from unintentional mistakes PE_M, studied in section 27.6.
6) Errors resulting from frauding and sabotage PE_F, studied in section 27.7.

Then, the total preparation error PE is :

$$PE \equiv PE_C + PE_L + PE_A + PE_P + PE_M + PE_F$$

28.1.10. Recapitulation :

Table 28.1. recapitulates the filiation of the components of the overall estimation error OE. Each of these components takes into account one and only one of the relevant properties of the material submitted to sampling and of the sampling operation. Their definition and properties are recapitulated in the following sections :

1) Fundamental error FE   Chapter 21 - Section 28.2.
2) Grouping and segregation error GE   Chapter 21 - Section 28.3.
3) Long-range quality fluctuation error QE_2.   Chapter 11 - Section 28.4.
4) Periodic quality fluctuation error QE_3 Chapter 12 - Section 28.5.
5) Weighting error WE   Chapter 13 - Section 28.6.
6) Increment delimitation error DE   Chapter 17 - Section 28.7.
7) Increment extraction error EE   Chapter 18 - Section 28.8.
8) Preparation errors PE   Chapter 27 - Section 28.9.
9) Analytical errors AE (not studied in this book).

Then, the overall estimation error OE is :

$$OE \equiv \sum_n TE_n + AE \qquad \text{with}$$

$$TE_n \equiv FE + GE + QE_2 + QE_3 + WE + DE + EE + PE$$

For each component of the total sampling error TE we shall recall :

- the definition,
- the properties of the mean or bias,
- the properties of the variance,
- the conditions of cancellation,
- the conditions of minimization.

When speaking of cancellation or minimization of a given error e, we refer to the cancellation or minimization of its mean square $r^2(e)$ with :

$$r^2(e) = m^2(e) + \sigma^2(e)$$

When the coefficient of bias $b(e) = |m(e)|/\sigma(e)$ is negligible, the mean square

Table 28.1. Recapitulation of the components of the overall estimation error OE.

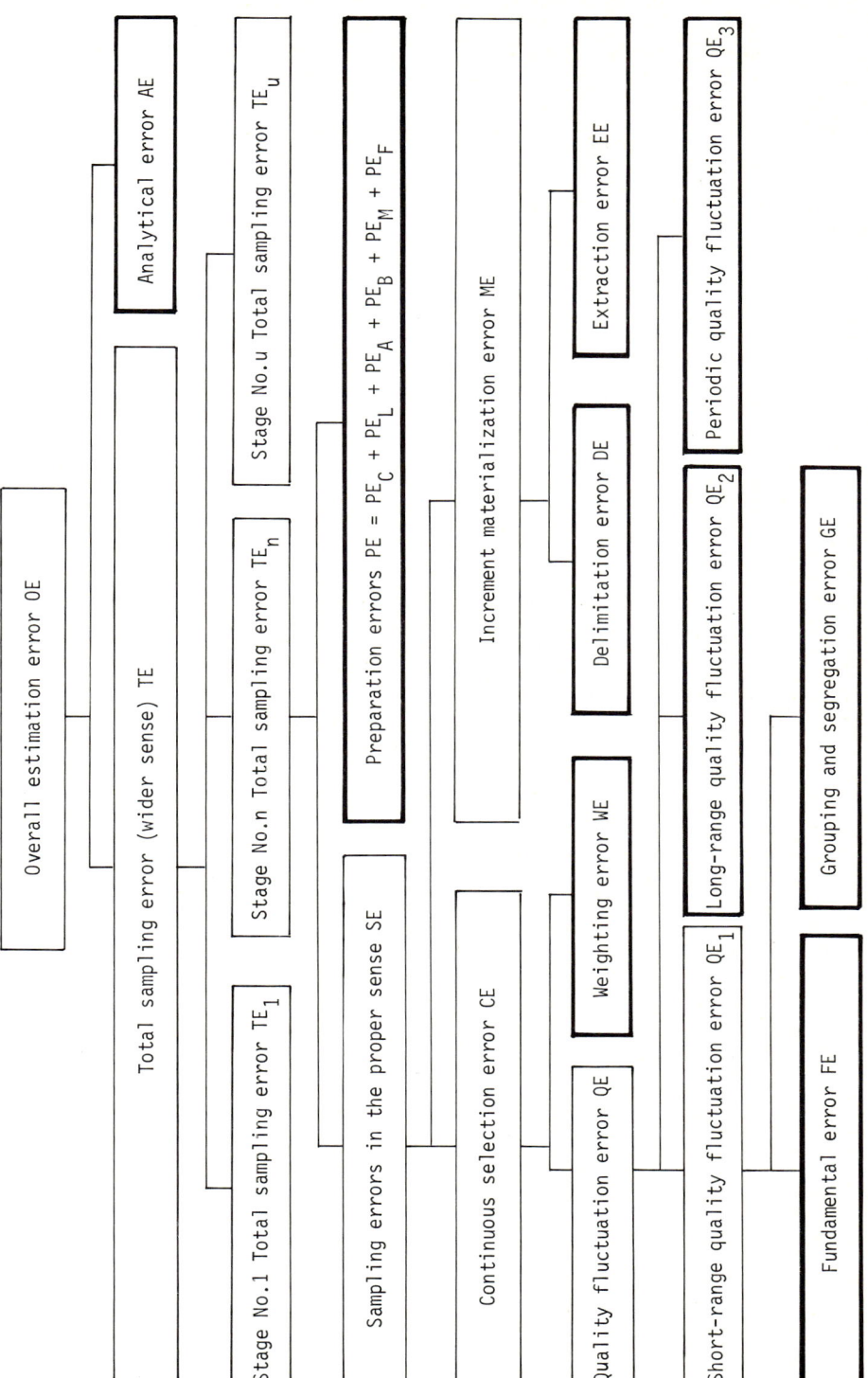

is practically equal to the variance. This is observed with all components of the continuous selection error CE, associated with the models of the selection process i.e. the errors FE, GE, QE_2, QE_3, WE.

When, on the contrary, the coefficient of bias is likely to be non-negligible, then the conditions of cancellation or minimization of the mean square practically amount to those of the mean m(e). This is observed with the errors associated with the material realization of the model, i.e. the errors DE, EE and PE.

We can reduce, seldom cancel the errors associated with the models. With these, the strategy consists in minimizing them. On the contrary we can cancel the errors associated with the realization of the models. With these errors, always the most dangerous of all errors, the strategy is simple : preventive elimination.

28.2. FUNDAMENTAL ERROR FE - Chapter 21 -

28.2.1. Definition :

The fundamental error has been defined and its properties reviewed in section 21.3.2. It results from the constitution of the set of particles submitted to sampling and takes into account all relevant properties of these particles.

28.2.2. Properties of the mean m(FE) :

The "fundamental bias" m(FE) is expressed as follows (second approximation) :

$$m(FE) = - \frac{1-P}{P a_L M_L^2} \sum_i (a_i - a_L) M_i^2 \qquad \text{with } i = 1, 2, \ldots N_F$$

P : selecting probability,
a_L : critical content of the lot L to be sampled,
M_L : weight of the lot L,
N_F : number of fragments in the lot L,
a_i : critical content of the fragment F_i belonging to L,
M_i : weight of the fragment F_i,

With the possible exception of very low-grade ores such as those of precious metals and minerals, this fundamental bias is always negligible.

28.2.3. Properties of the variance σ^2(FE) :

The "fundamental variance" σ^2(FE) is expressed as follows (same notations as above) :

$$\sigma^2(FE) = \frac{1-P}{P a_L^2 M_L^2} \sum_i (a_i - a_L)^2 M_i^2 \qquad \text{with } i = 1, 2, \ldots N_F$$

Though very important from a theoretical standpoint, this formula cannot be used in practical calculations where the following practically equivalent expression is particularly useful (derived in sections 22.4. and 22.6.) :

$$\sigma^2(FE) = \left[\frac{1}{M_S} - \frac{1}{M_L}\right] c \, \ell \, f \, g \, d^3 = \frac{C \, d^3}{M_S}$$

M_S : sample weight (in grams)
M_L : lot weight (in grams)
c : mineralogical constitution factor :

$$c = \frac{1 - a_L}{a_L} \{(1 - a_L) \lambda_c + a_L \lambda_g\} \qquad \text{with}$$

    a_L : critical content of the lot L (in decimal value : 10 % = 0.1)
    λ_c : specific gravity of the critical component (in g/cm^3)
    λ_g : specific gravity of the gangue materials (in g/cm^3)

ℓ : liberation factor of the critical component (dimensionless) :
    $0 \leq \ell \leq 1$. For all practical purposes, the following values will be retained :
    - if $d \leq d_\ell$: $\ell = 1$
    - if $d > d_\ell$: $\ell = \sqrt{\dfrac{d_\ell}{d}}$ with :

    d_ℓ : liberation size of the critical component
    d : maximum particle size in the lot L, mesh retaining a 5 % oversize,

f : particle shape factor (dimensionless). For all practical purposes, f is a constant : f = 0.5

g : size range factor (dimensionless) :
    $0 \leq g \leq 1$.
    - if d > 4d'   : g = 0.25
    - if 4d' > d > 2d' : g = 0.50
    - if 2d' > d > d' : g = 0.75
    - if d = d'   : g = 1.00   with :

    d : maximum particle size in the lot L (mesh retaining 5 % oversize)
    d' : minimum particle size in the lot L (mesh retaining 95 % oversize)

C : sampling constant characterizing the material to be sampled (in g/cm^3) :
    $C = c \, \ell \, f \, g$

The simplified formula :

$$\sigma^2(FE) = \frac{C \, d^3}{M_S}$$

is valid whenever the sampling ratio $\tau = \dfrac{M_S}{M_L}$ is smaller than 1/10. With splitting it is usually larger and the complete formula involving M_L must be utilized. This variance can be either calculated or estimated by means of the slide rule described in section 22.6.6.

28.2.4. Cancellation of the fundamental error FE :

As c, f, g, d are never nil, the fundamental variance would cancel either if $M_S = M_L$, which is not realistic, or if $\ell = 0$ which corresponds to a perfectly homogeneous constitution and is never achieved in practice.

Of all sampling errors, FE is the only error that never cancels.

28.2.5. Minimization of the fundamental error FE :

In order to reduce FE, the only factors on which we can act are :

1) <u>the sample weight</u> M_S : the larger the sample, the smaller the error.

2) <u>the particle size d</u> : within limits that vary from one case to the next, we can reduce the maximum particle diameter d. The smaller this maximum particle size and the smaller the fundamental error.

But the heavier the sample, the larger the comminution ratio and the higher the preparation and sample reduction cost. The solution to a given sampling problem is always a compromise between a suitable sampling representativeness and an acceptable sampling and sample reduction cost. FE does not depend on the selection scheme.

28.3. GROUPING AND SEGREGATION ERROR GE

28.3.1. Definition :

The grouping and segregation error GE has been defined and its properties have been reviewed in section 21.3.3.. This error takes into account the properties of the distribution of the particles throughout the domain (D_L) occupied by the lot L submitted to sampling. GE does not depend on the selection scheme.

28.3.2. Properties of the mean m(GE) :

For all practical purposes, the mean m(GE) may be regarded as negligible.

28.3.3. Properties of the variance σ^2(GE) :

By definition :

$\sigma^2(GE) = \gamma \, \xi \, \sigma^2(FE)$ with :

$\sigma^2(FE)$: variance of the fundamental error (see section 28.2. overleaf),

γ : grouping factor defined in section 19.3.5.3.

$$\gamma = \frac{N_{FL} - N_{GL}}{N_{GL} - 1} = N_{FG} \quad \text{with :}$$

    N_{FL} : number of fragments in the lot L,

    N_{GL} : number of groups of fragments (or number of potential increments) in the

N_{FG} : number of fragments in the average increment

ξ : segregation factor defined in section 19.3.5.6.

$0 \leq \xi \leq 1$

One never attempts to estimate $\sigma^2(GE)$ which in most cases remains smaller and even much smaller than $\sigma^2(FE)$. When designing a sampling plant, it is always safe to assume :

$\sigma^2(GE) = \sigma^2(FE)$ or in other words : $\sigma^2(QE_1) = 2\sigma^2(FE)$

28.3.4. Cancellation of the grouping and segregation error GE :

Since $\sigma^2(FE)$ never cancels, $\sigma^2(GE)$ cancels when one at least of the following conditions is fulfilled :

1) $\gamma = 0$: according to the mathematical definition of γ, this condition is fulfilled when $N_{FL} = N_{GL}$ or in other words when each increment is made of one and only one particle.

2) $\xi = 0$: the distribution of the particles is homogeneous.

The first condition is never achieved. The distribution tends to become homogeneous when the lot is carefully mixed prior to sampling. Three-dimensional homogenization is however difficult to realize and remains unstable due to the omnipresence of gravity, responsible for all vertical segregations. This point is studied in section 19.3.5.4.

28.3.5. Minimization of the grouping and segregation error GE :

We can reduce the three factors $\sigma^2(FE)$, γ and ξ.

1) minimizing $\sigma^2(FE)$: see section 28.2.5. overleaf,

2) minimizing γ : this is achieved by taking as small increments as possible, without transgressing the rules of correct delimitation and extraction which will be recalled in sections 28.7 and 28.8.

3) minimizing ξ : this is achieved by mixing or homogenizing the lot prior to sampling, whenever economically feasible.

Here again, the right solution results from a compromise between representativeness and cost.

28.4. LONG RANGE QUALITY FLUCTUATION ERROR QE_2

28.4.1. Definition :

The long-range quality fluctuation error QE_2 has been defined in chapter 9 and its properties have been reviewed in chapter 11. It takes into account the long-range non-periodic trends of the quality fluctuations of the material to be sampled. QE_2 depends on the selection scheme.

28.4.2. Properties of the mean $m(QE_2)$:

<u>Systematic and stratified random selection schemes</u> : the mean $m(QE_2)$ is zero if and only if the interval T_{sy} between increments or the strata extent T_{st} are sub-multiples of the extent T_L of the domain (T_L) occupied by the lot L on the time axis.

<u>Random selection scheme</u> : the mean $m(QE_2)$ is always zero.

When the mean is non-zero, it is likely to be negligible whenever the number of increments is large (say 30 or more).

28.4.3. Properties of the variance $\sigma^2(QE_2)$:

This variance can be expressed as follows :

1) <u>Systematic selection</u> with interval T_{sy} :

$$\sigma^2_{sy}(QE_2) = \frac{1}{T_L} \left[\frac{v'_{a2}}{6} T^2_{sy} \right]$$

2) <u>Stratified selection</u> with strata extent T_{st} :

$$\sigma^2_{st}(QE_2) = \frac{1}{T_L} \left[\frac{v'_{a2}}{3} T^2_{st} + \frac{v''_{a2}}{6} T^3_{st} \right]$$

3) <u>Random selection</u> with Q_{ra} increments :

$$\sigma^2_{ra}(QE_2) = \frac{1}{Q_{ra}} \left[\frac{v'_{a2}}{3} T_L + \frac{v''_{a2}}{6} T^2_L \right]$$

v'_{a2} and v''_{a2} : variographic parameters of the function $a(t)$. They can be experimentally determined (variographic experiment, see chapter 14).

28.4.4. Cancellation of the long-range quality fluctuation error QE_2 :

According to its definition, the error QE_2 cancels if its mean and variance are simultaneously zero.

1) <u>the mean</u> $m(QE_2)$ cancels, with systematic and stratified selection when the interval T_{sy} or the strata extent T_{st} is a sub-multiple of T_L.

2) <u>the variance</u> cancels if and only if the variographic parameters v'_{a2} and v''_{a2} are nil. This happens when the variogram is flat. The variogram of a given functio $a(t)$ becomes flat when the material to be sampled is blended in a bed blending system. Unfortunately the investments and operating cost of bed blending facilities is so high that they are never implemented for the sole purpose of cancelling the long-range quality fluctuation error.

28.4.5. Minimization of the long-range quality fluctuation error QE_2 :

The mean $m(QE_2)$ can be minimized by chosing T_{sy} or T_{st} among the sub-multiples of T_L. The variance $\sigma^2(QE_2)$ can be minimized in the following way :

1) By choosing the best selection scheme : when no periodic quality fluctuations are liable to take place (see section 28.5.), the most reproducible of all selection schemes (for a given tolerated cost) or the cheapest (for a given tolerated variance) is always the systematic scheme. When however periodic fluctuations are likely to occur, the overall reproducibility of the stratified scheme is often better than that of the systematic scheme. The random scheme is never better than the other two schemes. Its implementation is restricted to the sampling of zero-dimensional objects (see section 5.3.2.4.).

2) When the selection scheme has been decided upon, the shorter the interval T_{sy} or the strata extent T_{st} and the smaller the variance, but also the heavier the primary sample and the higher the sample reduction cost.

Once again, the solution lies in a compromise between reproducibility and cost.

28.5. PERIODIC QUALITY FLUCTUATION ERROR QE_3

28.5.1. Definition :

The periodic quality fluctuation error QE_3 has been defined in chapter 9 and its properties have been reviewed in chapter 12. It takes into account the periodic trends of the quality fluctuations of the material to be sampled.

28.5.2. Properties of the mean $m(QE_3)$:

1) Systematic and stratified selection : same observation as in section 28.4.2.

2) Random selection : same observation as in section 28.4.2.

28.5.3. Properties of the variance $\sigma^2(QE_3)$:

The expression linking up the variance with the properties of the material to be sampled, characterized by the variographic parameter v_{a3} and the period T_p of the phenomenon, and with those of the selection scheme is very complex. Furthermore, we seldom know with precision the values of v_{a3} and T_p and it would be impossible to estimate the variance in every case. In such conditions, the only relevant characteristic is the maximum of this variance. It can be expressed as follows :

$$\sigma^2_{sy}(QE_3)_{max} = v_{a3} \qquad \sigma^2_{st}(QE_3)_{max} = \frac{v_{a3}}{Q} \qquad \sigma^2_{ra}(QE_3)_{max} = \frac{v_{a3}}{Q}$$

where Q is the number of increments in the sample.

28.5.4. Cancellation of the error QE_3 :

The error QE_3 cancels when one at least of the following conditions is fulfilled :

1) $v_{a3} = 0$: the periodic fluctuations are naturally non-existent or have been artificially suppressed. This can be achieved, with restrictions, by bed blending but the latter is always regarded as too costly.

2) The interval T_{sy} of the systematic scheme and the period T_p of the phenomenon

are both sub-multiples of T_L BUT (and this is very important) T_{sy} is NOT a multipl[e] of the period T_p. This mathematical solution is unfortunately inapplicable in prac tice.

28.5.5. Minimization of the error QE_3 :

The periodic quality fluctuation error QE_3 can be minimized as follows :

1) <u>Choice of the right selection scheme</u> : the greatest danger when sampling a periodic function arises when the interval between increments of a systematic sche[me] is a multiple of the period T_p. Then, the variance $\sigma^2_{sy}(QE_3)_{max}$ reaches a maximum maximorum which is Q times larger than the corresponding maximum involved in a stratified selection.

When periodic fluctuations are liable to take place, a stratified selection sch[e]me always represents the safest of all solutions.

2) <u>Choice of the strata extent</u> T_{st} : the shorter T_{st}, the larger the number Q of increments, and the smaller the variance of QE_3.

28.6. WEIGHTING ERROR WE

28.6.1. Definition :

The weighting error WE has been defined in chapter 9 and its properties have been reviewed in chapter 13. It takes into account the fluctuations of the rate of flow of the stream to be sampled.

28.6.2. Properties of the mean m(WE) :

The weighting bias m(WE) can be estimated from the results of a variographic experiment. It is usually negligible in comparison with the standard deviation.

28.6.3. Properties of the variance σ^2(WE) :

The weighting variance σ^2(WE) can be calculated from the results of a variographic experiment and it is always interesting to try and estimate it. It is defined as the difference between the variance σ^2(CE) and the variance σ^2(QE).

28.6.4. Cancellation of the weighting error WE :

The error WE cancels if and only if the coefficient of correlation between the quality function a(t) (critical content) and the weighting function μ(t) (rate of flow) is nil. This absence of correlation may happen independently or when one of the two functions is uniform throughout (T_L).

1) $a(t) = a_o$ = constant : the material is homogeneous,

2) $\mu(t) = \mu_o$ = constant : the rate of flow is constant.

None of these conditions is ever strictly fulfilled.

28.6.5. Minimization of the weighting error WE :

When the rate of flow is not regulated prior to sampling, the weighting error WE can be and by far the largest of all components of the continuous selection error CE. It is therefore always advisable to regulate the rate of flow prior to any sampling operation.

Weigh feeders are much more efficient than simple volumetric regulating devices but are more expensive. As regulating devices are liable to introduce periodic fluctuations of the rate of flow about an average value (a sort of pendulum effect), a stratified random selection scheme is always recommendable.

The more efficient the regulation, the larger the number of increments and the smaller the weighting variance but also, the higher the investments and the higher the cost of sample reduction. As usual, we must seek a compromise between reproducibility and cost.

28.7. INCREMENT DELIMITATION ERROR DE

28.7.1. Definition :

The increment delimitation error has been defined in chapter 16 and its properties have been studied in chapter 17. It usually results from an incorrect design of the sampling device.

28.7.2. Properties of the increment delimitation error DE :

The delimitation bias m(DE) as well as the delimitation variance $\sigma^2(DE)$ cannot be expressed by mathematical formulas but can be experimentally estimated (costly procedure). The delimitation bias can be very important. It results from :

1) Non-uniform selecting probability of all elements of the cross-section of the stream undergoing sampling,

2) Correlation between the characteristics of a given particle (diameter, density, shape) and its position in the stream cross-section,

3) Resulting correlation between particle personality and selecting probability.

The only efficient strategy with delimitation errors is preventive or, in already operating facilities, corrective elimination according to the rules stated in chapter 17.

28.7.3. Cancellation of the increment delimitation error DE :

The delimitation error can be easily eliminated at the designing stage. Eliminating the delimitation bias from an operating device is always advisable but inevitably costly. There are on the market correct devices (respecting the rules that cancel the delimitation error) but also a number of incorrect devices introducing an often dangerous delimitation bias. The user must learn to recognize a correct from an incorrect sampler. A correct sampler respects the following conditions :

1) <u>Cutter geometry</u> : straight trajectory : the edges should be parallel. Circular trajectory : the edges should be radial. All other types are incorrect.

2) <u>Cutter velocity</u> : should remain constant during the crossing of the stream. Electric drive only can respect this rule. Hydraulic, pneumatic, magnetic and manual drives are always incorrect and therefore unadvisable.

3) <u>Cutter lay-out</u> : should be studied with care according to the rules detailed in chapter 17.

4) <u>Cutter maintenance</u> : the cutter should be regularly cleaned and checked for deformation and wear.

When these conditions are simultaneously fulfilled, the delimitation error is completely eliminated. The reader must remember that a sampler achieving a correct delimitation is in no way more expensive than an incorrect one, and does not result in an increase of sample weight and sample reduction cost. There is therefore no reason to implement incorrect samplers, except ignorance.

28.8. INCREMENT EXTRACTION ERROR EE

28.8.1. <u>Definition</u> : the increment extraction error has been defined in chapter 16 and its properties have been studied in chapter 18. It always result from an incorrect design or utilization of the sampling device.

28.8.2. <u>Properties of the increment extraction error EE</u> :

The extraction bias $m(EE)$ as well as the extraction variance $\sigma^2(EE)$ cannot be expressed by mathematical formulas but can be experimentally estimated (costly procedure). The extraction bias is the most important of all the potential sampling biases : there are known examples of 20 % extraction biases. The extraction bias results from a non-uniform selecting probability of the various size fractions and to a lesser extent of the various density or shape fractions of the material submitted to sampling.

The only efficient strategy with extraction errors is preventive or corrective elimination.

28.8.3. <u>Cancellation of the increment extraction error EE</u> :

The extraction error can be easily eliminated at the designing stage. Eliminating the extraction error from an already operating device is definitely advisable but inevitably costly. There are on the market correct devices (respecting the rules that cancel the extraction error) but a larger number of incorrect devices introducing an always dangerous extraction bias. The user must learn to recognize a correct from an incorrect device. A sampler must respect the following rules of correct extraction :

1) <u>Condition concerning the stream</u> : the stream should be sampled at a point where its trajectory is practically vertical.

2) <u>Condition concerning the cutter edges</u> : the cutter edges should be perpendicular to the stream, i.e. horizontal.

3) <u>Condition concerning the sampler lay-out</u> : The sampler should be set in such a way that the stream crosses the middle part of the area generated by the cutter edges during their travel through the stream.

4) <u>Condition concerning the cutter width</u> : with a maximum particle diameter d_M larger than or equal to 3 mm, the cutter width should be larger than or equal to three times the diameter d_M. With a maximum particle size smaller than 3 mm, the cutter width should be larger than 10 mm, irrespective of the maximum grain size.

5) <u>Condition concerning the cutter velocity</u> : the cutter velocity should not exceed $V_{on} = (1 + n)\ 0.3$ m/s if the actual cutter width $W = n\ W_o = 3\ n\ d_M$.

6) <u>economical optimum</u> : it is defined by $W = W_o$ and $V_C = 0.6$ m/s.

W : actual cutter width,
W_o : minimum cutter width,
V_C : actual cutter velocity,
V_{on}: maximum cutter velocity when $W = n\ W_o$.

7) <u>Condition concerning the cutter design</u> : the cutter should be so designed as to prevent any particle entering the cutter from bouncing out or overflowing.

When these conditions are simultaneously fulfilled, the extraction error is completely eliminated.

A sampler achieving a correct extraction is in no way more expensive than an incorrect sampler but designers, manufacturers and users are very often tempted to transgress the rules of extraction correctness, in the primary sampling stage at least, in order to reduce the primary sample weight and the sample reduction cost.

This amounts to forgetting the dangers of sampling bias which have been repeatedly pointed out in the preceding chapters. This point is particularly dangerous in commercial sampling to which chapters 30 to 32 are dedicated.

28.9. PREPARATION ERRORS PE

28.9.1. Definition :

The preparation errors have been defined and their properties have been studied in chapter 27. They are liable to take place in the course of all non-selective operations undergone by the lot and the successive samples : transfer, comminution, mixing, drying, etc.. They result from ignorance, awkwardness, carelessness, lack of qualification, dishonesty or malevolence of the sampling and preparation operators.

28.9.2. Properties of the preparation errors PE :

Preparation errors cannot be regarded as random errors with stationary properties such as mean and variance. The only efficient strategy with preparation error is preventive elimination.

28.9.3. Elimination of the preparation errors PE :

These errors belong to six categories :

1) <u>Errors by contamination PE_C</u> : the lot and the successive samples must not be contaminated by foreign materials,

2) <u>Errors by loss PE_L</u> : the integrity of the lot and of the successive samples should be respected,

3) <u>Errors by alteration of the chemical composition PE_A</u> : certain minerals or materials are liable to pick up oxygen, water or carbon dioxide from the atmosphere and to fix them in their molecules. Other minerals are liable to loose combined water, carbon dioxide, etc.. if dried at too high a temperature. Drying is always a critical operation and should always be carried out in a drying oven. Infrared or hot-plate dryers are very dangerous.

4) <u>Errors by alteration of the physical composition PE_P</u> : such errors are specific to moisture and size analysis. When sampling for a moisture estimation, no exchange of water (addition or substraction) should take place between the sample and its environment. When sampling for size analysis, free falls should be reduced as much as possible in order to prevent breakage and degradation.

5) <u>Errors resulting from unintentional mistakes PE_M</u> : the sampling and preparation operators should always be fully qualified for their jobs. They should have the qualities usually required from analysts, receive adequate teaching and follow very strict procedures prepared by a sampling specialist.

6) <u>Errors resulting from frauding or sabotage PE_F</u> : such errors are specific to commercial sampling. Commercial sampling operations should be automated with a minimum of human intervention. Sampling procedures as well as sales contracts should protect the interests of both parties in an equitable way.

28.10. CONCLUSIONS

Sampling and sample preparation should always be carried out by fully qualified operators placed under the supervision of the head of quality control. Sampling and sample preparation should be on an equal footing with assaying and given exactly the same consideration. Sampling has become a science and must be treated as such.

CHAPTER 29

SOLVABLE AND UNSOLVABLE SAMPLING PROBLEMS

29.1. DEFINITIONS

We regard a sampling problem as "solvable" if and only if the mean square $r^2(TE)$ of the total sampling error TE can be estimated. This estimation can simply consist of an inequality such as :

$r^2(TE) = m^2(TE) + \sigma^2(TE) \leq r^2(TE)_{max}$

In order to be consistent with this definition, the word "sample" should be restricted to fractions of the lot obtained at the cost of a total sampling error TE whose mean square can be estimated or is known to be smaller than a certain maximum $r^2(TE)_{max}$.

Reciprocally, we regard a sampling problem as "unsolvable" whenever the mean square of the total sampling error TE cannot be estimated and more specifically when unforeseeable biases are liable to take place. The word "specimen" should be used instead of the word "sample" to qualify fractions obtained in such conditions.

The failure of mining and metallurgical undertakings can nearly always be traced back to the confusion between a specimen on which no sane financial decision should ever be taken and a sample. In other words, the failure of what is aptly called a mining or metallurgical "venture" can nearly always be attributed to unaccounted for sampling errors. The main objective of the theory developed in this book is to give the responsibles of the mineral industries the possibility of transforming a risky "venture" into a safe "undertaking".

The purpose of this chapter is to review the various categories of sampling problems, to define whether they are solvable and whenever possible how they can be solved.

29.2. REPRESENTATIVENESS AND COST

From a theoretical standpoint, all sampling problems are solvable. From a practical standpoint however, even though we should assume that an unlimited budget is at our disposal (a very optimistic hypothesis), there are sampling problems that cannot be regarded as solvable. In a number of other cases, the notion of solvability is closely associated to that of acceptable cost and we shall say, again from a practical standpoint, that a sampling problem can be regarded as solvable if and only if a certain "acceptable representativeness standard" characterized by a tolerated mean square $r_o^2(TE)$ can be achieved at an "acceptable cost" expressed for instance in dollars per ton, or in percentage of the price of the commodity considered. This brings up two questions :

1) What is an acceptable representativeness standard ?
2) What is an acceptable cost ?

29.2.1. Notion of acceptable representativeness standard :

The overall estimation error OE (see section 28.1.) is the resultant of the total sampling error TE and of the analytical error AE. Assuming the delimitation bias m(DE), the extraction bias m(EE) and the analytical bias m(AE) to be negligible and the preparation errors PE to be kept within reasonable limits, which is the only sound basis of any estimating system and which we know how to achieve, then we can write :

$$r^2(OE) = \sigma^2(OE) = \sigma^2(TE) + \sigma^2(AE)$$

What do we know of $\sigma^2(TE)$?

Some contracts between manufacturers and users of sampling equipment fix a certain "expendable sampling variance" $\sigma_o^2(TE)$ disposable in one or several sampling stages. This case will be dealt with in chapter 33.

Certain commercial contracts fix the value of the "largest acceptable differenc a_o between independent estimates of the critical content a_L obtained by seller and buyer on a single shipment (single sampling + two independent assays or two independent sampling and assaying systems). Assuming a_o to be the 95 % probability confidence interval of this difference, the standard deviation of the difference is $a_o/2$ and its variance $a_o^2/4$. Assuming now both parties to carry out their estimations with equal precision, the relative overall estimation variance of each party should not exceed $\sigma_o^2(OE) = a_o^2/8a_L^2$.

Example : Seller and buyer of a 65 % Fe iron ore accept to split all differences between their estimates when they do not exceed a_o = 0.5 % Fe. Then :

$$\sigma_o^2(OE) = \frac{(0.5)^2}{8(65)^2} = 7.4 \times 10^{-6}$$

Allowing for an assaying variance $\sigma^2(AE)$ of about 1.0×10^{-6} ($\pm$ 0.13 % Fe), the expendable sampling variance is :

$$\sigma_o^2(TE) = \sigma_o^2(OE) - \sigma_o^2(AE) = 6.4 \times 10^{-6}$$

Now, in most cases, this variance is not imposed on us and we are at liberty to fix it but what should be our criteria ?

It would be meaningless to aim at a variance $\sigma_0^2(TE)$ significantly smaller than $\sigma_0^2(AE)$ since the relative reduction of the overall variance $\sigma^2(OE)$ would be small or even imperceptible. On the other hand, too high a variance $\sigma_0^2(TE)$ would make us loose the advantages of precise assaying. The arithmetical ideal would therefore seem to be to allow for a total sampling variance $\sigma_0^2(TE)$ equal to the analytical variance $\sigma_0^2(AE)$. But experience shows that it is usually cheaper to reduce $\sigma_0^2(AE)$ (for instance by repeating the assays) than to reduce $\sigma_0^2(TE)$ (for instance by handling heavier samples). For a given tolerated overall estimation variance $\sigma_0^2(OE)$ it is therefore economical to allow for a sampling variance larger but not much larger than the analytical variance. We shall therefore admit that the total expendable sampling variance is :

$$\sigma_0^2(TE) = K \, \sigma_0^2(AE) \quad \text{with} \quad 1 \leqq K \leqq 10$$

The value of K depends on several factors and more particularly on :

1) the cost of repeated assaying, which varies widely according to the element to be assayed and to the type of analysis carried out,
2) the cost we can (or we think we can) afford to spend on sampling.

The general idea being to minimize the overall estimation cost for a given value of $\sigma_0^2(OE)$ or to minimize the overall estimation variance $\sigma^2(OE)$ for a given cost.

29.2.2. Notion of acceptable cost :

This obviously depends on the use to be made of the figure delivered by the analyst. If the sample is assayed in order to estimate the iron content of a multi-million-dollar shipment of iron ore for settlement purposes, it may justify a larger expense than the sample routinely assayed in order to fill up the daily production sheet needed for administrative or internal accounting purposes. But how much are we going to spend and how are we going to justify this expense ?

To some people ranking from Presidents down to assistant-operators, sampling is all balderdash and the cost of sampling is acceptable if it does not exceed that of taking a handful of material at the nearest accessible point of the lot (or perhaps a scoopful if a scoop is available). Well, as things were in the recent past and as they still are today in some places, there may be something in their point but since those people are unlikely to read chapter 29 of a book such as this, we shall reluctantly admit that some money has to be spent on sampling.

The best way of looking at things is to assimilate the cost of representative sampling to that of an insurance policy against the risk of loosing money, the idea being on the one hand to prevent accidental big losses and on the other to be a winner in the long run. There is always indeed a risk of loosing money with a poorly representative or biased sampling but the difficulty arises when trying to assess this risk. We shall see now that the problem is set in different terms according as we are dealing with commercial, technical or administrative sampling.

29.2.2.1. Commercial sampling :

Various aspects of commercial sampling will be reviewed in chapters 30 to 32 but an analysis of the notion of equity (chapter 30) shows that if money is to be spent on commercial sampling, the first end to be achieved is a systematic eradication of all potential sources of sampling and preparation bias. If some money is left, it is not forbidden to spend a part of it in order to reduce the sampling variance but it is only of secondary importance (section 30.4.). Furthermore, we recall that the elimination of the delimitation and extraction biases always results in the suppression of the corresponding variances and in a significant reduction of the total sampling variance.

According to our experience, most mining or metallurgical Companies are ready to consider as acceptable a sampling cost ranging about 1 % of the market value of the commodity submitted to sampling. This round figure of 1 % does not seem to result from elaborate computations involving the calculus of probabilities nor from any kind of objective considerations but rather from the subjective idea that as individuals or as directors, we are ready to consider 1 % as a negligible quantity and, if we cannot help it, willing to spend or to loose 1 % of our private properties or productions. This rule seems to be true, regardless of the financial interest of accurate or reproducible sampling. In most cases, 2 % seems to be regarded as an intolerable burden. This observation could provide matter for a new Parkinson's law.

However unscientific it may appear and for want of a more objective approach, we shall therefore regard as acceptable a sampling cost representing no more than 1 % of the market value of the commodity to be sampled.

29.2.2.2. Technical sampling :

We call "technical sampling" a sampling operation providing informations necessary for controlling or improving a transformation process such as beneficiation or smelting or for computing mine reserves. Geological sampling falls also within the province of technical sampling.

When conducting a technical sampling, for instance in a processing plant, we aim at reaching an economical equilibrium corresponding for the Company to a maximum profit. Assuming that we can rely on an accurate model of the process involving a certain number of characteristics of the material fed to the plant, an estimation chain "sampling + analysis" is the only way to evaluate these characteristics. Then, theoretically at least, we should be able to compute, in terms of dollars per ton for instance, the losses incurred when departing from the economical equilibrium, which is however more readily said than done. The same holds true when sampling concentrates or tailings of a processing plant , metal and slags of a smelter.

In each particular case, one should therefore be able to establish an approximative relationship between sampling reproducibility and avoidance of certain losses and deduce therefrom an acceptable cost of sampling, an increase of the sampling cost being justified if and only if it suppresses the risk of a larger loss of money.

29.2.2.3. Administrative or internal accounting sampling :

We know of innumerable instances where the figures resulting of a costly sampling and assaying procedure are simply fed to daily, weekly or monthly production sheets (or to their up-to-date computer equivalents) with no apparent benefit to anyone. The only justification of such controls is that they have always been carried out that way and that nobody has ever wondered to what extent they are not useless, nor taken the decision to suppress them.

We do not mean to say that the control of the metallurgical balance in processing plants or smelters is useless. On the contrary, we believe it to be a priceless tool when performed with accuracy and motivation but the fact is that more often than not the relevant estimates are used only for administrative or internal accounting purposes.

Now, on what objective basis can we estimate the "acceptable cost" of such sampling operations ? We better leave the reader to answer this question.

Irrational though it may seem, we know a good many Companies spending more money on administrative sampling than on commercial sampling of their production.

29.2.3. Tentative conclusion :

As things are and if we allow for a few notable exceptions (such as that of the uranium industry for instance), there does not seem to be a general understanding of the necessity for an assessment of the relationship between the notion of acceptable representativeness standard and that of acceptable cost. The purpose of the preceding sections was to help the reader to gain consciousness of a problem which he and he alone can solve.

29.3. SAMPLING OF THREE-DIMENSIONAL OBJECTS

29.3.1. Definition - Examples (see also section 5.3.) :

Strictly speaking, all objects extend in a three-dimensional Euclidean space. We shall however restrict our definition to objects that simultaneously :

- cannot be efficiently represented by a two-dimensional model (see section 29.4), a one-dimensional model (see section 29.5. and 29.6.) or a zero-dimensional model (see section 29.7.).

- cannot be transformed into two-, one- or zero-dimensional objects.

Three-dimensional compact objects such as mineral deposits or metal ingots fall outside the scope of this book.

Examples : here are a few examples of three-dimensional objects made of particulate materials :

1) Certain mineral deposits made of unconsolidated materials,

2) Piles too heavy to be transferred or flattened for the sole purpose of their sampling. These may be made of off-grade materials cast aside for some reason or other, discarded overburden or old tailings that the increase of metal prices tends to transform into low-grade ores,

3) Loads of ship-holds, railway wagons or trucks, etc ...

29.3.2. Theoretical and practical solvability :

From a theoretical standpoint, the problem set by the sampling of three-dimensional particulate objects is in no way different from the problem set by compact objects to which the geostatistical theories are applicable (see for instance David, 1977).

From a practical standpoint, however, the problem is different. The sampling of consolidated orebodies is usually carried out by drilling (core drilling, percussive or rotary drilling). When the cores, drill cuttings or sludges are quantitatively recovered (which is sometimes a very optimistic hypothesis) such a sampling can be regarded as achieving an unbiased increment extraction (the delimitation is always correct and therefore unbiased), because the sample is practically made of the ideal cylinder taken into consideration by the geostatistical model. When drilling through unconsolidated materials on the contrary, an extraction bias is always likely to take place. Three phenomena can be considered as bias-generating :

- Vibrations or percussions transmitted to the sampled material by the drill are likely to upset the unstable equilibrium of fine particles when they do not fill up the interstices between lumps. The natural size distribution is altered and the vertical segregation is accentuated.

- The comminuting action of the drill adulterates the size analysis of the material and this can alter certain conclusions of the sample study.

- Under the pressure transmitted by the drill, the coarsest particles are likely to escape laterally. This phenomenon was responsible for a very important extraction bias during a drilling campaign carried out on a sedimentary phosphate deposit containing an important proportion of flint boulders, much larger than the drill diameter. The proportion of flint in the sampled was undervalued by several hundred percents but this bias was not discovered until the processing plant (erected according to the results obtained on the samples) was started. This bias resulted in huge unexpected expenditures and in a complete reconsideration of the whole project. Clearly, such a problem was to be considered a priori as unsolvable, due to the fact that there were no means of ascertaining that the samples actually extracted from the deposit were unbiased, which in fact they were not.

Practically, we must therefore conclude that the sampling of three-dimensional particulate objects is a generally unsolvable problem. One can always collect "specimens" which can be studied and assayed but one should never forget that they are nothing more than specimens and that unexpected biases are liable to take place that deprive the specimens of any value as "samples".

29.3.3. Possible solutions :

We shall obviously discard any non-probabilistic sampling method consisting in gathering for instance surface material for the sole reason that it is easily accessible and can be collected at practically no expense. The reader should always remember that superficial materials have usually been weathered by exposure to the elements. Rain as well as wind are likely to have washed away fine to medium particles. Oxidation or alteration of certain minerals is likely to be more important at the surface, etc ...

The only solution would consist in transforming the lot into a two-, one- or zero-dimensional object which is seldom materially achievable (e.g. the phosphate deposit), nearly always too expensive to be worth consideration and anyway excluded by the definition given in section 29.3.1.

29.4. SAMPLING OF TWO-DIMENSIONAL OBJECTS

29.4.1. Definition - Examples (see also section 5.3.2.1.) :

An object is said to be "two-dimensional" when it can reasonably be represented by a two dimensional model or in other words when one of its dimensions (usually the thickness) is relatively small as compared with the other two dimensions and relatively uniform. The two non-degenerate dimensions are usually those of the horizontal plane.

Two-dimensional compact objects such as certain mineral deposits or metal plates and sheets (cast iron, nickel matte, blister copper, etc ...) fall outside the scope of this book.

Examples : here are a few examples of two-dimensional particulate objects :

1) Certain flat mineral deposits made of unconsolidated materials,
2) Flat piles of materials that may for instance result from the flattening of three-dimensional objects,
3) To some extent at least, the loads of railway wagons or trucks when they are uniformly loaded or flattened.

29.4.2. Theoretical and practical solvability :

From a theoretical standpoint these particulate objects are in no way different from the two-dimensional compact objects to which the geostatistical theories are applicable (see David, 1977).

From a practical standpoint, however, we meet the same difficulties as with the three-dimensional objects but the fact that the thickness is relatively small (say two to three meters) makes it possible, though uneasy to sink uniform section vertical shafts with wood or metal lining preventing the external material from falling into the sample. Uniform section is one of the conditions of increment delimitation correctness. In order to achieve increment extraction correctness, the rule of the centre of gravity will have to be respected (see chapter 18). This method is costly indeed but it is the only direct "solution" to the problem of two-dimensional sampling. Any drilling method is likely to generate the errors pointed out in section 29.3.2.

29.4.3. Other possible solutions :

We discard superficial sampling as we did in section 29.3.3. and for the same reasons. There remains the solution which consists in transforming the two-dimensional object into a one- or zero-dimensional object. This is seldom materially achievable, nearly always too expensive to be worth consideration and anyway excluded from the general definition of two-dimensional objects.

29.5. SAMPLING OF ONE-DIMENSIONAL STATIONARY OBJECTS

29.5.1. Definition - Examples (see section 5.3.2.2.) :

An object is said to be "one-dimensional" when it can be reasonably represented by a one-dimensional model or in other words when two of its dimensions (usually those of a vertical cross-section) are relatively small as compared with the length and relatively uniform. It is said to be "stationary" as opposed to the flowing streams which will be dealt with in section 29.6.

Metal bars, such as copper wire bars, are an example of one-dimensional compact objects.

Elongated piles of ore such as those to be found in the bed blending systems are an example of one-dimensional particulate objects.

29.5.2. Theoretical and practical solvability :

From a theoretical standpoint, the continuous selection model developed in the second part of this book answers all the questions posed by the sampling of one-dimensional objects, whether stationary or flowing.

From a practical standpoint, there is no solution to the problem of sampling elongated piles, except when the section is small (e.g. less than 1 m^2) and when the maximum particle size does not exceed 10 mm. Then it is possible, though uneasy and costly, to force down vertical steel plates maintained by distance-pieces and to extract constant thickness slices made of the material contained between the plates. Each slice corresponds to an increment correctly delimited and extracted.

29.5.3. Other possible solutions :

In fact, one-dimensional piles are very often due to be reclaimed sooner or later, i.e. to be transformed into one-dimensional flowing stream or into a zero-dimensional object, and whenever possible, it is always advisable and much safer to sample them under this form.

29.6. SAMPLING OF ONE-DIMENSIONAL FLOWING STREAMS

This problem has been extensively studied in the preceding chapters. We know that it is always solvable and we know how to solve it. The reader can understand all the better now why this particular problem has received the best part of our attention : of all sampling problems, this is the only one that can always be regarded as solvable, nearly always at a very reasonable cost and with good representativeness.

All modern sampling facilities resort to this kind of sampling tending to be universally recognized as the only reliable one. We shall therefore conclude that whenever possible the sampling of a lot of particulate material must be carried out when the lot is under the form of a flowing stream.

29.7. SAMPLING OF ZERO-DIMENSIONAL OBJECTS

29.7.1. Definition - Examples (see section 5.3.2.4.) :

An object is said to be "zero-dimensional" when it is naturally divided into a large number of "units" of practically uniform weight and when its primary sampling consists of a selection of a certain number of these units. These are usually handling or transportation units such as series of rail wagons, truckloads, shovelfuls (either mechanical or hand shovel), drums, sacks, bags or any kind of container with a uniform capacity and more or less uniform loading ratio. Splitting methods transform three-dimensional objects into zero-dimensional objects prior to sampling which consists in selecting a certain number of units to form a sample. This problem has been dealt with in chapters 24 to 26.

Objects made of small numbers of units of uniform weight or of any number of units of non-uniform weight cannot be regarded as zero-dimensional objects and cannot be sampled as such by selection of a certain number of units.

29.7.2. Theoretical and practical solvability :

From a theoretical standpoint, two cases may arise :

1) The units are disposed in a natural order reflecting more or less the chronology of their production. This is the case for instance of a train of rail-wagons loaded from the same bin through the same feeder.

2) The units keep no mark of their original loading order. This happens with shipments of sacks or drums upon arrival at the buyer's facilities. For all practical purposes, these units are in "random order".

In the first case, the theoretical solution involves the variogram of the critical content of these units whereas in the second one it involves the variance of the population of units. The only problem is to estimate the variogram or the variance, which we know how to carry out, but which is a costly procedure.

In both cases, the solution is given in the second part of this book, any selection scheme carried out on a population in "random order" being equivalent to a random selection scheme carried out on a population of units in a non-random order.

From a practical standpoint, and as far as the primary sampling stage alone is concerned, the solution consists in selecting a certain number of units according to a certain selection scheme which can be systematic (units in chronological order with no risk of periodic quality fluctuations - see chapter 12) or stratified (units in chronological order when there is a risk of periodic quality fluctuations). With units in true random order, the selection scheme is irrelevant but as there nearly always remains something of the original chronological order, a systematic scheme cannot but result in a reduction of the sampling variance. Assuming the units to be actually separated, the problem can be solved on paper : you just have to decide that wagons No. 5 - 15 - 25 - 35 - etc... will be retained to make up a primary sample with a 10 % sampling ratio.

29.7.3. Further processing of the primary sample :

The difficulties usually arise with the secondary sampling stage. It is one thing to write on a sheet of paper that the primary sample is made of, say, 30 fifty-ton wagons and another one to carry out the processing of this primary sample.

As the case may be, each unit must be regarded as :

- a three-dimensional object : high capacity rail-wagons or trucks,

- a two-dimensional object : rail-wagons or truckloads after flattening of the surface,

- a one-dimensional flowing stream : the most satisfactory of all possibilities, when the increment-units are unloaded into a surge bin and reclaimed by means of a weigh feeder for secondary sampling and further processing (the sampling reject is then fed back to the wagons and forwarded to its natural destination).

- a zero-dimensional object : when each primary unit is made of a uniform number of secondary zero-dimensional units such as bags or drums primarily loaded in wagons or trucks (the primary units).

When handling large tonnages under the form of zero-dimensional objects in a routine way, the most accurate and the cheapest of all solutions consists in selecting for instance one unit out of 10 or 20 (primary sample) according to a systematic or stratified scheme, in discharging the increment-units into a surge bin, in feeding the reclaimed material to a cross-stream sampler (secondary sample) and in feeding back the sampling rejects to the empty units kept in stand-by. This solution requires capital expenditures but leads to very small operating costs (as practically no labour is involved) and very high reliability, of extreme importance in commercial sampling.

29.8. SAMPLING OF SMALL OR VALUABLE OBJECTS

29.8.1. Definition - Examples (see also chapters 24 to 26) :

As far as its sampling is concerned, an object is said to be "small" when its weight is small enough, or "valuable" when its value is large enough to justify its transformation into a one-dimensional or zero-dimensional object.

Valuable or small objects can range from 20,000 tons of a valuable commodity (high grade nickel ore for instance) to a 50 gram laboratory sample. They can be either split by one of the methods or devices described in chapter 24 (which as we know transform them into a zero-dimensional object prior to the selection) or submitted to increment sampling after transformation into a one-dimensional flowing stream.

29.8.2. Theoretical and practical solvability :

From a theoretical standpoint, these problems can always be solved as recalled in section 29.6. (flowing streams) and 29.7. (zero-dimensional objects).

From a practical standpoint, the fact that these objects are small or valuable warrants that the cost of careful and accurate handling and sampling will remain within reasonable limits. The emphasis should be put on care and accuracy, especially in the preparation stages (see chapter 27).

29.8.3. The Japanese slab-cakes :

Although small objects can always be easily put under the form of a one dimensional flowing stream or of a zero-dimensional object, of which we know that the sampling is always cheap and reliable, there are methods consisting in putting them under the form of two- or one-dimensional stationary objects. As pointed out in sections 29.4. and 29.5. these methods are likely to be incorrect and to generate a bias but the reader should be aware of their existence and of their defects.

29.8.3.1. Two-dimensional slab-cake :

The Japanese slab-cake is a laboratory sampling method applicable to lots weighing up to a few hundred kg. It can be reduced to a two-dimensional model and seems to be very popular in Japan though apparently unknown elsewhere.

The batch to be sampled is mixed and spread by hand on a concrete ground or a steel plate and put under the form of a flat cake, usually rectangular, 5 to 10 cm thick. Then, the rectangle is divided into a matrix of approximate squares such as those represented on fig. 29.1. The squares are delimited by a groove made by any available instrument and for instance the edge of a shovel.

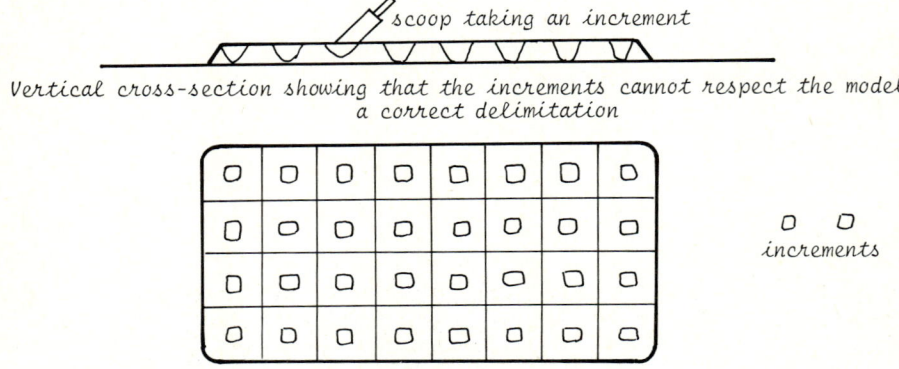

Vertical cross-section showing that the increments cannot respect the model of a correct delimitation

increments

Fig. 29.1. Two-dimensional Japanese slab-cake.

The size of the matrix (4 x 8 in our example) is always fixed with precision in the sampling procedure. A scoopful of material is collected about the middle of each square, or alternate squares, and constitutes an increment. All increments are then gathered and make up the sample. The method can be repeated if necessary.

Unfortunately, this method is subject to the defects enumerated in section 29.4.2. and is highly incorrect. It is likely to introduce a bias in the size distribution.

29.8.3.2. One-dimensional slab-cake :

This is a variant of the two-dimensional slab-cake. Instead of making a rectangular cake, the batch is made to form an elongated ribbon, a few meters long, about 10 to 20 cm wide and a few centimeters thick (fig. 29.2.).

Fig. 29.2. One-dimensional Japanese slab-cake.

Transversal cuts are made by means of a flat scoop and constitute increments which are gathered to form the sample. This method is subject to the same defects as the two-dimensional slab-cake and is always liable to introduce a bias.

29.8.3.3. Incorrectness of the Japanese slab-cakes :

Why should one implement an incorrect method when a correct one is available at the same or at a cheaper cost ? Fractional or alternate shovelling are correct by nature while slab-cakes are not. They require the same tools (shovels and scoops) and the same labour force. Why ? The answer is probably that of a French humorist : why should one make things in a simple way if they can be made in a complicated way ?

29.9. CONCLUSIONS

With practically no exception, we can state that the sampling of three-, two- and one-dimensional stationary objects is never solvable because one is never in a position to ascertain that the operation is unbiased : increment delimitation and extraction biases are always likely to take place. From such objects, specimens can be extracted, not samples.

On the contrary, the sampling of one-dimensional flowing streams or that of zero-dimensional objects is always solvable at an acceptable cost, with the exception of secondary sampling of zero-dimensional objects made of bulky units, which may prove costly as mentioned in section 29.7.3.

Of all the sampling problems, from a theoretical as well as from a practical or economical standpoint, the sampling of a flowing stream is by far the easiest, cheapest, most satisfactory and reliable. It can always be achieved at a reasonable cost. It can be completely automated, from the primary sampling of a 8,000 tons per hour stream of 200 mm material down to the reception of a 100 gram laboratory sample in a Mason jar, the only human intervention being the screwing on and sealing of the jar and the labelling of the sample. It requires little maintenance and is practically fool-proof. Not only can it be made correct but also the user can be certain that it is correct and therefore unbiased, and can rely upon its accuracy.

Mechanized sampling plants such as those which will be described in the ninth part of this book, achieving the primary sampling and the sample reduction require capital expenditure but involve a minimum of human labour and operate at an amazingly low cost.

Mechanized commercial sampling plants can be put under lock and key in order to protect the whole operation against any tampering. Here again, the absence of human intervention is a safety factor. Hand commercial sampling should be progressively replaced by mechanized sampling.

For the sampling of small or valuable objects, splitting methods are often advisable, especially when sampling for sales and when ensuring that they remain probabilistic and equitable, but should be integrated in automated plants as often as possible because they are labour consuming with all possible consequences of human unreliability.

When a collection of small objects are to be sampled in a routine way (e.g. reduction of drill cores or of any geological specimens) a small mechanized plant may save a lot of labour, time and money.

EIGHTH PART

PROBLEMS ASSOCIATED WITH COMMERCIAL SAMPLING

The trade of mineral commodities concerns huge tonnages and huge amounts of money that can be counted in billion dollars per year. All settlements are based on assays and all assays are based on samples. Unfortunately, the fact that sampling is an error-generating technique and is often overlooked and it may be worth mentioning that, as a general rule, technical sampling has progressed, during the last decades, much more than commercial sampling that remains very often primitive despite the fact that huge amounts of money are at stake. This is probably due to the fact that technical sampling is carried out under the responsibility of technicians who are probably quicker to follow the developments of a science such as sampling than administrative staff usually responsible for sales.

In the present conditions, selling or buying ores and concentrates remains very often a sort of gambling where the most astute has every chance to win and where cheating, one way or another, is not exceptional.

Discrepancies between seller and buyer are frequent and, probably not due to probabilistic causes, buyers nearly always find less valuable component than sellers are supposed to have shipped. Such discrepancies may result from errors in weighing, moisture sampling, moisture estimation, assay sampling and assaying, to say nothing of losses, accidental or not. Of all these operations, sampling is likely to be responsible for the largest part of the systematic differences recorded.

As a result of these discrepancies, disputes very often arise between parties to the sales agreement. The author was recently called as expert witness in a dispute arising between a group of South-American tin mines and a European smelter. The systematic differences recorded on a large number of shipments averaged up to about 9 % of the critical content and a considerable amount of money was involved.

A statistical study covering more than 200 shipments of tin concentrates delivered by two independent groups of tin mines M_1 and M_2 to two independent European tin smelters S_1 and S_2 proved beyond any possible doubt that :

1) M_1 and M_2 were in constant agreement with the smelter S_1,
2) M_1 and M_2 were in constant disagreement with the smelter S_2, both groups of mines recording systematic differences ranging between 8 and 9 %.

As the four parties involved performed independent estimations, by different methods on different continents, we had an overwhelming evidence, supported by a statistical analysis, that the estimates obtained by the smelter S_2 were heavily biased, due mainly to a biased hand sampling method.

This example is a good introduction to the following chapters :

1) it shows the importance of sampling biases in the trade of mineral commodities,

2) it shows the importance of a statistical analysis of the results obtained by both parties to the contract.

As, according to the agreement, the settlement price was calculated on the basis of the smelter's estimates, the mine was loosing 9 % of its production. This loss was due :

1) on the one hand to a faulty sampling procedure,

2) on the other hand to the dissymmetrical contract according to which the interests of the seller were not conveniently protected.

What mine can afford to loose 9 % of its production ?

This eighth part is made of three chapters :

Chapter 30 : Notion of equity

Chapter 31 : Testing the agreement between two series of estimates of a same characteristic - Discrepancies between sellers and buyers

Chapter 32 : Testing the agreement between an estimate and the true value - Check of sampling bias

We would like to point out the fact that, if chapters 31 and 32 are of extreme importance in commercial sampling, which justifies our decision to include them in this part of the book dealing with commercial sampling, they are also relevant in technical sampling and for instance when establishing a balance between :

1) the material (metal or mineral) fed to a processing plant or to a smelter,

2) the material (metal or mineral) recovered in concentrates or ingots and lost in tailings or slags.

CHAPTER 30

NOTION OF EQUITY

30.1. INTRODUCTION - DEFINITION

The notion of equity arises in commercial sampling where it is of extreme importance. The trade of mineral commodities nearly always consists of series of shipments from a same producer to a same consumer. Such shipments are usually governed by long-term agreements providing for tens, hundreds or even thousands of lots. According to these agreements, the purchase price of each lot or shipment is determined in terms of dry weights and critical contents evaluated on samples. In commercial sampling, we can define a "critical content" as the proportion of a component which is taken into account in the purchase price of the material object of the contract.

We shall assume in the following sections that relevant measurements such as gross weighing, tare weighing, moisture sampling and analysis, assaying are exact, the only residual errors being the sampling errors in the wider sense. These are undoubtedly very optimistic hypotheses but their discussion falls outside the scope of this book. It would anyway require a multi-volume textbook still to be written.

We shall reason on a single critical component, the main valuable component for instance, or in other words on a single term of the formula determining the purchase price. This is justified by the fact that all terms of the formula are usually built in the same way. In a given commodity, several components are usually critical such as, for instance :

   - main valuable component : iron in an iron ore, zinc in a zinc concentrate, barium sulphate in a barite concentrate, etc ...

   - secondary valuable component(s) : gold, silver, in a lead or copper concentrate, etc ...

   - penalized impurities : arsenic or zinc in a lead concentrate, silica in a barite concentrate, etc ...

   - penalized off-size material : oversize or undersize to a given mesh in an iron ore, etc ...

We shall call :

p_{Ln} : the true market price of a given lot L_n with critical content a_{Ln}

p_{Sn} : the settlement price of L_n as calculated when using the sample critical content a_{Sn} as a random estimator of a_{Ln}. The price p_{Sn} is a random estimator of p_{Ln}.

VE_n : the "relative settlement error" (V for value), relative difference between p_{Sn} and p_{Ln} :

$$VE_n = \frac{p_{Sn} - p_{Ln}}{p_{Ln}}$$

VE_n is a random variable.

$p_{L\Sigma}$: the cumulative true market price of a series of N lots L_n (n = 1, 2, ... N) :

$$p_{L\Sigma} = \sum_n p_{Ln}$$

$p_{S\Sigma}$: the cumulative settlement price of the series of N lots L_n :

$$p_{S\Sigma} = \sum_n p_{Sn}$$

VE_Σ : the cumulative settlement error defined as :

$$VE_\Sigma = \frac{p_{S\Sigma} - p_{L\Sigma}}{p_{L\Sigma}} = \frac{\sum_n p_{Ln} VE_n}{\sum_n p_{Ln}}$$

VE_Σ is the weighted mean of VE_n. As the shipments usually have comparable values, we shall admit that for all practical purposes this weighted mean is equal to the arithmetical mean :

$$VE_\Sigma = \frac{1}{N} \sum_n VE_n$$

Definition : A sampling procedure is said to be "equitable" when the mean of the random variable VE_n is zero : by definition of VE_n :

Equity : $\qquad m(VE_n) = 0$ which involves : $\qquad m(p_{Sn}) = p_{Ln}$

30.2. PROPERTIES OF THE SETTLEMENT PRICE ASSUMED TO BE A LINEAR FUNCTION OF THE CRITICAL CONTENT

With metal-bearing ores and concentrates as well as with industrial minerals, the settlement price p_{Sn} is nearly always a linear function of the critical content a_{Sn}, at least within the limits of a given range fixed by the contract. With valuable components, p_{Sn} is positive. With penalized components, p_{Sn} is negative.

Formulas linking up the price p_{Sn} to the critical content a_{Sn} may vary from one commodity or from one contract to the next but as a general rule, they can be expressed by the most general formula, valid for a single component :

$$p_{Sn} = M_{Ln}\left[(a_{Sn} - a_{On})(1 - b_n)(p_{On} - p_{1n}) - p_{2n}\right] \quad \text{with :}$$

M_{Ln} : dry weight of the lot L_n,

p_{On} : price of one ton of pure metal or mineral (positive) or penalty per ton of penalized metal, mineral or material (negative) in a material characterized by a critical content a_{Sn},

a_{On}, b_n, p_{1n} and p_{2n} : parameters accounting for smelting or processing losses, smelting and processing charges, etc ... Some of these parameters can be zero.

Whether these formulas are "equitable" is another proposition altogether : in the metal mining and metallurgical industries, they are usually imposed upon the producer by the consumer. As a general rule, they work in favour of the buyer but this is beyond the scope of this book.

We can calculate p_{Ln} according to the same formula :

$$p_{Ln} = M_{Ln}\left[(a_{Ln} - a_{On})(1 - b_n)(p_{On} - p_{1n}) - p_{2n}\right]$$

from which we deduce :

$$VE_n = \frac{(1 - b_n)(p_{On} - p_{1n})\, M_{Ln}}{p_{Ln}}(a_{Sn} - a_{Ln})$$

When, as assumed in this section (linear hypothesis), b_n, p_{On} and p_{1n} remain constant, the equity condition $m(VE_n) = 0$ reduces itself to :

$$m(a_{Sn}) = a_{Ln}$$

The sampling procedure is equitable if and only if it is unbiased. When the number N of shipments becomes very large, as it usually does with a long-range contract, it is a known result of mathematical statistics that VE_Σ tends towards $m(VE_n)$. Then, as N becomes infinite :

1) <u>if the sampling is equitable</u> : VE_Σ tends towards $m(VE_n) = 0$. The buyer has paid exactly for what he has received, the seller has been paid exactly for what he has shipped. The interests of both parties have been equally respected.

2) <u>if the sampling is inequitable</u> : VE_Σ tends towards $m(VE_n) \neq 0$. Two cases may arise :

- <u>$m(VE_n)$ is positive</u> : the settlement price is systematically higher than the true value. The sampling inequity works in favour of the seller. In the long run, the interests of the buyer are significantly jeopardized.

- <u>$m(VE_n)$ is negative</u> : the settlement price is systematically lower than the true value. The sampling inequity works in favour of the buyer. In the long run, the interests of the seller are significantly jeopardized.

As far as sampling is concerned and inasmuch as the formula linking up p_{Sn} to a_{Sn} is itself equitable, the settlement is equitable if and only if the samples are unbiased.

30.3. PROPERTIES OF THE SETTLEMENT PRICE ASSUMED TO BE A NON-LINEAR FUNCTION OF THE CRITICAL CONTENT

Formulas linking up p_{Sn} to a_{Sn} are linear only within limits defined in the contract. Outside these limits, the formula remains linear but with different parameters. An example will illustrate this point. According to a certain contract, a certain iron ore is paid as follows for one ton of dry ore :

- Normal range : 65 % Fe > a_{Sn} > 62 % Fe : 21 cents per 1 % Fe
- Low grade 1 : 62 % Fe > a_{Sn} > 61 % Fe : penalty of 42 cents per 1 % below 62
- Low grade 2 : 61 % Fe > a_{Sn} > 60 % Fe : penalty of 70 cents per 1 % below 61

This formula, which may seem complex, is illustrated in Fig. 30.1.

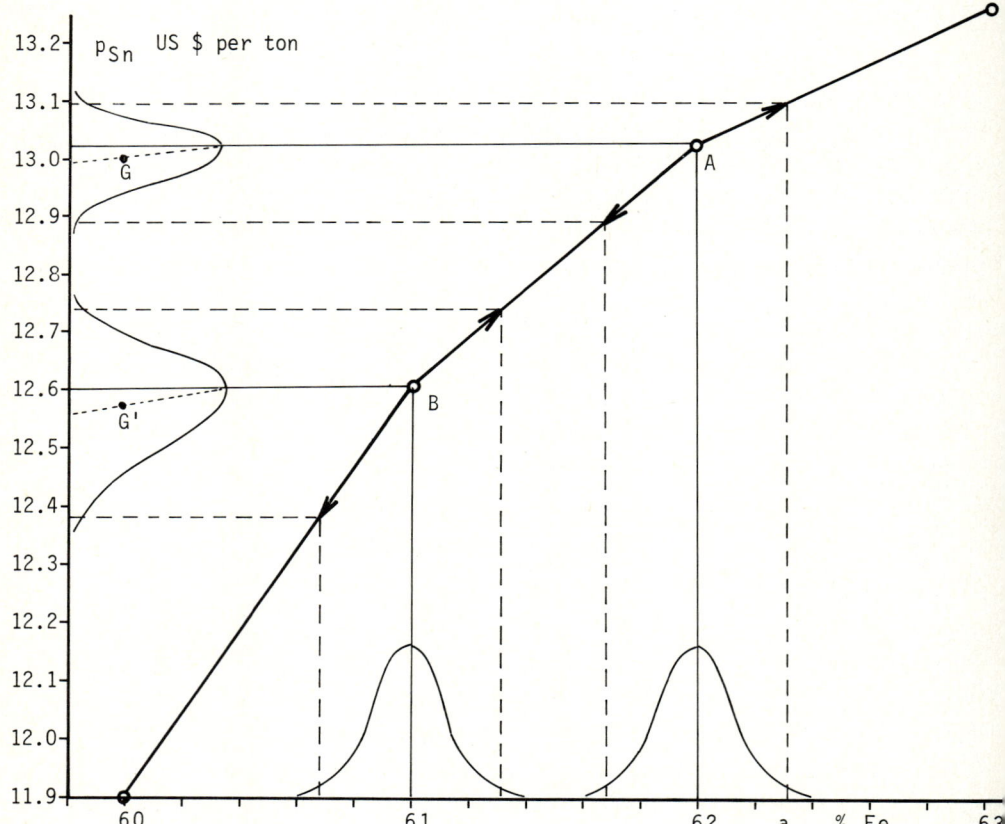

Fig. 30.1. *Example of non-linear relationship between the settlement price of an iron ore and its critical content Fe %.*

Let's suppose that the true content a_{Ln} corresponds with one of the apexes of the broken line, apex A for instance. The frequency distribution of the sample content a_{Sn} is, for the sake of convenience, represented along the axis of the abscissae. Assuming the sampling to be unbiased, the frequency curve is practically normal and symmetrical, both sides of a_{Ln} = 62 % Fe. The arrows on the broken line delimit the confidence interval ± 2 $\sigma(a_{Sn})$. The frequency distribution of the corresponding settlement price p_{Sn} is represented along the axis of the ordinates. Due to the change of slope at point A, the curve is made of two dissymmetrical halves, the lower half being obviously larger than the upper one. The true price p_{Ln} is the ordinate of point A (13.02 dollars), but the mean of p_{Sn}, the settlement price is the ordinate of the centre of gravity G of the frequency curve which, due to its dissymmetry, is necessarily smaller than p_{Ln}. The consequence is :

$m(p_{Sn}) < p_{Ln}$ which involves $m(VE_n) < 0$

Though the sampling is unbiased, the settlement is inequitable. The departure from equity is a consequence of the definition of p_{Sn}. Since in all examples in our possession the broken line is concave downwards like that of fig. 30.1., this lack of equity always works in favour of the buyer. In an equitable contract, this could and should be provided against by using a corrective factor larger than 1 to be applied to the sample content a_{Sn}. As the bias is of the order of 0.2 % in our example, the corrective factor should be 1.002 when a_{Sn} is near 62 %. It should be a little larger when a_{Sn} is near 61 % (point B).

Unfortunately, as far as sampling is concerned, we can do nothing about it : the best we can do is to achieve a correct and therefore unbiased sampling and even then the settlement is inequitable. As things are, the chances that this point will ever be taken into consideration seem very remote.

30.4. RELATIVE IMPORTANCE OF BIAS AND RANDOM ERROR IN COMMERCIAL SAMPLING

We shall call :

TE_n : the total sampling error committed on lot L_n. By definition :

$$TE_n = \frac{a_{Sn} - a_{Ln}}{a_{Ln}}$$

When lots of the same material are sampled under routine conditions, TE_n can be regarded as a random variable with a stationary mean $m(TE)$ and a stationary variance $\sigma^2(TE)$.

TE_Σ : the cumulative sampling error committed on the set of N lots L_n (n = 1, 2, ..N)

$$TE_\Sigma = \frac{a_{S\Sigma} - a_{L\Sigma}}{a_{L\Sigma}}$$

As the weights M_{Ln} of the lots are very often comparable, we may write :

$$m(TE_\Sigma) = m(TE) \quad \text{and} \quad \sigma^2(TE_\Sigma) = \frac{1}{N}\sigma^2(TE)$$

The 95 % probability confidence intervals are :

<u>For a single lot L_n</u> : $TE_n = m(TE) \pm 2\,\sigma(TE)$

<u>For a set of N lots</u> : $TE_\Sigma = m(TE) \pm \frac{2}{\sqrt{N}}\sigma(TE)$

In the linear range (see section 30.2.) the settlements errors VE_n and VE_Σ committed on a single lot L_n and on a set of N lots are random variables with stationary means and variances expressed as follows :

$$m(VE_n) = m(TE) \quad \text{and} \quad \sigma^2(VE_n) = \sigma^2(TE)$$

$$m(VE_\Sigma) = m(VE_n) = m(TE) \quad \text{and} \quad \sigma^2(VE_\Sigma) = \frac{1}{N}\sigma^2(VE_n) = \frac{1}{N}\sigma^2(TE)$$

The 95 % probability confidence intervals are therefore :

<u>For a single lot L_n</u> : $VE_n = m(TE) \pm 2\,\sigma(TE)$

<u>For a set of N lots</u> : $VE_\Sigma = m(TE) \pm \frac{2}{\sqrt{N}}\sigma(TE)$

Expressed in relative value, the confidence intervals of VE_n and VE_Σ on the one hand, those of TE_n and TE_Σ on the other hand are identical.

These formulas show that on long term contracts, with values of N likely to be very large, 100 or more, the term involving the random error, characterized by the standard deviation $\sigma(TE)$, becomes smaller and smaller whilst the term involving the bias, characterized by the mean $m(TE)$ remains unchanged. We shall show on one example how important can be the bias (section 30.4.1.) and on another how unimportant can become the random error.

30.4.1. Relative unimportance of random errors in commercial sampling :

By random errors, the reader must understand those errors with zero mean that are characterized by the variance $\sigma^2(TE)$. In order to illustrate this point, we shall present the example of uranium concentrates.

A contract concerning the shipment of 100 lots of calcined uranium oxide, each containing about 20,000 lbs of calcines, was signed between a given seller and a given buyer. The material contains an average of 93 % U_3O_8. The price of U_3O_8 is known to be 43 US $ per pound (December 1978). Each lot is to be represented by a single series of six twin-samples distributed to the parties or kept in reserve. We shall assume that the sampling bias is nil (hypothesis well supported by experience) that the total sampling variance associated with each of the twin-samples is $\sigma^2(TE)$ and that the lots have equal statistical weights (physical weights of the same order of magnitude). Let's assume now that the true (unknown) critical content a_{Ln} of lot L_n is 93 % U_3O_8. The true (unknown) market price of L_n is :

P_{Ln} = 20,000 x 0.93 x 43 = 799,800 $ or, in round figures 0.8 x 10^6 $

According to past experience, we know the 95 % probability confidence interval of the critical content to be about ± 0.4 % U_3O_8, from which we can estimate :

$\sigma(a_{Sn})$ = 0.2 % U_3O_8

$\sigma^2(TE) = \dfrac{\sigma^2(a_{Sn})}{a_{Ln}^2} = \dfrac{(0.2)^2}{(93)^2} = 4.62 \times 10^{-6}$ and $\sigma(TE) = 2.15 \times 10^{-3}$

As $\sigma(VE_n) = \sigma(TE)$ we can compute :

$\sigma(p_{Sn}) = P_{Ln}\, \sigma(TE) = 0.8 \times 10^6 \times 2.15 \times 10^{-3} = 1720$ $

Since we have assumed the sampling to be correctly carried out and therefore unbiased, the chances are, as well for the seller or the buyer, and for a single lot L_n :

0.13 % to gain or loose more than $3\sigma(p_{Sn})$ or : more than 5160 $
2.15 % to gain or loose between 2 and $3\sigma(p_{Sn})$ or : between 3440 and 5160 $
13.59 % to gain or loose between 1 and $2\sigma(p_{Sn})$ or : between 1720 and 3440 $
34.13 % to gain or loose between 0 and $1\sigma(p_{Sn})$ or : between 0 and 1720 $

This calls for two remarks :

1) As compared with the commercial value of 800,000 $, the maximum possible loss of some 5,000 $ does not seem to be excessive. It represents 0.65 % in relative value, and its probability is very small indeed.

2) Since the sampling is unbiased, it is therefore equitable which means that in the long run, nobody is likely to win or loose an appreciable proportion of the amount of money at stake.

Let's now consider the whole contract of 100 lots (N = 100). The relative standard deviation $\sigma(VE_\Sigma)$ of the cumulated error committed on the set of 100 lots is :

$\sigma(VE_\Sigma) = \dfrac{1}{\sqrt{N}} \sigma(VE_n) = \dfrac{1}{10} \sigma(TE) = 2.15 \times 10^{-4}$

The total price of the 100 lots is of the order of :

$P_{L\Sigma}$ = 100 P_{Ln} = 80 x 10^6 $

The absolute standard deviation of the cumulative price $p_{L\Sigma}$ is :

$\sigma(p_{S\Sigma}) = P_{L\Sigma}\, \sigma(VE) = 100\, P_{Ln} \times \dfrac{1}{10} \sigma(TE) = 10\, \sigma(p_{Sn}) = 17,200$ $

The relative standard deviation is ten times smaller for 100 lots than for a single lot, whilst the absolute standard deviation is ten times larger. If we accept $3\sigma(p_{S\Sigma})$ as the maximum possible loss of either party, it amounts to 51,600 $ or 0.065 % of the total commercial value of the concentrates object of the contract.

Such a maximum risk is absolutely trivial. Our conclusion is not founded on any unrealistic hypothesis : it so happens that uranium concentrates are very homogeneous materials and that their sampling is very reproducible but let's assume now a quite improbable ± 1 % U_3O_8 confidence interval. The maximum possible loss would not exceed 129,000 $ or 0.16 % of the total value and there is only a 0.13 % probability for such a loss to be reached or slightly exceeded. Furthermore, there is by hypothesis an equal risk (or chance) to gain the same amount of money since the whole procedure is assumed to be equitable.

Now, we pointed out on various occasions that biases ranging from 1 to 2 % could easily pass undetected (though we should emphasize the fact that the uranium industry is probably the only one where such biases are unlikely to be observed). What is the importance of a symmetrical risk to gain or loose 0.065 % as compared with the systematic risk to loose 1 or 2 % if the sampling is biased and if your party happens to be on the wrong side.

Some Companies are so sensitized to the possibility of loosing money through a biased sampling that they try to impose their own sampling conditions and sometimes to bias the sampling in their own favour. The truth of the matter is that they often succeed.

30.5. CONCLUSIONS

Who can afford to loose 9 % of his production like the tin mine mentioned in the introduction to this eighth part ? Can YOU ? In commercial sampling, bias is of a much more treacherous nature than random errors. According to our experience only a minority of sampling devices or facilities can be regarded as absolutely correct and apt to deliver unbiased samples, perhaps still less in commercial sampling than in technical sampling. A fair number of these introduce a 1 to 2 % bias usually too small to be clearly distinguished from the inevitable random errors but nevertheless significantly detrimental in the long run. A careful statistical analysis of the results (see chapter 31) can detect these biases if the number of data is large enough. A critical inspection of the sampling and preparation equipment and methods, when carried out by a specialist checking whether the conditions stated in chapters 17, 18 and 27 are actually respected, remains the best bias detector.

Larger biases are likely to arouse suspicion sooner or later but up to 5 %, experimental methods of checking for bias remain inefficient, especially when they are implemented in accordance with certain national or international Standards, a point which will be dealt with in chapter 32. Still larger biases (we had various opportunities to come across biases ranging between 10 and 20 %) are easily detected which does not mean that they can be easily corrected.

Our conclusion is that if money is to be spent on commercial sampling, the first concern should be to eliminate all possible sources of sampling and preparation bias (chapters 17, 18 and 27). There is no possible compromise with biases, either you destroy them or they will in the long run destroy you. There is only one possible strategy : preventive or curative eradication. If money is left, then but only then you can try and reduce random errors characterized by the variance $\sigma^2(TE)$ which is anyway likely to be significantly reduced when suppressing the delimitation, extraction and preparation biases.

Now, with the reader's permission, we would like to close this chapter with a few pieces of advice to buyers and sellers of mineral commodities all over the world.

1) Never rely on a non-probabilistic sampling method such as for instance the hammer and shovel method. You might loose much more money than you would spend with a correct, unbiased probabilistic sampling scheme.

2) Whether seller or buyer, never sign an agreement according to which the governing samples are those obtained by the other party, unless you are absolutely certain through an expert's valuation that the sampling facilities to be used are correctly designed and laid out and that your interests are not going to be jeopardized. Never forget that there are incorrect ways of using correct sampling devices and we have sufficient evidence that the right to be represented, in a foreign Country for instance, by an accredited sworn sampler is definitely no safeguard against bias.

3) The fact that the other party is a big Company trying to impose its conditions is no safeguard either.

The only way to make sure that your interests are safely protected is :

1) to have your own weighing and sampling facilities, to have these designed or evaluated by a sampling expert and corrected if necessary,

2) never to sign agreements where your rights are not strictly equivalent to those of the other party,

3) to agree with the other party on an acceptable systematic difference between his estimates and yours of one of the following magnitudes :

- critical content as estimated on twin-samples obtained from a unique primary sample : check of the difference between assays.

- critical content as estimated on samples obtained from two independent sampling systems : check of the differences between samples and assays.

- weight of critical component as estimated by both parties independently : check of the differences between gross and tare weights, moisture sampling and analysis, assay sampling and assaying.

- settlement price : check of the differences between all relevant factors including secondary valuable components and penalized impurities.

A number of contracts provide for a certain acceptable difference between assay carried out on twin-samples obtained from a same primary sample by the laboratories of both parties (first case of our list). Such a procedure is definitely insufficient because it can only detect lot by lot assaying discrepancies which, according to our experience, are only minor as compared with sampling discrepancies.

4) to implement the statistical method described in chapter 31 to decide on a sound basis :

- whether a systematic difference exists between both estimates,
- that, if a systematic difference exists, it is smaller than the maximum agreed upon (see point 3 above).

5) if the systematic difference is smaller than the maximum agreed upon, decide to split the difference and to adopt the arithmetic mean of both estimates as the governing value of the critical content, weight of critical component or settlement price as the case may be.

6) if the systematic difference is larger than the maximum agreed upon, the contract should provide first for an arbitration out of court by an independent expert agreed upon by both parties. We know at least one contract written in such terms and working for a few years with the complete satisfaction of both parties. If for some reason or other no amicable agreement can be reached between seller and buyer and if you are practically certain that your interests are jeopardized, ask for the arbitration of the International Chamber of Commerce in Paris.

Such a procedure may seem costly but the relevant expenses must be considered as an insurance policy against much more important losses : the cost of such a control, which obviously varies widely from one case to the next, is not likely to be larger than 1 % of the price of your production or consumption whereas much more important biases are liable to take place.

30.6. EQUITY - LOUIS-LE-DEBONNAIRE'S SPLITTING METHOD

Unlikely though it may appear, the ninth century French history provides a very clever model of splitting equity or more accurately of "overequity". Louis-le-Débonnaire, Charlemagne's son and heir, ruled over the so-called "Holy Roman Empire" from 814 to 840. This was neither holy, nor roman but was nevertheless a mighty empire covering a large part of today's western Europe. Towards the end of his reign, Louis conceived the project of splitting his empire between two of the most restless of his male offsprings : Lothaire, the eldest, and Charles-le-Chauve. In order to disarm the jealousy and mistrust of his sons, the old Emperor invited Lothaire to draw the line that would split the empire, knowing that Charles would then have the privilege of choosing his portion.

It does not need a huge stretch of imagination to transpose this model to commercial sampling of mineral commodities. One of the parties to the contract, each of them by turns for instance, would perform the splitting operations in the presence of the other and the latter would have the privilege of choosing the potential sample that would become the actual sample.

As already pointed out in chapter 24, the idea remains to separate the splitting operation from the selection but suppose now that for some reason or other the operator attempts to bias the sample : in a probabilistic approach such as ours, there remains a 50 % probability for the bias to work in favour of the cheat whereas in Louis-le-Débonnaire's approach, assuming the representatives of both parties to be equally experienced and clever, the bias can work only to the detriment of the cheat which, from a moral standpoint, is highly satisfactory. But who cares about morals in the trade of mineral commodities ?

CHAPTER 31

TESTING THE AGREEMENT BETWEEN TWO SERIES OF INDEPENDENT ESTIMATES OF A SAME CHARACTERISTIC - DISCREPANCIES BETWEEN SELLER AND BUYER

31.1. INTRODUCTION

In the mineral industries, whether for technical, accounting or commercial purposes it may happen that two independent estimates are obtained of a same characteristic and it is often interesting to test their agreement on a sound scientific basis. Here are a few examples :

1) Sampling : critical content of a certain lot L estimated on two independent samples obtained by means of a device to be checked and of a reference method respectively. The technical part of this problem will be dealt with in chapter 32.

2) Processing plant or smelter : incoming and outgoing weights of a certain commodity or element.

3) Internal accounting : critical content of the material delivered by a mine to the adjoining processing plant belonging to the same Company, weight of concentrates delivered by a processing plant to the adjoining smelter, etc..

4) Trade of mineral commodities : dry weight of solids, critical content, weight of critical component, settlement price of a given shipment, independently estimated by seller and buyer (see section 30.5.).

The statistical method which will be presented in the next sections makes it possible to detect two kinds of detrimental anomalies :

- measurement or sampling biases : e.g. gross weighing, tare weighing, moisture sampling, moisture analysis, assay sampling, assaying, etc..

- inconspicuous losses of material : e.g. losses of flotation concentrates stored in the open in a dry and windy Country (we know many examples of such losses), losses under the form of dust or fumes, under the form of fine concentrates in a thickener overflow, etc...In commercial operations, theft or loss of bags of valuable concentrates.

We chose to present this method, which is based on well known statistical results but which has not been described in a practical way as we do here, in the part of this work dedicated to commercial sampling because it is certainly in this field that the method can render the most valuable service.

31.2. NOTATIONS AND DEFINITIONS

We shall define :

L_n : the n th lot of a certain commodity in a certain series (n = 1, 2, ... N). The lot L_1 corresponds with the beginning of our test.

X : a certain critical characteristic to be estimated : weight, assay, price, etc.

x_n : the true (unknown) value of X in lot L_n.

Y and Z : two independent estimation systems of the characteristic X.

y_n and z_n : estimates of x_n obtained by Y and Z respectively.

d_n : the algebraical difference : $d_n = y_n - z_n$

D : true (unknown) mean of the distribution of the random variable d_n. It is the systematic difference between the estimates obtained by Y and Z. When both parties work in a routine way, we may assume that D remains constant.

σ^2 : true (unknown) variance of the distribution of d_n.

D_N : the estimate of D after a series of N trials. By definition :

$$D_N = \frac{1}{N} \sum_n d_n \qquad \text{with} \qquad n = 1, 2, \ldots N$$

s_N^2 : unbiased estimate of σ^2 after a series of N trials :

$$s_N^2 = \frac{1}{N-1} \sum_n (d_n - D_N)^2$$

The statistical analysis of such a sequence of data is carried out by means of the Student-Fisher (SF) test where a certain hypothesis is checked.

31.2.1. Positive and negative tests :

We shall say that a given SF test is "<u>positive</u>" when the hypothesis checked must be rejected. It is said to be "<u>negative</u>" when the hypothesis cannot be rejected, which does not mean, as is too often believed, that it must be blindly accepted.

31.2.2. Certainty, presumption and uncertainty :

The notion of "<u>certainty</u>" is alien to mathematical statistics. We must therefore be prepared to draw from our tests conclusions that are not absolutely sure. We shall however speak of "<u>practical certainty</u>" when the risk of being wrong is smaller than 1 % (this negligible quantity already referred to in section 28.2.2.1 and of "<u>serious presumption</u>" when this risk does not exceed 5 %. This is purely conventional but corresponds with subjective feelings widely agreed upon. We shall speak of "<u>uncertainty</u>" when neither a practical certainty nor even a serious presum tion can be reached. Whenever possible, we should always proceed with a progressive test until a practical certainty or at least a serious presumption is attained to.

31.2.3. Agreement and disagreement between two series of estimates :

One can never draw any conclusion from a single set of data (y_n, z_n, d_n). We shall speak of "agreement" between two series of estimates when there is a practical certainty that the systematic difference D is (in absolute value) smaller than or equal to a certain tolerated systematic difference D_A, and of "disagreement" when there is a practical certainty that D is (still in absolute value) larger than D_A. We shall assume that both parties involved in the test, seller and buyer for instance, have agreed upon a certain value of D_A.

31.2.4. Kinds of risk and level of risk :

When implementing a statistical test, we can incur two opposite "kinds" of risk called α and β respectively.

1) The risk alpha is the risk incurred when the hypothesis tested is in conformity with the unknown truth, if the test is positive and leads us to reject this hypothesis.

2) the risk beta is the risk incurred when the hypothesis tested is wrong, if the test is negative and if we accept this hypothesis as representing the truth.

The distinction between those two kinds of risk is capital but is not always clearly pointed out, even in textbooks of mathematical statistics. Even standards seem to ignore this difference. We shall see an example of the danger incurred when confusing α and β in chapter 32.

Each of these two risks can be accepted at different probability levels and for instance the 1 % and 5 % levels retained for the definition of practical certainty and presumption.

31.3. TESTING THE HYPOTHESIS $H \equiv \{D = 0\}$

When checking the agreement between two series of estimates, the first question we have to answer is : are we entitled to conclude that D is or is not zero ? The technique consists in implementing a SF test in order to check the hypothesis $H \equiv \{D = 0\}$. Assuming the unknown truth to be in conformity with H, we know that the random variable w_N :

$$w_N = \frac{D_N \sqrt{N}}{s_N}$$

follows a Student-Fisher's "t distribution" with $\nu = N - 1$ degrees of freedom (df). We shall call $t_{\nu\alpha}$ the value of the SF t function with (N - 1) df which is exceeded with a symmetrical probability equal to 2α. Then, assuming H to be true :

Prob $\{w_N > t_{\nu\alpha}\} = \alpha$ and Prob $\{w_N < - t_{\nu\alpha}\} = \alpha$

In our demonstrations and examples we shall always retain $\alpha = 1\ \%$.

The value of $t_{\alpha 1}$ can be found in any statistical tables under the value 0.99 of the cumulative probability. Table 31.1 gives the values of t_ν for a risk $\alpha = 1\ \%$ and for different values of $\nu = N - 1$.

Table 31.1. Value of t_ν for a risk $\alpha = 1\ \%$

$\nu = N - 1$	t_ν	$\nu = N - 1$	t_ν
1	31.821	21	2.518
2	6.965	22	2.508
3	4.541	23	2.500
4	3.747	24	2.492
5	3.365	25	2.485
6	3.143	26	2.479
7	2.998	27	2.473
8	2.896	28	2.467
9	2.821	29	2.462
10	2.764	30	2.457
11	2.718	40	2.423
12	2.681	50	2.403
13	2.650	60	2.390
14	2.624	70	2.379
15	2.602	80	2.370
16	2.583	90	2.365
17	2.557	100	2.363
18	2.552	110	2.361
19	2.539	120	2.358
20	2.528	∞	2.326

The principle of the test consists in :

1) Collecting the data y_n and z_n
2) computing the algebraical difference $d_n = y_n - z_n$
3) computing the mean D_N of the population of N differences d_n
4) computing the standard deviation s_N of the same population
5) computing the variable $w_N = D_N \sqrt{N}/s_N$
6) comparing w_N with t_ν and $- t_\nu$.

After any number N of trials (N > 1), two possibilities may arise :

(a) : $|w_N| > t$: <u>the test is positive</u> : there is a practical certainty that D is non-zero. The sign of D is that of D_N. Our conclusion is attained to with a risk $\alpha < 1\ \%$ of being wrong. But as we are usually not satisfied with a mere qualitative conclusion, our next objective is to ascertain whether or no $|D|$ is smaller than the tolerable difference D_A. In other words, we must now test the complementary hypothesis $H' \equiv \{D' = |D| - D_A\} = 0$. This point is developed in the next section.

(b) : $|w_N| \leq t_\nu$: <u>the test is negative</u> : we remain in a state of uncertainty and it would be erroneous to conclude that D = 0, as one too often does. Assume, for instance, that D = ε, a very small, non-zero systematic difference. The test is obviously unable to discriminate the two hypotheses :

$H \equiv \{D = 0\}$ and $H_\varepsilon \equiv \{D = \varepsilon\}$

The risk incurred when concluding that D = 0 while D = ε ≠ 0 is not the risk α = 1 % but the risk β = 1 - α = 99 %, which in terms of risks makes a serious difference. So many people, encouraged by so many authors and standards, believe that they are entitled in such a case to conclude that D = 0, that the point was worth emphasizing. When the test is negative, we have no right to draw any conclusion whatsoever, because the present situation may result from two opposite causes :

- either D is actually zero
- or D is non-zero but the number N of trials is too small to show it up. To remove this uncertainty a larger number of data would be required.

What we can do at that point is to check whether we have not yet reached a practical certainty that $|D|$ is smaller than D_A. This point is dealt with in section 31.4.

31.4. TESTING THE HYPOTHESIS : $H' \equiv \{D' = |D| - D_A = 0\}$

Irrespective of the result of the preliminary test described in the preceding section, we have found that this complementary test had to be carried out. The problem is now to ascertain :

(a) If the answer to the preliminary test is positive : whether $|D|$ is larger than D_A or not. Is the systematic difference acceptable ?

(b) If the answer to the preliminary test is negative : whether $|D|$ is smaller than D_A or not. If there is a systematic difference, is it acceptable ?

With : $D' = |D| - D_A$, we shall now test the complementary hypothesis :

$H' \equiv \{D' = 0\}$

exactly as we did with hypothesis H in the preceding section. Assuming hypothesis H' to be in conformity with the unknown truth, we know that the random variable :

$$w'_N = \frac{D'_N \sqrt{N}}{s_N} \qquad \text{with} \qquad D'_N = |D_N| - D_A$$

follows a SF distribution with $\nu = N - 1$ degrees of freedom. Then, after point 6) of the preliminary test we shall add the following points :

7) computing the difference $D'_N = |D_N| - D_A$

8) computing the variable w'_N defined overleaf : $w'_N = D'_N \sqrt{N}/s_N$

9) comparing w_N' with t_ν and $-t_\nu$.

Here again, two possibilities may arise :

(a) $|w_N'| > t_\nu$: <u>the complementary test is positive</u> : there is a practical certainty that D' is non-zero and the sign of D' is that of D_N'.

- If $D_N' > 0$: this possibility can arise only when the preliminary test is positive. We already know that $|D| > 0$. We are now entitled to conclude that :

$$D' = |D| - D_A > 0 \qquad \text{which involves} \qquad |D| > D_A$$

We have gained a practical certainty that the disagreement between the estimates y_n and z_n is significantly larger than the acceptable difference D_A.

. If $D_N > 0$: Y's estimates are systematically higher than those of Z by more than D_A.

. If $D_N < 0$: Y's estimates are systematically lower than those of Z by more than D_A.

We have reached a practical certainty that there is something wrong somewhere and a specialist should inspect the two estimation circuits Y and Z, in order to ascertain the causes of over- or under-valuation of one or several of the relevant data. As already mentioned, these causes may belong to two categories :

° sampling or measurement bias (including preparation biases),
° loss of material.

The specialist will have to discriminate these possible causes and to propose solutions in order to suppress the sources of unacceptable errors.

- If $D_N' < 0$: this possibility can arise whether the preliminary test is positive or negative. We are entitled to conclude that :

$$D' = |D| - D_A < 0 \qquad \text{which involves} \qquad |D| < D_A$$

We have gained a practical certainty that, if there is a systematic difference between Y's and Z's estimates, it is smaller than the acceptable difference D_A. It is then common practice to split the difference and to adopt the arithmetical mean $x_n' = (y_n + z_n)/2$ as the best available unbiased estimate of the true unknown quantity x_n, this for the N lots tested so far. But another question is now worth being taken into consideration : the statistical properties of D may be stationary over shorter or longer periods, there is no guarantee that they will remain so for ever. As long as data such as y_n and z_n are collected, the test must be carried on. Special techniques exist making it possible to disclose if and when the systematic difference fluctuates. These must be implemented by a specialist.

(b) $|w_N'| < t_\nu$: <u>the complementary test is negative</u> : we remain in a state of uncertainty as regards the sign of $D' = |D| - D_A$. In this case, the conclusions of the complete test (preliminary and complementary) can be summarized as follows :

- If the preliminary test was positive :

$$\frac{t_\nu s_N}{\sqrt{N}} < |D| \leq \frac{t_\nu s_N}{\sqrt{N}} + D_A \quad \text{(confidence interval of the systematic difference)}$$

- If the preliminary test was negative :

$$|D| \leq \frac{t_\nu s_N}{\sqrt{N}} \quad \text{(maximum of the systematic difference likely to be observed).}$$

31.5. PRACTICAL IMPLEMENTATION OF THE COMPLETE TEST

From a practical standpoint, any desk or pocket computer, programmable or not, can be used to implement this test.

The procedure to be followed is :

1) Enter the first set of data y_1 and z_1,
2) compute the difference $d_1 = y_1 - z_1$,
3) enter the second set of data y_2 and z_2,
4) compute the difference $d_2 = y_2 - z_2$,
5) compute the mean $D_2 = (d_1 + d_2)/2$
6) compute the standard deviation s_2 (defined in section 31.2. with N = 2),
7) compute the random variable w_2 (defined in section 31.3. with N = 2),
8) compute the difference $D_2' = |D_2| - D_A$
9) compute the random variable w_2' (defined in section 31.4. with N = 2)
10) compare w_2 with t_1 and $-t_1$ (defined in table 31.1 with $\nu = N - 1 = 1$). If w_2 falls between t_1 and $-t_1$, as it usually does at the beginning of a test, and as long as w_N does, collect new data and proceed to step 11). If it does not, proceed to step 21) or 22), whichever is applicable.
11) compare w_2' with t_1 and $-t_1$. If w_2' falls between t_1 and $-t_1$, and as long as w_N' does, collect new data and proceed to step 12). If it does not, proceed to step 23) or 24), whichever is applicable.
12) enter the Nth set of data y_N and z_N,
13) compute the difference $d_N = y_N - z_N$,
14) compute the mean D_N of the N values of d_N
15) compute the standard deviation s_N of the population of N values of d_N
16) compute the difference $D_N' = |D_N| - D_A$
17) compute the variables w_N and w_N' (see overleaf)
18) compare w_N with t_ν and $-t_\nu$
19) compare w_N' with t_ν and $-t_\nu$
20) if both $|w_N|$ and $|w_N'|$ are smaller than t_ν and as long as they are, proceed back to step 12) and enter the (N + 1)th set of data.

21) If w_N is larger than t_ν, conclude that Y is systematically higher than Z and proceed to step 23).
22) If w_N is smaller than $-t_\nu$, conclude that Z is systematically higher than Y and proceed to step 23).
23) If w_N' is larger than t_ν, conclude that $|D|$ is larger than D_A : there is a systematic difference between Y and Z and this is larger than the acceptable difference D_A. This may happen only if $|w_N|$ is itself larger than t_ν (steps 21) or 22). It is up to the trouble-shooting teams of the control departments of Y and Z, or to the expert mentioned in the contract, to look out for a measurement, sampling or preparation bias or for inconspicuous losses of material.
24) If w_N' is smaller than $-t_\nu$, conclude that if there is a systematic difference D between the estimates of Y and Z, it is smaller than the acceptable difference D_A. This may happen only if $|w_N|$ is itself smaller than t_ν. Then, proceed to step 25).
25) Compute all values of $x_n' = \dfrac{y_n + z_n}{2}$ and retain these as the best available unbiased estimates of the unknown quantities x_n. In this procedure, x_n' is likely to have been retained anyway for the provisional settlement.

Remarks : When following this procedure, we draw no conclusion unless we have reach a practical certainty that :

- either there is a systematic difference D which is larger than the acceptable difference D_A,

- or, if there is a systematic difference $D \neq 0$, this is smaller than the acceptable difference D_A,

which is exactly what we want to know.

Examples given in section 31.7 and in chapter 32 will illustrate the implementation of this method.

31.6. GRAPHICAL PRESENTATION OF THE RESULTS OF THE TEST

Remark : This section is slightly different from the corresponding section of the first printing of this book (1979).

The graphical method we describe below consists in plotting :

$$W_N = \frac{D_N \sqrt{N}}{s_N t_\nu} = \frac{w_N}{t_\nu} \quad \text{and} \quad W_N' = \frac{D_N' \sqrt{N}}{s_N t_\nu} = \frac{w_N'}{t_\nu} \quad \text{against N}$$

The "intervals of uncertainty" of W_N and W_N' are ± 1, irrespective of N, which defines the "area of uncertainty" as the region delimited by the ordinates $+1$ and

In the central part of the diagram (area of uncertainty) : As long as the experimental points remain in this area, we are not entitled to draw any conclusion with a practical certainty. We must carry on the test and collect new data.

<u>In the upper part of the diagram</u> :

- as soon as W_N enters this area, we are entitled to conclude that D is larger than zero. Y is therefore systematically higher than Z.

- as soon as W'_N enters this area, we are entitled to conclude that $|D|$ is larger than D_A. The systematic difference already detected is unacceptable.

<u>In the lower part of the diagram</u> :

- as soon as W_N enters this area, we are entitled to conclude that D is smaller than zero. Z is therefore systematically higher than Y.

- as soon as W'_N enters this area, we are entitled to conclude that $|D|$ is smaller than D_A. If there is a systematic difference between Y and Z, irrespective of its sign, it is definitely acceptable and may be split.

This method is illustrated in section 31.7 and in fig 31.1 and 31.2.

31.7. EXAMPLES

Uranium concentrates are shipped by a producer Y to a consumer Z under the form of lots made of 24 drums each, containing an average of 475 kg of concentrate.

After a period of reasonable agreement between the gross weights estimated by both parties, a discrepancy arose, the feeling being that the weight of certain lots tended to be overvalued by Y and we were asked to carry out a statistical analysis of the data. Two typical lots will exemplify our demonstration. The corresponding tests are represented in fig. 31.1 (lot 520) and 31.2 (lot 522). We computed the values of w_N and W_N, as well as those of w'_N and W'_N (for D_A = 0.5 kg) and those of w''_N and W''_N (for D_A = 1.0 kg).

<u>Lot 520 (fig. 31.1)</u> : With the tenth drum we may conclude that Y is systematically higher than Z. With the 22nd drum we are practically certain that the systematic difference is larger than 0.5 kg per drum (0.1 % relative). The broken line of W''_N remains in the area of uncertainty and has not been represented. We are not entitled to conclude that the systematic difference is larger than 1 kg. The best estimate available for D is D_{24} = 1.0 kg

<u>Lot 522 (fig. 31.2)</u> : All points W_N remain in the area of uncertainty. We are not entitled to conclude that D is different from zero which, as already pointed out, does not mean that we are entitled to conclude that D is zero. The 24th point of the curve W'_N falls just short of the lower border of the area of uncertainty (W'_{24} = - 0.96). We are practically certain that if there is a systematic difference, this is smaller than 0.5 kg, which is acceptable. When studying the curve of W''_N

we see that with drum No. 15 it enters the lower part of the diagram. If there is a systematic difference, it is definitely smaller than 1 kg.

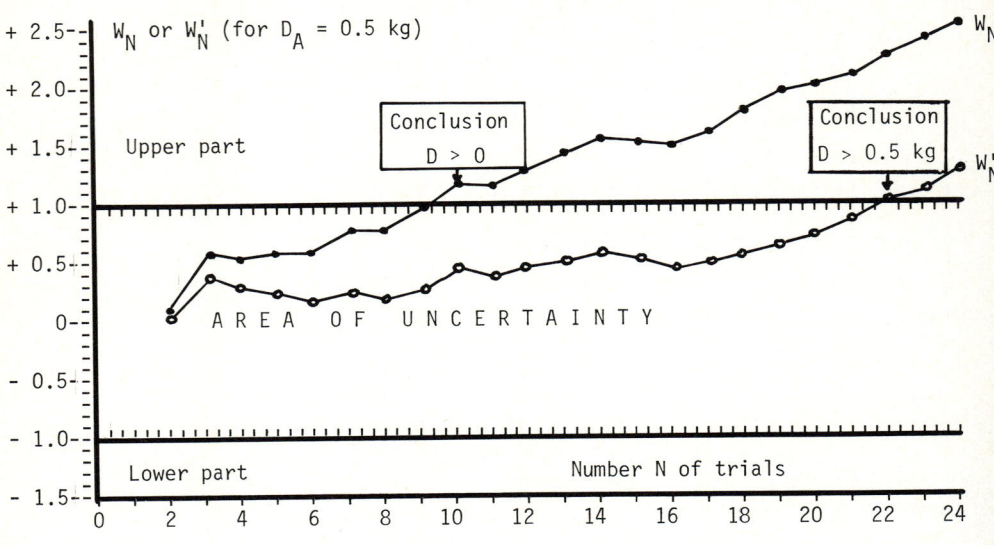

Fig. 31.1. Graphical representation of the test - First example - Lot 520

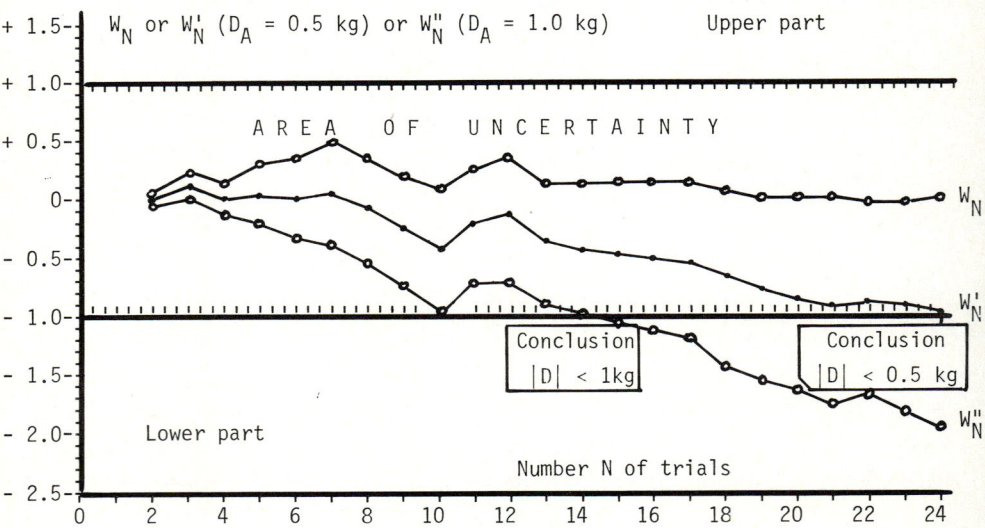

Fig. 31.2. Graphical representation of the test - Second example - Lot 522

Remark : the weighing of the drums making up the lots 520 and 522 was carried out by both parties under routine conditions a few days apart and we might have expected some stability in the properties of D. On the contrary we reached opposite conclusions on various lots belonging to the same shipment :

Lot 519 : N = 5 : conclusion D > 0
         N = 6 : conclusion D > 0.5 kg
         N = 7 : conclusion D > 1.0 kg

There is a systematic difference and it is certainly larger than 1 kg. The average D_{24}, which remains the best available estimate for D is 1.54 kg.

Lot 520 (see above) : N = 10 : conclusion D > 0
                      N = 22 : conclusion D > 0.5 kg

There is a systematic difference and it is larger than 0.5 kg. The best estimate of D is D_{24} = 1.00 kg.

Lot 521 : N = 9 : conclusion |D| < 0.5 kg
         N = 10 : no significant conclusion
         N = 11 : no significant conclusion
         N = 12 : conclusion |D| < 0.5 kg confirmed
         N = 13 : no significant conclusion
         N = 14 : no significant conclusion
         N = 15 through 24 : conclusion |D| < 0.5 kg definitely confirmed.

If there is a systematic difference it is smaller than 0.5 kg. The same test carried out with D_A = 0.25 kg remains inconclusive. We are not entitled to conclude that D is smaller than 0.25 kg. The best estimate of D is D_{24} = 0.08 kg

Lot 522 : N = 15 : conclusion |D| < 1.0 kg
         N = 24 : conclusion |D| < 0.5 kg practically certain (probability 98 %)

If there is a systematic difference it is smaller than 0.5 kg. D_{24} = 0.00 kg.

Practical consequences : A detailed inspection showed that Y's scales needed maintenance. They were repaired accordingly and the discrepancies disappeared.

31.8. AVERAGE NUMBER N_o OF TRIALS NECESSARY TO SHOW UP A SYSTEMATIC DIFFERENCE D or D'

The quantities W_N and W_N' are random variables with means :

$$m(W_N) = \frac{D\sqrt{N}}{\sigma t_\nu} \quad \text{and} \quad m(W_N') = \frac{D'\sqrt{N}}{\sigma t_\nu}$$

As an average, the broken lines joining the points W_N or W_N' will cross the borderline of the area of uncertainty when :

$$\frac{t_\nu^2}{N_o} = \frac{D^2}{\sigma^2} \quad \text{and} \quad \frac{t_\nu^2}{N_o} = \frac{D'^2}{\sigma^2}$$

The relevant factor is therefore the ratio D/σ or D'/σ. Table 31.2 gives the value of the average number of trials N_o required to show up a certain systematic difference D or D' when the ratio D/σ or D'/σ ranges from 4.0 to 0.1.

Table 31.2. Average number N_o of trials necessary to show up a systematic difference D or D', when the standard deviation of the difference d_n is σ.

$\dfrac{D}{\sigma}$	N_o
4.0	3
3.0	4
2.0	5
1.0	9
0.75	13
0.50	25
0.40	40
0.30	64
0.25	90
0.20	135
0.15	240
0.10	540

This table shows that as soon as the ratio falls below 0.5, a large number of trials is necessary to show it up. For this reason, whenever devising an experiment of this type involving measurements (weighings, etc..), estimations (assays, etc..) or samplings, it is of the utmost importance to precise the experimental procedure in the greatest detail and to implement it with the greatest care, so as to reduce the standard deviation σ to a minimum.

CHAPTER 32

TESTING THE AGREEMENT BETWEEN AN ESTIMATE AND THE TRUE VALUE
CHECK OF SAMPLING BIAS

32.1. INTRODUCTION - NOTATIONS

From a statistical standpoint, this is but a particular case of the general problem dealt with in chapter 31. We intend to develop in the present chapter the very important though ill-known subject of the experimental check of a sampling bias. Whereas chapter 31 was mainly concerned with the statistical aspects of the problem, this chapter will emphasize the practical side of the operation. According to the notations retained in the preceding chapter, we shall define :

L_n : the n th lot or sub-lot of a series submitted to the test (n = 1, 2, .. N),
S_n : the sample extracted from L_n by the method or device to be tested,
X : the critical characteristic to be estimated. It can be the percentage of a given element or mineral, of a given size fraction, etc..
x_n : the true value of X in L_n. Contrary to the case dealt with in chapter 31, x_n is assumed to be known, at least within the limits of a certain confidence interval.
y_n : the estimate of x_n obtained by means of the device to be tested,
d_n : the algebraical difference :

$$d_n = y_n - x_n$$

D : the true unknown mean of the distribution of d_n. D is the absolute sampling bias we want to detect.
σ^2 : the true unknown variance of the distribution of d_n. It is the sum of the sampling and analysis variances. D and σ^2 are assumed to be stationary, at least during the period covered by the test.
D_N : the estimate of D obtained after a series of N trials :

$$D_N = \frac{1}{N} \sum_n d_n \quad \text{with } n = 1, 2, .. N.$$

s_N^2 : the unbiased estimate of σ^2 obtained after a series of N trials :

$$s_N^2 = \frac{1}{N-1} \sum_n (d_n - D_N)^2$$

D_A : a standard of accuracy : the sampling is regarded as accurate if and only if we reach a practical certainty that : $|D| \leq D_A$

From a theoretical standpoint, the problem of ascertaining whether $|D|$ is smaller or larger than D_A has been dealt with in chapter 31. We shall simply observe that :

When X is a metal or mineral content, we check the <u>accuracy</u> of the sampling method or device with respect to this critical content.

When X is the proportion of one of the size fractions, we check the accuracy of the method or device with respect to its size analysis which amounts to checking its <u>correctness</u> or uniformity of the selection probability with respect to the size fractions. This is a consequence of the observation made in chapters 17 and 18 that the main sources of sampling bias were the result of a non-uniform selecting probability of the various size fractions and more specifically of the coarser or finer fractions. A sampler which is not accurate with respect to the size analysis is necessarily incorrect.

This remark is important from a practical standpoint, since it is always easier to show up a lack of accuracy with respect to the size analysis than with respect to the mineralogical composition : the latter is only a random consequence of the former. The proof of accuracy towards the size analysis is a proof of correctness which is an intrinsic property of the sampling device (for a material of given maximum particle size). Such a proof is more convincing than the proof of accuracy towards a given metal or mineral content which may result from favourable circumstances such as chance compensation of two opposite sources of incorrectness.

The problem we are going to examine in the next sections is that of ascertaining the true value x_n of X in L_n or at least an unbiased estimate z_n of x_n. We shall describe three different methods in sections 32.2. to 32.4.

32.2. METHOD No.1 - ABSOLUTE METHOD INVOLVING A SYNTHETIC LOT

This absolute method is applicable to any sampling method or device. It uses a synthetic material made up by mixing known proportions of pure components. The synthetic material should have approximately the same size analysis and definitely the same maximum particle diameter as the material to be sampled in routine conditions by means of the device being tested. These pure components can be minerals (when testing the accuracy towards the mineralogical composition) or size fractions (when testing the correctness). When using "pure" minerals, their composition must however be checked by careful assaying, assuming that their constitution heterogeneity is small enough to result in a negligible sampling error. When using "pure" size fractions, we recommend to build up a discontinuous size analysis by reunion of non-adjacent size fractions covering the whole size range to be tested and for instance : 100-50 mm, 20-10 mm, 4-2 mm, 0.8-0 mm (all sieve openings mentioned here belong to the French AFNOR standard series based on the following basic figures :

1.00 - 1.25 - 1.60 - 2.00 - 2.50 - 3.15 - 4.00 - 5.00 - 6.30 - 8.00 - 10.00 - etc...

This allows for the well known lack of reproducibility of the screening operations and reduces the errors committed in the estimation of x_n and y_n to practically zero. The fraction 20-10 mm, for instance, has passed a 20 mm screen and is unlikely to contain elongated fragments coarser than 30 mm. On the other hand, the fraction 100-50 mm has been retained on a 50 mm screen. It may contain a few fragments finer than 50 mm but certainly not finer than say 40 mm. By screening the samples on a 31.5 mm screen (mid-way between 50 and 20 mm) we are certain to carry out a very clean work and to achieve a very high degree of reproducibility. The + 31.5 mm will be attributed to the 100-50 mm fraction, the + 6.3 mm to the 20-10 mm fraction, the + 1.25 mm to the 4-2 mm fraction and the - 1.25 mm to the 0.8-0 mm fraction.

When carefully carried out, this method is very accurate. Special care will be taken in order to prevent preparation errors and more specifically breakage of fragments. The synthetic lot must be carefully mixed in order to reduce the grouping and segregation error.

According to our experience, this is the only efficient and foolproof method for testing the sampling accuracy and correctness, because it is the only method involving the true value of x_n (or the best possible estimate of a "true" value which is never exactly known). It is a variant of this method, involving a single size fraction that has been implemented in the experiment related in section 18.5.

In order to check an eventual breakage of fragments, the sampling reject can be screened on the same sieves as the sample and the size analysis of the lot can be reconstituted by computation. If the size analysis has slipped downwards, the reconstituted size analysis will be substituted for the original one.

32.3. METHOD No.2 - RELATIVE METHOD INVOLVING A REFERENCE METHOD OR DEVICE

This method amounts to testing the difference between two estimates of a same unknown quantity x_n, the reference estimate z_n being assumed, rightly or wrongly, to be unbiased. This method is sometimes used for checking cross-stream samplers operating at the discharge of a belt conveyor. In this case, it consists in :

1) taking an increment I_n by means of the mechanical sampler to be tested,
2) stopping the belt immediately after the increment I_n has been taken,
3) taking a reference increment J_n of about the same weight on the stopped belt, usually by means of a couple of steel plates maintained parallel by a frame and driven through the material at right angle with the belt axis and surface, as near the head pulley as possible,
4) preparing, analysing or assaying the couple of increments I_n and J_n by means of the same procedure carried out by the same operator, in order to reduce the standard deviation s_n of the difference d_n.
5) recording the values of y_n (I_n) and z_n (J_n) and carrying out the test described in chapter 31, proceeding with the trials until a practically certain conclusion

can be attained at or, whenever more convenient, carrying out series of 5 or 10 trials at one time.

An example of this method is presented in section 32.5.

When checking a cross-stream pulp sampler, the bias usually results from an extraction error (chapter 18) due to too narrow a cutter. The reference increment J_n can be obtained by means of a cutter wide enough to be presumed correct and therefore unbiased, joined side by side to the cutter to be tested. This method has been implemented in the experiments reported in section 18.4.5.1.

Due to the fact that this method is based on a comparison with a reference method or device, it is valid only inasmuch as the latter has been proved to be unbiased. We have seen a correct sampler provisionally discarded after a test involving a flap sampler as reference device. The flap sampler being definitely incorrect as pointed out in chapter 17, the conclusion was obviously wrong. This shows that the "judgment errors" should be included among the sampling errors.

32.4. METHOD No. 3 - RELATIVE METHOD INVOLVING A COMPARISON BETWEEN SAMPLE AND SAMPLING REJECT

In this method, the reference increment J_n is nothing else than the sampling reject left after the extraction of the increment I_n. This method is recommendable for instance with splitting methods or devices. It has been implemented for instance in various experiments carried out on riffle splitters and reported in chapter 24. It is applicable also to the testing of cross-stream samplers, either as a variant of the absolute method described in section 32.2. or according to the following procedure. Its implementation requires a minimum of 15 to 20 meters of accessible belt just up-stream of the sampler to be tested, which is not always very easy to achieve.

32.4.1. Variant No.1 :

1) Load 4 to 6 meters near the lower accessible end of the stopped conveyor belt feeding the sampler, as uniformly as possible. This is the lot L_n.

2) Start the belt,

3) At a time carefully computed (or determined by trial and error) start the sampler and check that the sampler is completely out of the stream both at the beginning and at the end of the flow.

4) Stop the belt as soon as it is empty,

5) Recover the increment I_n and the reject, used as reference increment J_n.

6) Proceed as recommended in section 32.3., points 4) and 5).

32.4.2. Variant No.2 :

1) Stop the loaded belt and its feeder : the lot L_n is made of the material present on the belt. This belt can be much shorter than in the first variant.

2) Start the belt, the feeder remaining stopped.

3) Proceed as above, the sample being taken about the middle of the load.

4) Recover the increment I_n and the sampling reject J_n and follow the procedure described in section 32.3. points 4) and 5).

32.4.3. Interest of this method :

It consists in the amplifying effect observed especially when the sampling ratio is high. With a riffle splitter, for instance (ratio 1/2), if the critical content of L_n is x_n = 4.0 % and that of the sample I_n is y_n = 4.2 %, the content of the sampling reject J_n will be z_n = 3.8 %, since I_n and J_n have practically equal weights. If we compare y_n to x_n, we have to show up a systematic difference d_n = 0.2 %. If we compare y_n to z_n, we have to show up a systematic difference d'_n = 0.4 %, which is much easier and requires a smaller number of trials.

32.5. EXAMPLE OF APPLICATION OF METHOD No.2

This example is worth mentioning because it shows the danger of certain Standards and the limitations of the experimental method when the statistical analysis is not conducted as recommended in chapter 31.

A certain lumpy iron ore was sampled at a transfer point upon loading aboard sea-going vessels carrying up to 150,000 metric tons of ore at a rate of 4,000 to 5,000 tons/hour.

32.5.1. Critical inspection of the sampler :

The primary sampler was of the swing-arm type and is schematized in fig. 32.1.

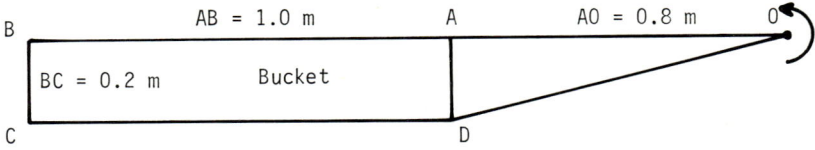

Linear velocity : 0.93 m/s in A and 2.1 m/s in B

Fig. 32.1. *Swing-arm type sampler object of the test.*

A critical inspection of the sampler showed that :

1) Though of the circular-path type, the cutter edges were parallel instead of being radial as they should have been (chapter 17 - delimitation error). The sampling ratio is 2.3 times larger in A than in B.

2) Though it was used to sample lumpy ores up to 120 mm, the cutter width was only 200 mm, or 1.67 times the diameter of the coarsest particles instead of three times (Chapter 18 - First extraction error).

3) The cutter velocity was 0.93 m/s at point A and 2.1 m/s at point B, i.e. more than three times the critical speed of 0.6 m/s (Chapter 18 - Second extraction error).

4) The capacity of the bucket was insufficient. As soon as the rate of flow exceeds 4 500 t/h, the bucket overflows (Chapter 18 - Third extraction error).

5) When operating during the dry season, the excessive cutter speed generates a dust cloud (Chapter 17 - Second delimitation error and chapter 27 - Preparation error).

32.5.2. Experimental bias testing :

As the producer of this iron ore had observed important discrepancies between his own iron assays (obtained on samples taken by this device) and the buyer's, he first asked for the advices of a firm of professional samplers justly held in good repute and this implemented a bias test recommended in the draft of an ISO standard. This test is based on the method described in section 32.3 and involves a comparison with a reference sample taken on the stopped belt. The test was carried out on a 35 mm rubble ore, much finer than the coarsest ore to be sampled, which was an error : the test should have been run in the most critical conditions i.e. with a 120 mm ore. According to the procedure described in the draft standard the number N of trials was limited to 10 : second error, the number N should not be fixed in advance. Anyway, it should be much larger than 10 as shown in chapter 31. According to the draft standard, the device was to be regarded as unbiased if a SF test performed with a risk α = 2.5 % (or 2α = 5%) did not show up a significan difference : third error, 5% is too large a risk, fourth error and probably the most dangerous of all errors reviewed so far, as already pointed out in chapter 31, the fact that the test does not disclose a significant difference may not be interpreted as an absence of bias. The standard confuses "absence of proof of bias" with "proof of absence of bias", which is a very different proposition.

1) <u>Bias test on iron assays</u> : we shall use this example to show the reader a new way of interpreting the results of a SF test. The confidence interval of the average difference D_N can be expressed as follows :

$$D_N - \frac{t_\nu s_N}{\sqrt{N}} < D < D_N + \frac{t_\nu s_N}{\sqrt{N}}$$

(a) When $|W_N| > 1$: D is significantly different from zero. Its sign is that of D_N and its minimum absolute value is :

$$|D|_{MIN} = |D_N| - \frac{t_\nu s_N}{\sqrt{N}}$$

(b) When $|W_N| < 1$: D is not significantly different from zero. We can however estimate its maximum absolute value which is :

$$|D|_{MAX} = |D_N| + \frac{t_\nu \, s_N}{\sqrt{N}}$$

The results of the present test are given in table 32.1. When $|W_N| < 1$, the table gives the value of $|D|_{MAX}$. When $|W_N| > 1$, it gives the value of $|D|_{MIN}$.

Table 32.1. Bias testing of an iron ore sampler - Method No. 2 - Fe %.

n N	y_n Fe %	z_n Fe %	d_n Fe %	D_N Fe %	s_N Fe %	W_N	$|D|_{MIN}$ Fe %	$|D|_{MAX}$ Fe %
1	62.79	63.97	-1.18	-1.18	-	-		-
2	62.76	63.38	-0.62	-0.90	0.40	-0.10		9.81
3	62.56	62.73	-0.17	-0.66	0.51	-0.32		2.69
4	61.06	61.34	-0.28	-0.56	0.45	-0.55		1.59
5	61.17	61.93	-0.76	-0.60	0.40	-0.90		1.28
6	62.04	62.40	-0.36	-0.56	0.37	-1.10	0.048	
7	61.93	61.47	+0.46	-0.42	0.52	-0.68		1.03
8	62.93	62.03	+0.92	-0.25	0.67	-0.35		0.96
9	61.98	62.27	-0.29	-0.25	0.63	-0.41		0.86
10	61.79	61.70	+0.09	-0.22	0.60	-0.41		0.76

The analysis of the results given in this table shows that for N = 6 we are entitled to conclude with a 99 % probability to be right that D is non-zero, negative (the sampler under-values the iron content) and that the absolute value of this bias is larger than 0.048 (say 0.05) % Fe. But this conclusion is not confirmed by the following trials and the test remains inconclusive.

The only conclusion that may be drawn after the 10th trial is that, if the sampler introduces a bias, this is likely to be smaller than 0.76 % Fe in absolute value and negative (sign of D_{10}). There remains however a 1 % probability for the bias to be larger than this value.

The table very well shows how the confidence interval of the bias D narrows when the number N of trials increases. A larger number of trials would have led to a more precise conclusion.

Anyway, the rightful conclusion we may draw at the end of the test is much less encouraging than the erroneous conclusion resulting from the strict application of the standard recommendation, namely that the sampler is unbiased and may be utilized in routine operation.

Fortunately, all increments had been carefully screened before reduction and analysis, which makes it possible to perform the same test for the size analysis.

2) <u>Bias test on the percentage of - 5 mm</u> : This test is very important for, according to the contract, the percentage of - 5 mm is a critical characteristic liable to be penalized when too high. We therefore carried out the same test as in the preceding section with :

y_n : percentage of - 5 mm in the n th increment I_n taken by the sampler,
z_n : percentage of - 5 mm in the n th reference increment J_n (stopped belt).
The results of this test are given in table 32.2.

Table 32.2. *Bias testing of an iron ore sampler - Percentage of - 5 mm.*

n N	y_n %	z_n %	d_n %	D_N %	s_N %	W_N	$\|D\|_{MIN}$ %	$\|D\|_{MAX}$ %
1	12.7	11.5	+1.2	-	-	-	-	-
2	11.0	12.9	-1.9	-0.35	2.19	-0.01		49.67
3	15.0	13.0	+2.0	+0.43	2.06	+0.05		8.72
4	5.7	5.9	-0.2	+0.28	1.71	+0.07		4.16
5	10.3	5.9	+4.4	+1.10	2.37	+0.28		5.07
6	9.4	7.0	+2.4	+1.32	2.18	+0.44		4.31
7	11.4	9.7	+1.7	+1.37	2.00	+0.58		3.74
8	13.6	9.2	+4.4	+1.75	2.14	+0.77		4.01
9	12.7	9.8	+2.9	+1.88	2.04	+0.96		3.84
10	10.2	11.1	-0.9	+1.60	2.11	+0.85		3.48

In the whole range of N, $|W_N|$ remains smaller than 1. After the 9th trial, however, $W_9 = 0.96$ and we reach a 98 % probability conclusion that D is non-zero and positive. The sample contains more fines - and less coarse fragments - than the reference sample, which confirms that the selection probability of the coarse fragments is significantly reduced. At the end of the test, we may conclude that if there is a bias, its absolute value is smaller than 3.48 %, i.e. 36 % relative.

32.5.3. Conclusions of the test (see also figures 32.2. and 32.3.) :

1) <u>According to ISO standard</u> the procedure of which was strictly followed by the operators : the sampler is unbiased and may be used for routine operation.

2) <u>According to our own complete statistical analysis</u> of the results obtained in the same test : the number N = 10 of trials is insufficient to show up any bias either on the Fe content or on the percentage of - 5 mm. The complementary tests show that the only conclusions we are entitled to draw are :

- If there is a bias on the Fe-content, it is likely to be smaller than 0.76 % Fe or 1.3 % relative.

- If there is a bias on the size distribution and more specifically on the percentage of - 5 mm, it is likely to be smaller than 3.48 % or 36 % relative.

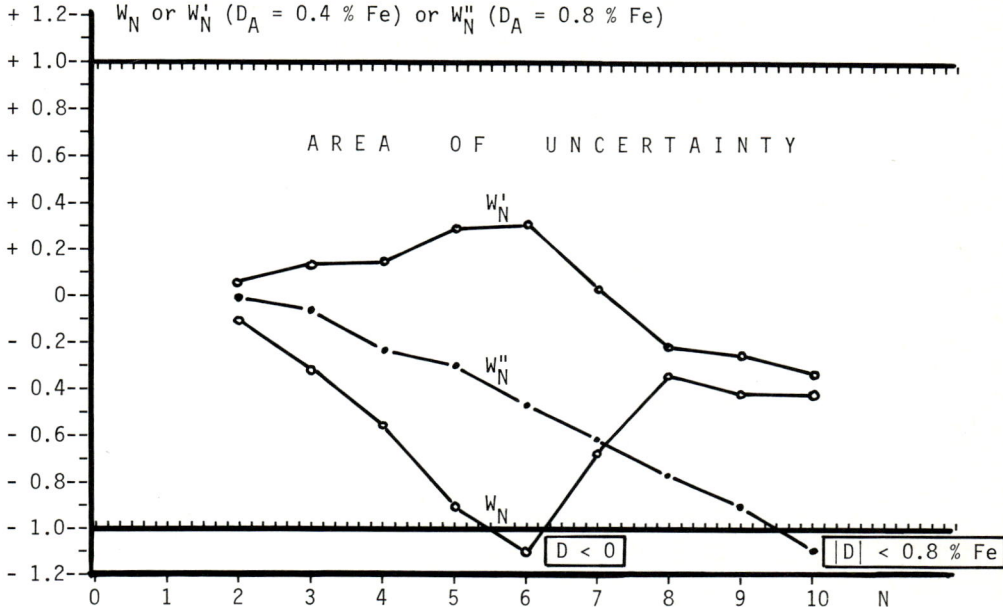

Fig. 32.2. Bias testing of an iron ore sampler - Graphical representation Fe %.

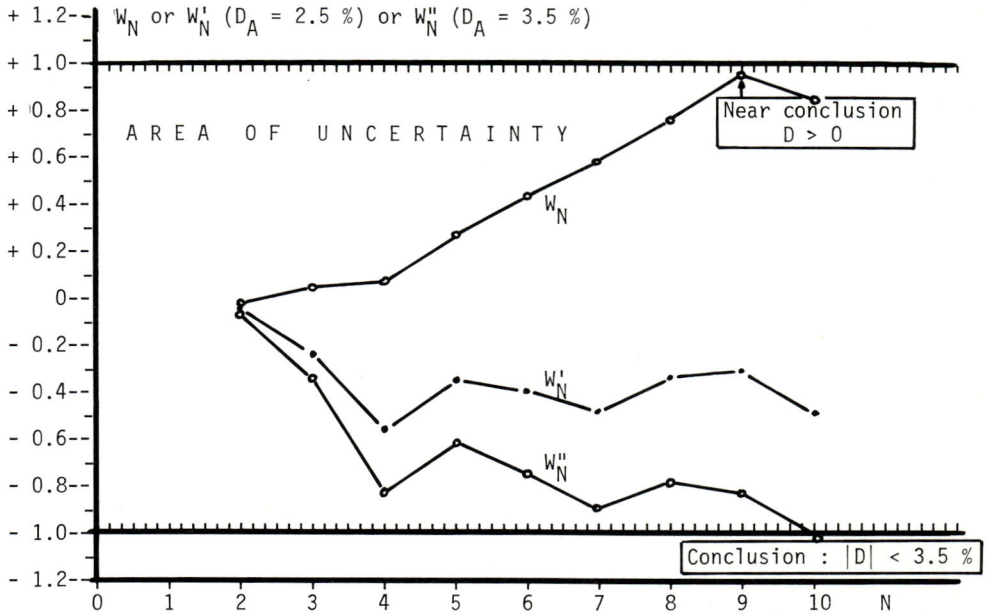

Fig. 32.3. Bias testing of an iron ore sampler - Graphical representation - Percentage of - 5 mm.

3) According to our critical inspection : the sampler introduces at least five sources of bias and is likely to sample the coarse fragments with a selection probability much smaller than that of the fines. The sampler is highly incorrect and should never be used though it was built by a world-famous manufacturer.

This example is typical : a careful critical examination which took less than one hour led us to the above conclusions, it pointed out the five sources of incorrectness and enumerated the transformations to carry out in order to suppress all potential sources of bias. On the contrary, an experimental test conducted with great care and efficiency by a team of excellent professional samplers following a standardized procedure led to the wrong conclusion that the sampler was clean and could be utilized for routine operation : the only mistake of these professional samplers had been to follow the recommendations of a standard.

Now, when carrying out a complete statistical analysis of the test results we reach different if not opposite conclusions : it is one thing to say that a sampler is harmless and delivers unbiased samples and another to conclude as we did that if the sampler introduces a bias on the Fe-content, this bias is likely to be smaller than 0.8 % Fe (1.3 % relative). On the percentage of - 5 mm the bias is likely to be smaller than 3.5 % (36 % relative). We are not entitled to conclude for instance that the bias on the iron content is smaller than 0.6 % Fe nor that the bias on the percentage of - 5 mm is smaller than 3 %. These conclusions are certainly more disquieting than the conclusions drawn in conformity with the standard that the sampler was foolproof.

This bias test took the best part of a couple of days during which the loading of the vessel was seriously disturbed and delayed, to the great displeasure of the ship's Captain. The preparation of the 20 increments took another two days and the whole operation cost the Mining Company a lot of money for practically nothing, except a wrong conclusion highly detrimental to the Company's interests. Our inspection which took place a few months later lasted one day including the drawing up of our report. It did not disturb the loading operations and required no labour. Our report pointed out the defects of the sampler and described how to correct them Including fee, air fare (the sampler was some 8 000 km from Europe) and accomodation it cost about five times less than the bias test.

32.6. ACCURACY vs. REPRODUCIBILITY - EXAMPLE OF APPLICATION OF METHOD No. 3

So many people muddle up the notions of accuracy and reproducibility (already defined in section 1.5.) that it is not useless to recall their definitions :

1) Accuracy is a property of the mean D of the distribution of d_n. A sampling system is said to be accurate if and only if $|D| \leq D_A$ (with D_A tolerated bias).

2) Reproducibility is a property of the variance σ^2 of the distribution of d_n. A sampling system is said to be reproducible if and only if $\sigma^2 \leq \sigma_0^2$ (with σ_0^2 tolerated variance).

3) A test involving repeated measurements of the sole variable y_n can provide (under certain conditions) an estimate of σ^2 but no information concerning the value of the bias D. This will be illustrated by an example :

A potential buyer of a certain sampler, advised by the manufacturer, conducted a test meant for checking an eventual bias and following the procedure we are going to describe now.

A 260-kg lot of - 5 mm iron ore was fed to the sampler achieving a 1 % sampling ratio. This first operation generated a sample S_1 and a reject R_1. The latter was fed again to the same sampler, which produced a sample S_2 and a reject R_2. This process was repeated seven times until obtention of the sample S_7 and the reject R_7.

At this stage our analytical laboratory was asked to prepare and assay the samples S_1, S_2, ... S_7 and we obtained the following results (each content is the mean of two assays) :

S_1 = 51.450 % Fe
S_2 = 51.435 % Fe
S_3 = 51.485 % Fe
S_4 = 51.470 % Fe
S_5 = 51.485 % Fe
S_6 = 51.550 % Fe
S_7 = 51.460 % Fe

Mean = 51.476 % Fe
Standard deviation : 0.037 % Fe or 0.72×10^{-3} relative
Variance : 0.0014 or 0.52×10^{-6} relative

Everybody was satisfied until we asked to prepare and assay the last sampling reject R_7 which fortunately had not yet been thrown away. We obtained :

R_7 = 50.855 % Fe

Knowing the weights of the seven samples and of the seventh and last sampling reject, we were able to calculate the Fe contents of R_6, R_5, ... R_1 and that of the original lot L. By calling y_n the contents of the sample S_n and z_n those of the rejects R_n we were in a position to implement the method No. 3 described in section 32.4. The results of this test are presented in table 32.3 (next page). We shall test only two hypotheses :

$H \equiv \{D = 0\}$: Computation of W_N
$H' \equiv \{D' = |D| - 0.45 \% \text{ Fe} = 0\}$: Computation of W'_N

and compare W_N and $W_N^!$ with their interval of uncertainty ± 1.

As early as the second trial we know that the sampler is biased and after the fifth trial we reach a practical certainty that this bias is larger than 0.45 % Fe. The best estimate we have of this bias is D_7 = + 0.567 % Fe or 1.11 % relative.

Table 32.3. Example of application of method No. 3 - Sampling of an iron ore

n	y_n	z_n	d_n	D_N	s_N	W_N	$W_N^!$	$\|D\|_{MIN}$
1	51.450	50.961	+ 0.489	+ 0.489	-	-	-	-
2	51.435	50.946	+ 0.489	+ 0.489	0.000	∞	∞	+ 0.489
3	51.485	50.926	+ 0.559	+ 0.512	0.040	+ 3.18	+ 0.39	+ 0.348
4	51.470	50.913	+ 0.557	+ 0.524	0.040	+ 5.77	+ 0.81	+ 0.432
5	51.485	50.895	+ 0.590	+ 0.537	0.046	+ 6.95	+ 1.13	+ 0.460
6	51.550	50.873	+ 0.677	+ 0.560	0.070	+ 5.84	+ 1.15	+ 0.463
7	51.460	50.855	+ 0.605	+ 0.567	0.066	+ 7.22	+ 1.49	+ 0.487

Thanks to good advertising and using the first part (but the first part only) of the test, forgetting the sampling reject and the obvious bias, keeping up the confusion between accuracy and reproducibility, this manufacturer was able to sell hundreds of devices of this type before the users found out that their samples were heavily biased. A few of these are still in operation 25 years after the test.

32.7. CRITICAL INSPECTION vs. EXPERIMENTAL TESTING OF ACCURACY

The problem of the bias test is a false one. Bias testing was justified when the rules and the notion itself of sampling correctness were unknown. Manufacturers and utilizers of sampling equipment were blind people trying to progress in the dark. Standards organizations had to devise some safeguard against the worst, a good illustration of the blind leading the blind, and they proposed the bias test.

But in the mean time, sampling has become a science, even though some people, rooted in the past, are reluctant to admit the fact. Times have changed but mentalities have not yet adapted to the present situation. The following considerations should induce all interested parties : users, manufacturers, engineering companies, standards organizations, inspection companies, etc.. to reconsider their position :

1) The sampling theory developed in this book shows that the best we can do is to give all fragments of the lot to be sampled a uniform probability of being selected. In other words the best we can do is to perform a correct sampling.

2) Even in this case, sampling is usually biased (section 20.5.). This bias results from the properties of the material being sampled, not from those of the sampling operation. No mechanical implement will ever suppress it and no bias test, however carefully it may be conducted, will ever be capable of detecting it. The question is therefore not to check whether the sampling is biased - we know it is - but to estimate to what extent it is biased.

3) Theoretical considerations, backed up by experimental estimations, show however that when the sampling is correct, the resulting "fundamental bias" is always negligible. To all intents and purposes, correct sampling may therefore be regarded as always accurate and we need no experimental proof of it, no more than we need to walk (or fly) around the earth to check that it is not flat as some people believe.

4) The rules of sampling correctness have been exhaustively reviewed in chapters 16-18. When these rules are not strictly respected, a correlation is generated between the selection probability of a fragment and its "personality" (sections 17.8 and 18.8) which in turn generates an "incorrectness bias". Our theoretical analysis (chapter 20) and the examples of chapter 23 show that, contrary to the fundamental bias, it is never safe to assume that the incorrectness bias is negligible.

5) Another conclusion of chapters 20 and 23 is that the delimitation and extraction errors are characterized not only by a non-negligible bias D but also by an important standard deviation σ, which entails two detrimental consequences :

a) the delimitation and extraction errors affect the reproducibility as well as the accuracy of sampling.

b) the ratio D/σ is likely to be small, even with an unacceptable bias, which makes it all the more difficult to disclose it by experimental means, which is well illustrated by table 31.2., page 390.

6) A critical inspection of a sampling system detects all deviations from delimitation and extraction correctness.

a) Either the sampling system is correct and we are satisfied that it will provide accurate samples as long as it is correctly operated and maintained.

b) Or it is not, in which case our inspection detects the deviations from correctness and recommends what should be done in order to fix the sampler.

7) Now, assuming it is conducted according to the recommendations of chapters 31 and 32, and pursued until a practically certain conclusion has been arrived at, what kind of conclusion can we come to at the end of a bias test ?

a) Either the test shows there is a 99 % probability that if there is any bias, this is acceptable. A careful analysis of the problem shows that such a situation can in fact result from two opposite causes :

- the sampler is correct and genuinely delivers unbiased samples. We know it will always do so.

- the sampler is incorrect but through an intricate network of correlations (see sections 17.8 and 18.8) between the selection probabilities and the personalities of the fragments making up the lot the resulting bias is negligible. But what sort of guarantee do we have that it will always remain so ? On the contrary, experience has repeatedly shown that one of the most disconcerting properties of biases is to shift from one day to the next, following the properties of the material to be sampled with which they are closely correlated. They may very well escape our test, reappear to-morrow and vanish again. One thing is certain : we have no right to generalize our conclusion that the sampler is unbiased.

b) Or the test shows there is a 99 % probability that there is a bias and that this bias is definitely unacceptable. But the test does not say why the sampler is biased nor how we could fix it. The only thing we can do in order to improve things is to carry out a critical inspection of the sampler. Now, on what basis are we going to establish our recommendations ? What criterion are we going to use if not the respect of the rules of sampling correctness ?

8) Our conclusion follows at once from the preceding paragraphs : the bias test is useless for it never results in a clear, indisputable conclusion. It is not only useless it is dangerous since it is too often ill-conducted. In any case, the last word belongs to the critical inspection of the sampling system and to a check that the rules of sampling correctness are respected. And when we know they are, we need no test to be satisfied that the sampler is unbiased.

NINTH PART

AUTOMATIC SAMPLING PLANTS

Whether for technical or for commercial sampling, it is definitely cheaper and more reliable to carry out all sampling operations, from the primary sampling stage of streams ranging up to 8,000 tons per hour down to the reception of the finely pulverized laboratory sample in a small Mason jar, in a completely mechanized and automated sampling plant than to implement manual methods such as those described in the textbooks of the first half of this century.

Our purpose, in this ninth part, is not to describe machinery and equipment but to help designers of automatic sampling plants to solve the theoretical part of their problems.

This ninth and last part of our book is made of two chapters :

Chapter 33 : Design of automatic sampling plants
Chapter 34 : Typical flow-sheets of automatic sampling plants.

CHAPTER 33

DESIGN OF AUTOMATIC SAMPLING PLANTS

33.1. INTRODUCTION

A "sampling plant" is a plant achieving automatically :

1) the primary sampling stage of a flowing stream of particulate material,
2) the reduction of this primary sample to the bulk and particle size of the required laboratory sample.

The word "plant" suggests the handling of large tonnages of coarse materials but we shall use it with the general meaning of a sequence of sampling, comminution and other preparation devices working automatically without human intervention, irrespective of the tonnage handled.

When designing such a plant, one usually aims at obtaining a representative sample or in other words a sample which is at the same time accurate and reproducible (definition of these words in section 1.5.). We shall review in turns the conditions of achievement of sampling accuracy and reproducibility.

33.2. ACHIEVEMENT OF SAMPLING ACCURACY

This amounts to eliminating all possible sources of selection and preparation bias. We shall assume that the designer has well assimilated the recapitulative chapter 28, that he has carefully read chapters 17 and 18 in order to remember how to eliminate the always treacherous delimitation and extraction biases and chapter 27 recalling the various sources of preparation bias. If he respects all the conditions stated in these chapters and if he is conscious of the risks incurred if he does not (the re-reading of chapter 23 may prove instructive), if furthermore he does not forget what has been said in chapter 24 of the possible splitting biases and if he succeeds in avoiding the pitfalls of commercial sampling detailed in chapter 30, then, his sampling plant has every chance to deliver accurate samples.

We would like to remind the reader that there is no possible compromise with the rules of sampling correctness. The only efficient strategy with sampling and preparation biases is one of preventive elimination AT THE DESIGNING STAGE. Later, when the plant is operating it is usually too late to carry out a good work. The designer knows better than anybody that transformations are costly and unwelcome.

When selecting sampling equipment, especially for the primary sampling stage where the largest errors are likely to occur, the designer should be very critical. There are on the market a good 50 % of primary sampling devices that do not respect the rules of delimitation and extraction correctness stated in chapters 17 and 18. If he objects to these defects, he will probably be answered by the manufacturer that such devices have been working satisfactorily for quite a number of years without any complaints. He will perhaps be tempted to allow himself to be convinced : he should not. There are, all over the world, thousands of improper samplers working "satisfactorily" for the sole reason that nobody has ever cared to check whether or no they were correct and delivered unbiased samples. The fact that a manufacturer claims that there have never been any complaints may be true (or not) it is definitely no proof that his equipment is correct.

33.3. ACHIEVEMENT OF SAMPLING REPRODUCIBILITY

Chapter 28 has recapitulated the properties of all sampling and preparation errors. If the designer has carefully followed the pieces of advice dispensed in the preceding section, he has already disposed of three major sources of error :

- Increment delimitation error   DE $\equiv$ 0
- Increment extraction error   EE $\equiv$ 0
- Preparation errors   PE $\equiv$ 0

and he can concentrate his attention on the remaining errors :

- Fundamental error FE
- Grouping and segregation error GE
- Long-range quality fluctuation error QE_2
- Periodic quality fluctuation error QE_3
- Weighting error WE.

The practical implementation of the results of our theory (chapter 15) is based on the results of a variographic experiment (chapter 14) which are practically never available at the designing stage of a sampling plant. The designer must therefore rely on his own experience or on that of a competent adviser and reason by analogy.

He will first decide to regulate the rate-of-flow prior to all sampling stages, thus eliminating the best part of the weighting error WE, remembering however that he will not be able to cancel it out completely, especially at the primary sampling stage where an efficient regulation is not always feasible.

Then, in order to reduce the errors GE, QE_2 and QE_3 that he cannot estimate for want of the results of a variographic experiment, he will :

1) Primary sampling : opt for a random stratified selection scheme which is the only safeguard against large periodic quality fluctuation errors QE_3, remembering

that there are on the market patented time switches delivering electrical impulses at random stratified intervals (US patent, Gy, 1968). Then, he will select a uniform strata extent of no more than 5 mn and make sure that the number of increments Q making up the sample is larger than or equal to 30. These provisional conditions being satisfied, he will design the primary sample reduction flow-sheet according to the rate-of-flow of primary sample so calculated, knowing that after a variographic experiment has actually been carried out, during the running in of the plant for instance, the user will probably be able to increase the strata length to 10 or even 15 mn (the number Q remaining however larger than 30). Such a flexibility is necessary : if the reduction flow-sheet is designed on the basis of four primary increments per hour (15 mn strata extent) and if the variographic experiment shows that a minimum of 12 increments per hour is necessary to achieve the standard of reproducibility required by the customer, the complete sampling plant will have to be transformed and probably rebuilt. According to our experience, twelve increments per hour with a minimum of 30 increment per sample is always safe.

2) Further sampling stages : one can usually implement a systematic scheme with intervals ranging from 2 mn for the secondary stage to a fraction of one second for the last two sampling stages usually carried out by means of sectorial dividers such as those described in chapter 24. It is always a safe rule to select shorter and shorter intervals. The purpose of selecting short intervals is to reduce :

- Grouping and segregation error GE
- Long-range quality fluctuation error QE_2
- Periodic quality fluctuation error QE_3
- Weighting error WE.

GE is an increasing function of the increment bulk : short intervals ensure a large number of increments and, when aiming at a given sample weight, a small increment weight. QE_2 and QE_3 are by definition increasing functions of the interval between increments and so is WE.

There remains the fundamental error FE and the residual fraction of GE, QE_2, QE_3 and WE. This problem will be dealt with in the next section.

33.4. ALLOTMENT OF THE EXPENDABLE SAMPLING VARIANCE

The notion of expendable sampling variance has been introduced in section 29.2.1. It is the total sampling variance that can be allotted to the various sampling stages. We shall call it σ_o^2. It is a dimensionless relative variance and ranges usually between 10^{-4} and 10^{-6}. Two points must be taken into consideration :

1) We can never completely suppress the variances of GE, QE_2, QE_3 and WE. As a whole, we shall regard them as negligible when their sum is smaller than the fundamental variance $\sigma^2(FE)$. If the recommendations suggested in the preceding sections

are followed, this condition is likely to be fulfilled. Then, in our computations, we shall allot one half of the expendable sampling variance σ_o^2 to these residual variances, leaving the other half for the sum of the fundamental variances of all sampling stages.

2) Then, if the subscripts 1, 2, ... specify the rank of the sampling stages, we must solve the inequality :

$$\sigma^2(FE_T) = \sigma^2(FE_1) + \sigma^2(FE_2) + \ldots \leq \frac{\sigma_o^2}{2}$$

The variance of the ultimate sampling stage which consists in taking the assay portion from the laboratory sample is usually taken into account by the assaying variance and does not have to be considered here.

It is now up to the designer to decide what fraction of the expendable variance is to be allotted to each sampling stage. Reproducibility being more expensive at the primary sampling stage than at any further stage, it is advisable to allot one half of $\sigma^2(FE_T)$, i.e. $\sigma_o^2/4$ to the primary sampling stage. The allotment of the other half is a question of convenience and can be left at the designer's initiative. A rough rule of thumb which is not worse than any other is to allot :

- $\sigma_o^2/2$ to the residual fraction of GE, QE_2, QE_3 and WE (all sampling stages),
- $\sigma_o^2/4$ to the primary fundamental error FE_1
- $\sigma_o^2/8$ to the secondary fundamental error FE_2
- $\sigma_o^2/16$ to the tertiary fundamental error FE_3 etc...

knowing that, irrespective of the number of sampling stages, the sum converges towards σ_o^2. At each sampling stage, the problem is then solved according to the procedure developed in chapter 22.

From a practical standpoint, the whole problem depends upon the available crushing, grinding and pulverizing equipment. There are no general rules : every problem is a new problem and must be solved with due consideration to the characteristics of the material to be comminuted. A few typical examples will be presented in chapter 34.

33.5. SAMPLING LARGE TONNAGES OF COARSE MATERIALS

The most delicate of all sampling problems is definitely the problem set by the sampling of large tonnages of coarse materials. The notion of "minimum increment weight" M_{Io} has been introduced in section 18.7. It is the the minimum increment weight that can be obtained by means of a correct device. It is defined by :

$$M_{Io} = \mu_o \frac{W_o}{V_o} \quad \text{with :}$$

μ_o : average rate-of-flow
W_o : minimum cutter width which, for coarse materials, is equal to three times the

diameter d of the coarsest fragments.

V_o : maximum cutter velocity which, when the actual cutter width W is equal to its minimum W_o, is equal to 0.6 m/s.

The reader must understand that this is the only condition to be respected by the increment weight. Contrary to the specifications of nearly all Standards, there is no such thing as an "<u>absolute</u> minimum increment weight". The minimum increment weight is proportional to the flow-rate. An example will illustrate this point.

A -100 mm bauxite ore is to be sampled at the discharge of a belt conveyor at a rate of 3,000 t/h. Let's first define a consistent system of units : we recommend the CGS system based on centimeter (cm), gram (g) and second (s). We compute :

$$\mu_o = \frac{3,000 \times 10^6}{3,600} = 0.833 \times 10^6 \text{ g/s} \qquad d = 10 \text{ cm}$$

$$W_o = 3 d = 30 \text{ cm} \qquad V_o = 60 \text{ cm/s}$$

$$M_{Io} = \frac{0.833 \times 10^6 \times 30}{60} = 0.416 \times 10^6 \text{ g or 416 kg.}$$

For this particular flow-rate, the minimum increment weight is 400-450 kg. Assuming an interval of 5 mn between increments as recommended in section 33.3., the primary sample represents some 5 t/h. If a sample is collected at the end of every 8-hour shift, the primary sample will weigh about 40 tons. As often as not, at this point of his calculations, the designer will throw this book away, and select "more reasonable" values such as, for instance, $W = 2 d = 20$ cm , $V_C = 200$ cm/s and compute a "much more satisfactory" increment weight :

$$M_I = \frac{0.833 \times 10^6 \times 20}{200} = 83 \times 10^3 \text{ g} \quad \text{or} \quad 83 \text{ kg},$$

which amounts to forgetting the dangers of extraction bias pointed out in chapter 18. We saw such an example in section 32.5 .

NO ! The answer does not consist in reducing W below W_o nor in increasing V_C above V_o, it simply consists in performing a secondary sampling stage without any intermediate comminution. We shall therefore design our primary sampler with due respect to the notion of minimum increment weight, collect 12 increments per hour, receive these in a surge bin and feed them at a constant rate of 5 t/h to a secondary sampler similar to the primary. Like the primary, the secondary sample will be perfectly correct and each secondary increment will weigh :

$$M'_{Io} = \mu'_o \frac{W_o}{V_o} = \frac{5 \times 10^6 \times 30}{3,600 \times 60} = 700 \text{ grams.}$$

At this point of the project, at the original top particle size d, the only relevant factor is the "<u>absolute minimum sample weight</u>" M_{So}. Let's denote by :

M_{S1} and M_{S2} : the primary and secondary sample weights,
C : the sampling constant that in the present case can be estimated at 10^{-2} g/cm3
σ^2 : the sum of the primary and secondary fundamental variances : by definition :

$$\sigma^2 = \left[\frac{1}{M_{S1}} - \frac{1}{M_L}\right] C d^3 + \left[\frac{1}{M_{S2}} - \frac{1}{M_{S1}}\right] C d^3 = \left[\frac{1}{M_{S2}} - \frac{1}{M_L}\right] C d^3 = \frac{C d^3}{M_{S2}}$$

As far as the fundamental variance is concerned, the secondary sample weight alone is relevant. If, according to the rule suggested in section 33.4. we allot a variance $\sigma^2/4$ to the sequence primary + secondary sampling, we can define an "absolute minimum sample weight M_{So} at the original particle size" with :

$$M_{S2} \geq M_{So} = \frac{C d^3}{\sigma_o^2/4}$$

If we accept for instance a relative confidence interval $\pm 2\sigma_o = 10^{-2}$, then $\sigma_o^2 = 25 \times 10^{-6}$ and we readily compute :

$$M_{So} = \frac{4 \times 10^{-2} \times 10^3}{25 \times 10^{-6}} = 1.6 \times 10^6 \text{ g} = 1.6 \text{ tons}$$

From which we can compute a minimum number of increments Q_o :

$$Q_o = \frac{M_{So}}{M_{Io}} = \frac{1.6 \times 10^6}{700} = 2\ 285$$

If these increments must cover an eight-hour shift, we can compute the constant interval T_{sy} or the average interval T_{st} between consecutive increments :

$$T_{sy} \text{ or } T_{st} = \frac{8 \times 3\ 600}{2\ 285} = 12.6 \text{ seconds}$$

Accordingly, the secondary sampler will be designed so as to run continuously, taking an increment every 12 seconds. This is the economically optimum solution but all solutions respecting the conditions of minimum cutter width, maximum cutter speed and minimum sample weight are technically acceptable and for instance for the secondary stage : W' = 60 cm (W_o = 30 cm), V'_C = 30 cm/s (V_o = 60 cm/s) and M'_{S2} = 2 tons (M_{So} = 1.6 t), from which we can compute :

$$M'_I = \frac{5 \times 10^6 \times 60}{3\ 600 \times 30} = 2\ 800 \text{ grams} \qquad Q' = \frac{2 \times 10^6}{2\ 800} = 714$$

$$T'_{sy} \text{ or } T'_{st} = \frac{8 \times 3\ 600}{714} = 40.3 \text{ seconds.}$$

In any case, the secondary sample must imperatively be comminuted prior to the tertiary sampling stage, for instance by means of a jaw crusher reducing the top particle size to about 25-30 mm.

CHAPTER 34

TYPICAL FLOW-SHEETS OF AUTOMATIC SAMPLING PLANTS

34.1. INTRODUCTION

This chapter does not intend to exhaust the subject nor to give a ready-made solution to any sampling problem. The reader should remember that it is always dangerous to copy the neighbour's sampling plant. Our purpose is to present five examples of flow-sheets of sampling plants actually operating on various materials in various parts of the world and collecting correct samples. One at least of these is old and would be designed differently today. All but one concern coarse materials :

Bauxite ore (200 to 250 mm) : section 34.2.
Nickel ore (100 to 150 mm) : section 34.3.
Iron ore (45 to 50 mm) : section 34.4.
Raw mix of a cement plant (40 to 50 mm) : section 34.5.
Uranium concentrates (10 microns) : section 34.6.

We have mentioned between brackets the normal maximum particle size (first figure) and the exceptional particle size (second figure). We shall use the following notations:

d : maximum particle size
μ : rate of flow
W : cutter width
V_C : cutter speed
T_I : duration of an increment : $T_I = W/V_C$
T_S : interval T_{sy} between increments or strata extent T_{st}
T_L : duration of the flow to be sampled
Q : number of increments making up a sample : $Q = T_L/T_S$
M_L : weight of the batch (lot or sample) submitted to the sampling operation
M_I : average increment weight : $M_I = \mu\, T_I$
M_S : average sample weight : $M_S = Q\, M_I$

The data are presented under tabular form, expressed in usual units and whenever necessary converted into CGS units, to be used in the calculations. Some of the data have been rounded off. We shall use only metric tons.

34.2. SAMPLING OF A BAUXITE ORE

Shipments of 1,200 tons of a lumpy bauxite ore are sampled upon unloading at the processing plant. The unloading lasts two hours and two laboratory samples are prepared in the sampling plant : one for moisture estimation, the other for assay.

Table 34.1. Sampling of a bauxite ore.

Sampling stage		Primary	Secondary	Tertiary (moisture)	Tertiary (assay)	Quaternary (assay)
d	(mm)	250	50	10	10	2
	(cm)	25	5	1	1	0.2
μ	(t/h)	600	6	0.125	0.125	–
	(g/s)	170×10^3	1,700	35	35	17
W	(mm)	750	200	25	40	50
	(cm)	75	20	2.5	4	5
V_C	(cm/s)	32	20	20	20	20
T_I	(s)	2.34	1	0.125	0.2	0.25
T_{sy}	(s)	240	48	4.2	4.2	2.4
T_L	(hours)	2	2	2	2	–
	(s)	7,200	7,200	7,200	7,200	720
Q		30	150	1,710	1,710	300
M_L	(tons)	1,200	12	0.25	0.25	–
	(g)	$1,200 \times 10^6$	12×10^6	250×10^3	250×10^3	12×10
M_I	(kg)	400	1.7			–
	(g)	400×10^3	1,700	4.4	7	4
M_S	(kg)	12,000	250	7.5	12	1.2
	(g)	12×10^6	250×10^3	7,500	12×10^3	1,200

The characteristics of the samplers and of the sampling operations are summarized in table 34.1. (overleaf) and the flow-sheet of the complete sampling scheme is represented in fig 34.1. (opposite page).

One of the interesting features of this plant is the tertiary sampler equipped with two chute-type cutters joined side by side and inclined in opposite directions.

The first chute takes a 7 kg moisture sample at 10 mm and the second one a 12 kg assay sample. Both samples are dispatched to the sampling and assaying laboratory for further processing.

If this plant was to be built today, it could have been completely automated until obtention of the six twin-samples, without any human intervention.

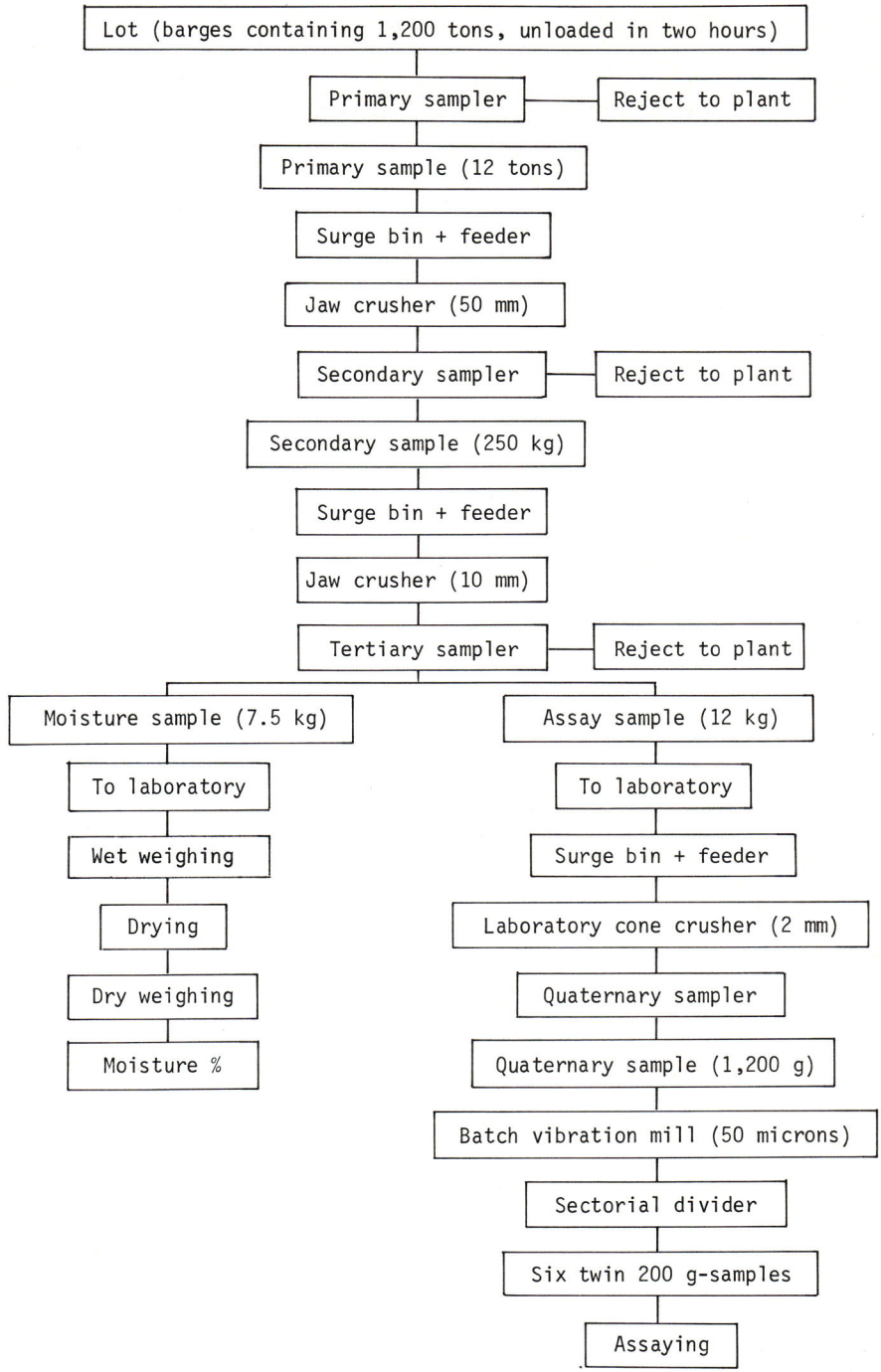

Fig. 34.1. Flow sheet of a bauxite sampling plant (including laboratory processing).

34.3. SAMPLING OF A NICKEL ORE

Sampling of 15,000 ton shiploads, broken up into six 2,500 ton sub-lots, upon loading aboard sea-going vessels at a rate of 800 t/h. The ore (100-0 mm) is very wet (20 to 35 % moisture) and sticky. Primary and secondary cutters are heated in order to prevent clogging. The characteristics of the samplers and of the sampling operations are summarized in table 34.2 and the flow-sheet is represented in fig. 34.2. (opposite page).

Table 34.2. Sampling of a wet and sticky nickel ore.

Sampling stage →	Primary	Secondary	Tertiary (assay)
Type of sampler	Straight path	Straight path	Sectorial splitter
d (mm)	100 (150)	30	5
(cm)	10	3	0.5
μ (t/h)	800	4	-
(g/s)	220×10^3	1,100	12.5
W (mm)	600	180	Angle 60°
(cm)	60	18	
V_C (cm/s)	75	20	10 rpm
T_I (s)	0.8	0.9	1
T_{sy} (s)	600	60	6
T_L (hours)	3.13	0.85	0.85
(s)	11,250	3,060	3,060
Q	19	51	510
M_L (tons)	2,500	3.4	-
(g)	2.5×10^9	3.4×10^6	38×10^3
M_I (kg)	180	1	-
(g)	180×10^3	10^3	12.5
M_S (kg)	3,400	51	6.4
(g)	3.4×10^6	51×10^3	6,400
Sampling ratio	1.4×10^{-3}	1.5×10^{-2}	0.167

One of the original features of this sampling scheme is that at the secondary sampling stage, one increment out of four is automatically diverted towards the moisture sample container by means of a reciprocating belt. The three remaining increments form the secondary assay sample. The moisture sample is for instance made of increments 2, 6, 10 etc.. and the assay sample of increments 1, 3, 4, 5, 7, 8, 9, 11, etc..This is perfectly correct from a theoretical standpoint : the assay sample is in fact the sum of two correct samples made of increments 1, 3, 5, 7, 9, 11, etc .. and 4, 8, 12, 16, etc .. respectively.

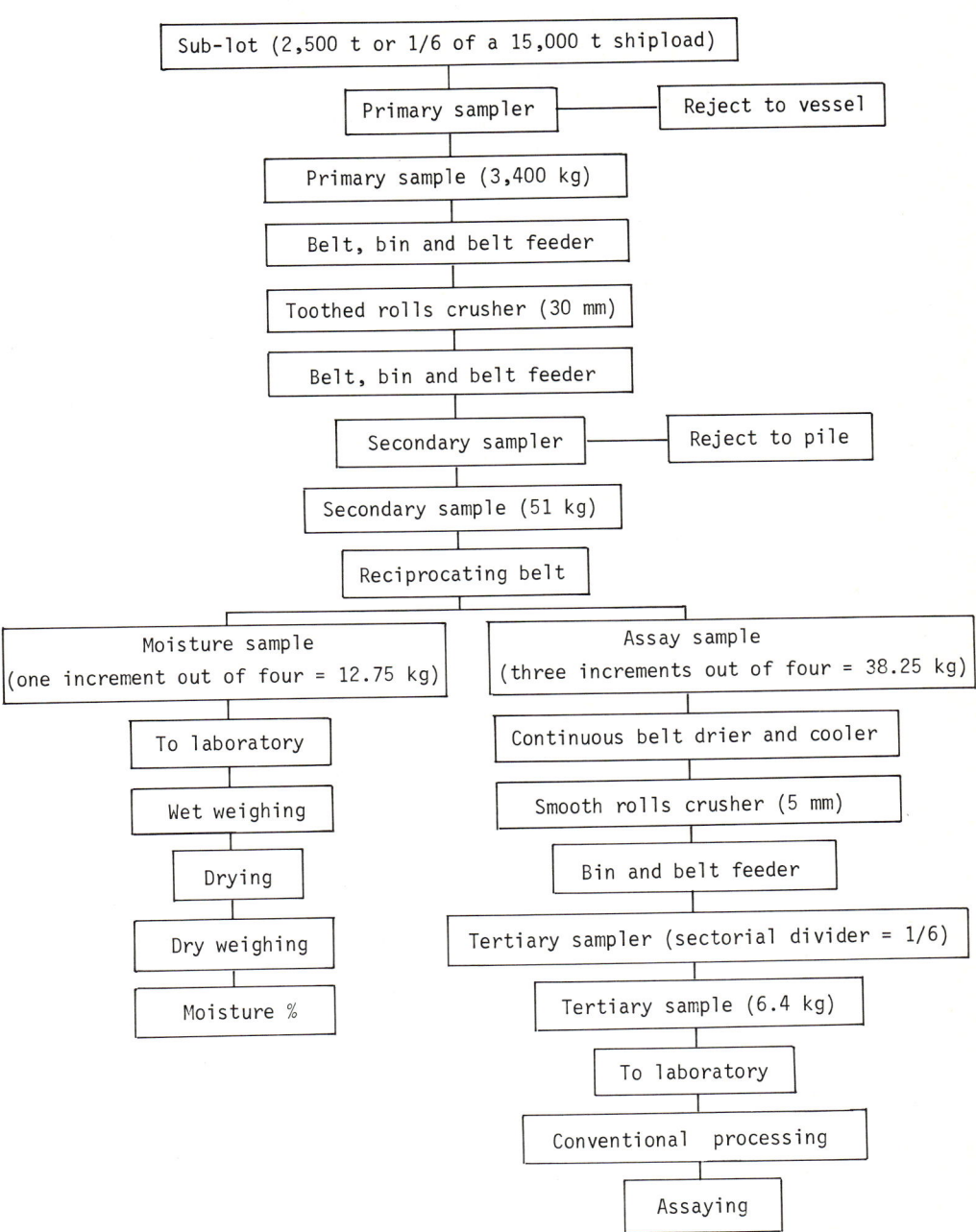

Fig. 34.2. Flow-sheet of a nickel ore sampling plant

34.4. SAMPLING OF AN IRON ORE

Lots of 1,200 to 3,600 tons of a calibrated 45-8 mm iron ore are sampled prior to stacking. The sampling plant has been designed in order to provide a 20 kg, 15 mm moisture sample and a 4 kg, 3 mm assay sample, irrespective of the lot weight.

The characteristics of the sampling equipment are given in table 34.3. for a 3,600 ton lot. For smaller lots, the constant interval T_{sy} between increments obtained at the secondary sampling stage is proportionally reduced. For a 1,200 ton lot, for instance, T_{sy} is 36 seconds instead of 108 seconds.

The corresponding flow-sheet is given in fig.34.3. The secondary sampler is equipped with two identical chute cutters inclined in opposite directions. These cutters deliver secondary moisture and assay samples respectively.

Table 34.3. Sampling of an iron ore.

Sampling stage →	Primary	Secondary	Tertiary (assay)
d (mm) (cm)	45 (50) 4.5	15 1.5	3 0.3
μ (t/h) (g/s)	600 167×10^3	1.2 333	0.04 11.1
W (mm) (cm)	140 14	60 6	65 6.5
V_c (cm/s)	40	20	11
T_I (s)	0.35	0.3	0.59
T_{sy} (s)	175	108	3
T_L (hours) (s)	6 21,600	6 21,600	0.5 1,800
Q	123	200	600
M_L (tons) (g)	3,600 3.6×10^9	7.2 7.2×10^6	- 20×10^3
M_I (kg) (g)	58.5 58.5×10^3	0.1 100	- 6.7
M_S (kg) (g)	7,200 7.2×10^6	20 20×10^3	4 4,000

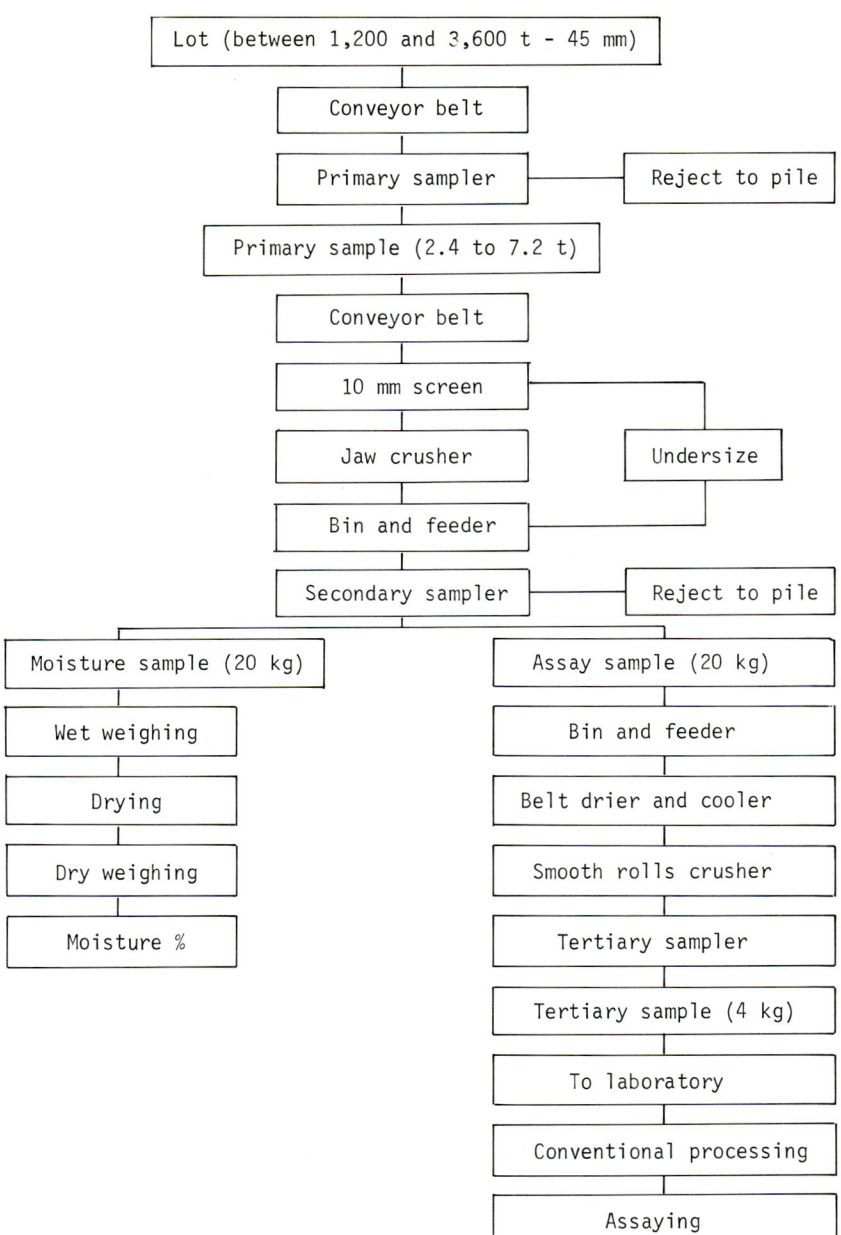

Fig.34.3. Flow-sheet of an iron ore sampling plant.

34.5. SAMPLING OF THE RAW MIX FED TO A CEMENT FACTORY

The raw mix is made of 80-90 % limestone (40 mm) and of 20-10 % slate (50 mm) and is sampled prior to being fed to a bed-blending system. Hourly samples are taken, immediately reduced and assayed in order to know with accuracy at every moment the average composition of the pile. The assays obtained by X-ray fluorescence are automatically fed to a computer that eventually corrects the proportioning of the two main components in order to obtain a pile of predetermined composition.

The characteristics of the sampling equipment are summarized in table 34.4 and the complete flow-sheet is presented in fig.34.4.

Table 34.4. Sampling of the raw mix fed to a cement plant.

Sampling stage	Primary	Secondary	Tertiary	Quaternary
d (mm)	50	20	5	0.16
(cm)	5	2	0.5	0.016
μ (t/h)	900	8	-	-
(g/s)	250×10^3	2,222	46.3	1.45
W (mm)	200	100	25	22
(cm)	20	10	2.5	2.2
V_c (cm/s)	20	20	11	11
T_I (s)	1	0.5	0.23	0.20
T_{st} (s)	112	24	7.3	2.95
T_L (hours)	1	1	1	1
(s)	3,600	3,600	3,600	3,600
Q	32	150	493	1,220
M_L (tons)	900	8	-	-
(g)	9×10^8	8×10^6	167×10^3	5,200
M_I (kg)	250	1.1	-	-
(g)	250×10^3	1,100	10.6	0.29
M_S (kg)	8,000	167	5.2	-
(g)	8×10^6	167×10^3	5,200	350

This sampling plant presents three interesting features :

1) The primary and secondary samplers are equipped with random stratified timers in order to prevent errors due to eventual periodic quality fluctuations,

2) The tertiary sample is ground in a peripheral discharge rod mill, as a pulp containing about 50 % water, prior to quaternary sampling,

3) The quaternary sampling reject is collected in a stirred tank the content of which has an average composition identical with that of the pile. The pulp being very thick is practically homogeneous and can be easily sampled if a check is necessary.

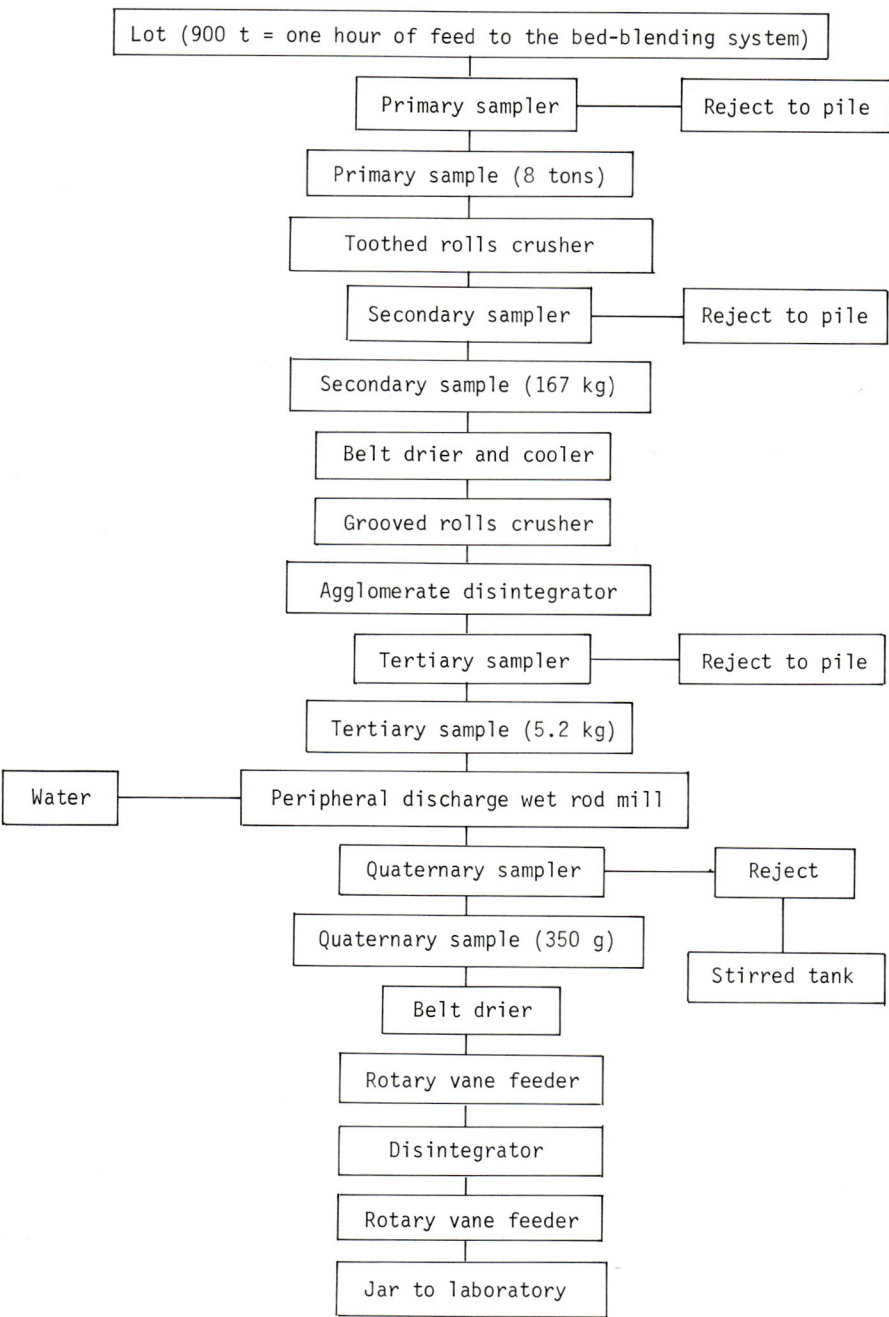

Fig. 34.4. Flow-sheet of the sampling plant of the raw mix fed to the blending pile of a cement factory.

34.6. COMMERCIAL SAMPLING OF URANIUM CONCENTRATES

Dry and pulverulent uranium concentrates made of micron size particles are sampled upon loading into 1,800 kg containers. Each lot is made of 10 containers and contains about 18 tons of concentrates. In order to avoid loss of valuable material (43 $ per pound U_3O_8 in March 1979), air pollution and pick up of moisture in a tropical atmosphere, the first sampling stage is put under a slightly low pressure. Secondary and tertiary stages are carried out in an air-condioned laboratory.

Secondary and tertiary samplers are stationary-feeder sectorial dividers such as those represented in fig. 24.10 and 24.11 respectively. Both achieve a correct though degenerate splitting. The secondary splitter is equipped with a time-switch but the tertiary splitter works continuously.

Table 34.5. Sampling of uranium concentrates.

Sampling stage →	Primary	Secondary	Tertiary
Type of sampler	Straight path sampler	Stationary feeder sectorial divider	Stationary feeder sectorial divider
d (microns)	10	10	10
µ (t/h) (g/s)	6 1,670	- 42	- 3.1
W (mm) (cm)	50 5	Four ten degree sectors	Twelve 18 degree sectors
V_C (cm/s) (rpm) (°/s)	22 - -	- 10 60	- 10 60
T_I (s)	0.23	4 x 0.167	12 x 0.3
T_{sy} (s)	30	16	6
T_L (hours) (s)	3 10,800	0.91 3,260	0.25 900
Q	360	204	150
M_L (tons) (g)	18 18 x 10^6	- 136 x 10^3	- 2 x 1,400 = 2,800
M_I (g)	378	7	0.93
M_S (kg) (g)	136 136 x 10^3	4 x 1.4 4 x 1,400	- 12 x 140

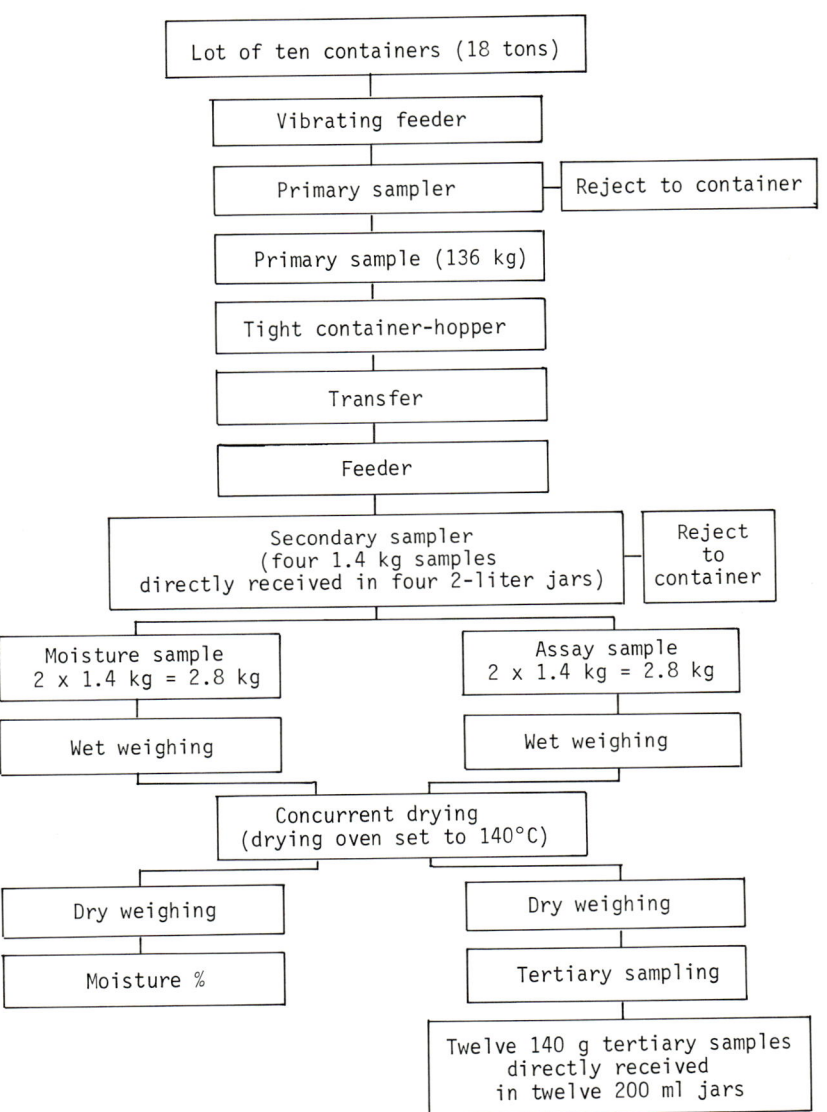

Fig. 34.5. *Commercial sampling of uranium concentrates*

REFERENCES

Armstrong-Smith, George,(1974). Sampling and sample preparation of copper concentrator products. In : The Institution of Mining and Metallurgy, London (Editor), Geological, Mining and metallurgical sampling: 232-246 and discussion 259.
Bartlett, M.S. (1960). An introduction to stochastic processes. Cambridge University Press, 312 pp.
Bastien, M. (1960). Loi du rapport de deux variables normales. R. Stat. Appl. 8: 45-50.
Becker, R.M. (1964-66). Some generalized probability distributions with special reference to the mineral industries. US Bureau of Mines RI 6329 : 53 pp, RI 6552 : 101 pp, RI 6598 : 79 pp, RI 6627 : 57 pp, RI 6768 : 60 pp.
British Standards Institution (1960). The sampling of coal and coke. BS 1017. Part 1 : Coal : 124 pp. Part 2 : Coke : 106 pp.
Brunton, D.W. (1895). The theory and practice of ore sampling. Trans. AIME, 25 : 826.
Cochran, W.G. (1953). Sampling Techniques. John Wiley, New-York, 330 pp.
Colijn, H. (1975). Weighing and Proportioning of Bulk Solids. Trans Tech Publications, Clausthal, Germany.
David, Michel (1977). Geostatistical ore reserve estimation. Elsevier Scientific Publishing Company, Amsterdam, Oxford, New-York. 364 pp.
Demond, C.D. and Halferdahl, A.C. (1922-23). Mechanical sampling of ores. Eng. & Min. J. 114: 280-284 and 115: 525-527.
Desnoes, André (1965). Utilisation des suspensions denses de mercure dans le bromoforme au laboratoire. R. Ind. Min. 47: 33-38.
Geary, R.C. (1930). The frequency distribution of the quotient of two normal variables. J. Roy. Stat. Soc. 93: 442.
Gy, Pierre (1953). Erreur commise dans le prélèvement d'un échantillon sur un lot de minerai. Congrès des Laveries des Mines Métalliques Françaises. R. Ind. Min. 36: 311-345.
Gy, P. (1955). Erforderliche Probemenge - Kurventafeln. Internationales Kongress für Erzaufbereitung. Erzmetall 8B: 199-220.
Gy, P. (1956). Nomogramme d'échantillonnage, Sampling Nomogram, Probenahme Nomogramm. Minerais et Métaux, Paris (Editor).
Gy, P. (1957). Sampling - The error committed in size distribution. Jamshedpur Congress. Ind. Min. J. (1957).
Gy, P. (1957). A new Theory of ore sampling. AIME annual meeting, New-Orleans.
Gy, P. (1964). Le principe d'equiprobabilité. Ann. Min. (Dec. 1964): 779-794.
Gy, P. (1965). Sampling of ores and metallurgical products during continuous transport. Trans. IMM 74: 165-199.
Gy, P. (1965). Calculateur d'échantillonnage - Sampling slide rule. Soc. Ind. Min. Saint-Etienne, France.
Gy, P. (1965). Procédé pour améliorer la précision de l'échantillonnage des matériaux. Brevet français No.1,455,938. July 6, 1965.
Gy, P. and Minerais et Métaux, Paris (1966). Process and apparatus for improving the precision of the sampling of materials. British Patent No.1,110,724. July 6, 1966.
Gy, P. and Minerais et Métaux, Paris (1968). Process for reducing error in the sampling of materials. US Patent No. 3,392,587. July 16, 1968.
Gy, P. (1967). L'échantillonnage des Minerais en vrac. Tôme 1 : Théorie générale. Numéro spécial Rev. Ind. Min. 15 janvier 1967 : 188 pp.
Gy, P. (1971). L'échantillonnage des Minerais en vrac. Tôme 2 : Théorie générale (fin) - Erreurs opératoires. Rev. Ind. Min. 15 septembre 1971 : 280 pp.

Gy, P. (1972). Contribution à l'étude de l'hétérogénéité d'un lot de matière morcelée. Thèse de Docteur-Ingénieur. Université de Nancy.
Gy, P. (1972). Die Probenahme bei stückigen Erzen. Aufbereitungs-Technik 13: 687-69
Gy, P. (1973). The sampling of broken ores - A review of principles and practice. In : The Institution of Mining and Metallurgy, London (Editor), Geological, Mining and Metallurgical Sampling : 194-205. and discussion 261-263.
Gy, P. (1974). El control minero y metalurgico. Lecture, Instituto de Ingenieros de Minas, Santiago, Chile. Rev. Minerales 24: 7-11.
Gy, P. (1975). Théorie et pratique de l'échantillonnage des matières morcelées. Editions PG, Cannes, France. 597 pp.
Gy, P. (1976). The sampling of particulate materials - A general theory. Symposium on Sampling practices in the mineral industries. The Australasian Institute of Mining and Metallurgy, Victoria, Australia : 17-33.
Gy, P. (1976). The sampling of particulate materials - A general theory. Int. J. Miner. Process. 3: 289-312.
Gy, P. and Marin, L. (1978). Unbiased sampling from a falling stream of particulate material. Int. J. Miner. Process. 5: 297-315.
Gy, P. (1978). Teoria matematica generale della campionatura dei materiali granulati in trasporto continuo. Lecture, University of Trieste, Italy. To be publish
Gy, P. (1978). Sampling particulate materials: a general theory. In: Sampling systems for on-line analysers. SIRA Institute and Warren Spring Laboratory. A seminar organized at the City University, London. To be published.
Hassialis, M.D. (1945). Sampling. Section 19 in : Taggart, A.F. Handbook of mineral dressing. John Wiley, New-York. 71 pp.
Marin, Lucien (1978). see Gy and Marin (1978).
Matheron, G. (1965). Les variables régionalisées et leur estimation. Masson, Paris, 212 pp.
Matheron, G. (1969). Le krigeage universel. Cahiers du centre de morphologie mathématique, Ecole Nationale Supérieure des Mines de Paris, Fasc. No.1: 82 pp.
Matheron, G. (1970). La théorie des variables régionalisées et ses applications. Cahiers du centre de morphologie mathématique, Ecole Nationale Supérieure des Mines de Paris, Fasc. No.5 : 212 pp.
Matheron, G. (1973). The intrinsic random functions and their applications. Adv. in Appl. Prob. 5: 439-468.
Richards, R.H. (1908). Ore dressing. Sampling : Vol. 2: 843-852. Vol. 3: 1571-1578. Vol. 4: 2031-2033. Mac-Graw Hill, New-York.
Warwick, A.W. (1903). Notes on sampling. Denver.

A more complete list of references can be found in Gy (1971).

Some of the Author's publications are available. Please apply to :

                        Pierre Gy
                        Consulting Engineer,
                        Résidences de Luynes,
                        14, Avenue Jean-de-Noailles,
                        06400 CANNES
                        FRANCE

INDEX

Abrasion, 203, 326
absolutely correct selection 96
absolute minimum increment weight 411
absolute minimum sample weight, 412
acceptable cost, 353
acceptable representativeness standard, 352
acceptable systematic difference, 375,381
accounting sampling, 355
accuracy, 17, 400
active components, 13
actual extended increment, 160
actual fragmental increment, 163
addition of critical component, 330
administrative sampling, 355
agreement between two series of estimates, 379, 381
agreement between estimate and true value, 391
alpha (risk), 381
alteration of the chemical composition, 27, 328
alteration of the physical composition, 27, 330
alternate shovelling, 298
alternation rule (riffling), 296, 316
analysis, 13, 14
analysis error, 15, 23, 335
analysis sample, 12
a posteriori qualities, 17
a priori qualities, 16
Armstrong-Smith, George, 199
assay portion 12
assaying, 13
automatic sampling plants, 405, 407 413

Bastien, M., 98, 230
batch vibration mill, 312
Becker, R.M., 4
bed-blending systems, 222
beta (risk), 381
bias 15
biased selection, 17
blending, 323
breakage of fragments, 330
breaking up of :
 continuous selection error CE, 105,336
 materialization error ME, 337

breaking up of :
 overall estimation error OE, 335
 quality fluctuation error QE, 336
 sampling error SE, 336
 short-range quality fluctuation error QE, 336
 total preparation error PE, 337
 total sampling error TE, 335
Brunton, D.W., 3
bucket-type cutters, 203

Central limit theorem, 98,230
certainty, 380
check of sampling bias, 391
chemical analysis, 13
chemical components, 13
chute cutters, 202
circular path cutters, 171
coefficient of bias, 16
 of correlation, 15
 of cubicity, 256
 of variation (Pearson), 15
collision between fragment and cutter edge, 183
collision of the first type, 188
 of the second type, 195
commercial sampling, 290, 354, 365
comminution, 323
compromise reproducibility-cost, 249,315
coning and quartering, 293
constant term of a(t), 56, 105
constant tonnage sampling systems, 122
constitution heterogeneity CH, 26, 218, 237
 homogeneity, 219
contamination, 27, 203, "é(
content, 14
continuous selection error CE, 25, 85, 99, 106, 336
continuous selection model, 23,24,41,53, 85
continuous variogram, 63
core halving, 311
core sample reduction, 311
Cornish quartering, 293
correct selection, 236, 251
 increment delimitation, 160
 increment extraction, 162
correctness, 16
correctness bias, 237
corrected variogram, 65,
correlation bias, 97, 101, 121
corrosion, 203,326
cost minimization, 202
cost of correct delimitation, 182

cost of correct extraction, 211
    of sample reduction, 211, 249, 342
creation of critical component, 330
crenellated weighting functions, 125
critical component, 14
    content, 14, 29, 367
    diameter, 196, 205, 284
    plane, 187
    point, 184, 194
    positions, 188
    speed, 205
    width, 196, 198, 200, 205
cutter capacity, 202
    construction, 202
    depth, 202
    geometry, 167
    lay-out, 175, 202
    speed, 173, 194, 200
    width, 194
cyclic phenomena, 105

David, Michel, 1, 34, 356
decimal minute, 66
definitions, 11
degenerate models, 43
degenerate splitting processes, 292
degree of representativeness, 272
delimitation correctness, 179
    error, 25, 347
Demond, C.D., 3
density of selection probability, 49, 75, 165
design of automatic sampling plants, 405, 407
Desnoes, André, 252
destruction of critical component, 331
deterministic selection, 16
diameter of a particle, 233
disagreement between two series of estimates, 381
discrete selection error, 26
    model, 23, 25, 42, 50, 213, 227
discrete variogram, 63
distribution heterogeneity DH, 26, 218 238
distribution homogeneity, 222
domain, 11, 16
drill core sample reduction, 311
dry fines sampling, 177, 203
dust, 178, 203, 325

Economical optimum (cutter width and speed), 201
electric cutter drive, 173, 175
elimination of carbon dioxide, 329
    combined water, 329
    critical component, 331
equiprobable sampling model, 4, 243
equity, 17, 27, 292, 367
errors, 15

estimate, 15
estimator, 15
Euclidean division, quotient, remainder, 76, 79
exactness, 17
expendable sampling variance, 352, 409
experimental determination of critical cutter characteristics, 204
experimental estimate, 15
experimental variogram, 64
extended function, 48
    increment, 35, 157
    sample, 35, 157
extension domain, 48
extent of a domain, 16
extraction error, 25, 183, 348
    probability, 165
    ratio 204

Field of application of selection schemes, 83
filtration, 323
first critical position, 188
fixation of carbon dioxide, 329
    water, 329
flap samplers, 169
flat variogram, 58, 62
flexible hose pulp samplers, 170
flowing streams (model), 46
flow-sheets of sampling plants, 413
fraction dealing-out, 36, 291
    delimitation, 36, 291
    separation, 36, 291
fractional shovelling, 297
fragment, 11
fragmental functions, 48
    increment, 35, 157
    sample, 35, 157
frauding, 332
free parameter of a selection scheme, 82
fundamental error FE, 26, 241, 246, 337, 340

Gangue minerals, 257
Geary, R.C., 98, 230
General Superintendence Co(GSC), 210
geometrical fractions, 36, 291
geostatistics, 1, 34
grade, 14
gravity segregation, 222
group of particles, 11
grouping and segregation error GE, 26, 202, 241, 247, 337, 342
grouping factor 221

Halferdahl, A.C., 3
hammer-and-shovel method, 16
handling in a damp atmosphere, 329
Hassialis, M.D., 4
heterogeneity carried by a group of

particles, 218
heterogeneity carried by a particle, 217
heterogeneity of a continuous set, 24, 55
heterogeneity of a discrete set, 26, 215
heterogeneity of constitution, 218
           of distribution, 218
homogeneity of a continuous set, 55
           of a discrete set, 216
homogenisation, 220
hydraulic cutter drive, 174

inclination of cutter edges, 192
incorrectness, 16
incorrectness bias, 97, 104, 232
           curve, 283
incorrect selection, 229
increment, 11
increment delimitation, 35, 161
        delimitation correctness, 179
        delimitation error DE, 25, 160, 162, 167, 337, 347
        extraction, 35
        extraction correctness, 189, 198
        extraction error EE, 25, 163, 183, 337, 348
        integrity, 202
        sampling process, 23
inspection of sampling equipment, 402
interstitial fluid, 14
intervals of uncertainty, 386

Japanese slab-cakes, 361
judgment errors, 394

Krige, D.G., 1

laboratory sample, 12
largest acceptable difference, 352
level of risk, 381
liberation diameter (or size), 262
           factor, 257, 262
local phenomena, 105
logical approach, 23
long-range quality fluctuation error QE_2, 25, 107, 111, 337, 343
long-range term of a(t), 56, 105
long-wave weighting functions, 125
losses, 27, 202, 327
lot, 11
Louis-le-Debonnaire's splitting method, 376
lower size limit, 263

Magnetic cutter drive, 174, 175
manual cutter drive, 174, 175
Matheron, G., 1, 4, 56
material fractions, 36, 291

Marin, Lucien, 198, 200, 204
materialization error ME, 25, 155, 336
materialization of the punctual increments, 155
maximum maximorum of the periodic quality fluctuation error, 117
maximum particle diameter (or size), 205, 256, 263
maximum distribution heterogeneity, 223
mean, 15
mean square, 15
Minemet-Industrie, 198, 200, 267, 301
mineralogical composition factor, 257
minimum increment weight, 211, 250, 410
minimum sample weight, 211, 412
mixing, 220, 250, 312, 323
model, 39
models of the increment sampling process, 39
model variogram, 65
model extended increment, 158
model fragmental increment, 162
moisture analysis, 13
        content, 14
movable batch, 33, 289

Natural distribution homogeneity, 220
negative Student-Fisher test, 380
nominal particle diameter (or size), 256
non-periodic variogram, 140
non-probabilistic selection, 16, 33
           sampling schemes, 23
notations, 11, 19
numerical sampling ratio, 234

One-dimensional distribution homogeneity 222
one-dimensional geometrical space, 16
           model, 45
           temporal space, 16
           temporal model, 46
open-discharge cutters, 202
overall estimation error OE, 23, 335
overdrying, 329
oxidation of sulphides, 328

Parabolic variogram, 62
particle, 11
particle shape factor, 263
particulate structure, 241
passive components, 13
Pearson, 15
percentage, 14
perfect distribution homogeneity, 219
periodic quality fluctuation error, 25, 107, 113, 337, 345
periodic term of a(t), 56, 105
periodic variogram, 131
physical components, 13
pneumatic cutter drive, 174, 175
point selection, 35

positive Student-Fisher test, 380
potential increments, 242
    samples, 36, 291
practical certainty, 380
precision, 17, 400
preparation, 12, 23, 27, 323
preparation errors, 23, 27, 325, 337, 349
presumption, 380
primary sample, 12
probabilistic selection, 16, 33
    sampling schemes, 23
pulp concentration, 13
    sampler, 193
    sampling (high volumes), 200
punctual functions, 42
    increment, 35, 148
    sample, 35, 75, 157
purposive selection, 16

Quadratic mean, 16
qualitative function, 24
qualities of a selection process, 16
quality fluctuation error QE, 25, 106, 336
quantitative function, 24

Random selection, 24, 75, 82
    stratified selection, 24, 79
    variable, 15
range of a variogram, 58
rate-of-flow, 46
rate-of-flow regulation, 127
rebounding range, 196
    rule, 162, 183
recapitulation of sampling errors, 335
rectilinear variogram, 62
reference selection scheme, 75
    sampling method, 285, 393
region of uncertainty, 386
reject, 12
relative standard deviation, 15
    variance, 15
representativeness, 17
representativeness and cost, 351
reproducibility, 17, 400
residual bias, 97, 103
resolution of sampling problems, 27, 51, 264, 333
revolution distribution homogeneity, 223, 293
revolving feeder sectorial splitters, 300
Richards, R.H., 3
riffle splitter, 294
riffling bias, 295
risks, 381
rule of the centre of gravity, 162, 183
rules of delimitation correctness, 179
    extraction correctness, 198, 210

Sabotage, 332
safe reduction scheme, 318
safety line, 314
    rule, 314
salting, 327
sample, 11
sample preparation, 323,
    reduction, 13, 313
    selection, 37, 292
sampling, 12, 13
sampling constant, 258
    error, 15, 23, 40, 232, 336
sampling of :
    one-dimensional flowing streams, 359
    one-dimensional stationary objects, 358
    small objects, 361
    three-dimensional objects, 355
    two-dimensional objects, 357
    valuable objects, 361
    zero-dimensional objects, 359
sampling plants, 407, 413
    probability, 166
    processes, 33
    ratio, 90
    reject, 12
    slide rule, 264
    stage, 12, 23
screening, 323
second critical point, 194
    position, 188
    situation, 195
secondary sample, 12
sectorial splitters, 300
segregation error, 247
    factor, 224
selection error, 15, 336
    probability, 165
selective process, 16
selecting condition, 16
    probability, 16
    results, 16
settlement error VE, 368
    price, 368
SF = Student-Fisher, 380
shape factor, 256, 263
short-range quality fluctuation error, 25, 107, 109, 242? 337
short-range term of a(t), 56, 105, 241
Sichel, H.S., 1
simply correct selection, 96
size analysis (sampling for), 13, 254, 2
size-density fractions, 233
    analysis, 269
slide rule (sampling), 264
Société Générale de Surveillance, 210
solvable sampling problems, 27, 351
sources of selection bias, 97
specimen, 11
splitting errors, 305

splitting process, 23, 27, 35, 289
    291, 305, 311
standard deviation, 15
standard of accuracy, 17
            representativeness, 17, 272
            reproducibility, 17
stationary feeder, sectorial splitter,
    301
statistical definitions, 15
            notations, 15
sticky materials, 203
stirring of pulps, 323
straight-path cutter, 167
straightness of cutter edges, 191
strata, 79
stratified selection, 24, 79
Student-Fisher (SF) test, 380
systematic selection, 24, 76
synthetic lot, 392

t distribution, 381
technical sampling, 354
thickness of cutter edges, 191
threshold of a variogram, 60
three-dimensional heterogeneity, 223
three-dimensional distribution homo-
    geneity, 222
three-dimensional model, 43
                space, 16
time sampling ratio, 90
tolerated systematic difference, 375,
    381
total sampling error TE, 23, 335
total selection bias, 97
transfer, 323
true market price, 368
true splitting processes, 292
twin-samples, 12, 37, 292
two-dimensional distribution homo-
    geneity, 222
two-dimensional model, 44
                space, 16
                sampling, 180

Uncertainty, 380
unbiased selection, 17
uniform weighting,
unintentional mistakes, 27, 331
units, 46
unsafe reduction scheme, 319
unsolvable sampling problems, 27, 351
unmovable batch, 33
unmovable objects (sampling of), 34
unweighted lot, 106
            sample, 106
useful domain of a variogram, 62
upper size limit, 263

Variance, 15
variogram, 24, 56
variographic experiment, 25, 63, 129
            parameters, 63, 73
Vezin, H.A., 3

Warwick, A.W., 34, 289
weighting bias, 121
        error, 25, 106, 121, 336, 346
        function, 43
        fluctuation factor, 122
de Wijs, H.J., 1

zero-dimensional lots, 73
            model, 46